Earth Observation for Monitoring Mangrove Ecosystems

Sanjeev Kumar Srivastava
Prashant K. Srivastava
Tapas K. Biswas • Mukunda Dev Behera
Sharif A. Mukul
Editors

Earth Observation for Monitoring Mangrove Ecosystems

Springer

Editors
Sanjeev Kumar Srivastava
Geospatial Analytics for Conservation and Management, School of Science Technology and Engineering
University of the Sunshine Coast
Sippy Downs, QLD, Australia

Tapas K. Biswas
CSIRO
Canberra, ACT, Australia

Sharif A. Mukul
Department of Environment and Development Studies
United International University
Dhaka, Bangladesh

Prashant K. Srivastava
Institute of Environment and Sustainable Development
Banaras Hindu University
Varanasi, Uttar Pradesh, India

Mukunda Dev Behera
Centre for Ocean, River, Atmosphere and Land Sciences (CORAL)
Indian Institute of Technology (IIT)
Kharagpur, West Bengal, India

ISBN 978-3-032-16667-8 ISBN 978-3-032-16668-5 (eBook)
https://doi.org/10.1007/978-3-032-16668-5

This Springer imprint is published by the registered company Springer Nature Switzerland AG
The registered company address is: Gewerbestrasse 11, 6330 Cham, Switzerland

This book is respectfully dedicated to the cherished memory of my beloved daughter, the late Ayushee Srivastava.
It is also devoted to my wife, Abha Srivastava, and my son, Sarthak Srivastava, whose love and support have been a constant source of strength.
Lastly, to my poodle, Sanoi, whose companionship has brought comfort and joy throughout this journey.

Foreword

Mangrove ecosystems are among the most valuable coastal environments on Earth. They protect shorelines, nurture biodiversity, store vast amounts of carbon, and support millions of livelihoods. Yet, they remain highly vulnerable to human pressures and the accelerating impacts of climate change. Safeguarding these ecosystems requires tools that can capture their complexity, monitor their health, and inform sustainable management at multiple scales.

Earth Observation (EO) technologies have emerged as indispensable instruments in this endeavour. Multispectral and hyperspectral imagery enables precise mapping of mangrove species, vegetation health, and canopy density. LiDAR provides accurate measurements of forest structure, height, and biomass, while microwave sensors offer the unique ability to track mangrove extent and inundation patterns in all weather conditions, day and night. Integrated within Geographic Information Systems (GIS), these datasets form the backbone of long-term monitoring, carbon stock assessment, restoration planning, and conservation strategies—from local communities to global frameworks.

This volume, authored by leading experts in mangrove ecology and conservation, demonstrates the power of combining ecological knowledge with cutting-edge technologies. Through case studies and methodological advances, it showcases how field surveys, advanced satellite systems, Synthetic Aperture Radar (SAR), LiDAR, and UAV-based sensors can provide unparalleled insights into mangrove structure, species composition, and ecosystem dynamics. More importantly, it illustrates how these technologies, when linked with GIS-based analysis and decision-support tools, can translate complex data into actionable strategies for restoration, sustainable use, and policy development.

By bridging science, technology, and socio-economic perspectives, this book makes a vital contribution to the growing body of knowledge on mangrove ecosystems. It not only documents their current status but also highlights pathways for monitoring change, guiding interventions, and shaping effective policies.

Ultimately, this work is an invitation: to researchers, practitioners, and policy-makers alike, to harness Earth observation and ecological science in the service of conserving and restoring one of the planet's most vital ecosystems.

Chief Executive Officer & Managing Director Australasian Space Innovation Institute
Adelaide, SA, Australia
Professor Emeritus, Adelaide University
Adelaide, SA, Australia

Andy Koronios

Foreword

Mangroves line many coasts of Central and South America, Tropical Africa, India, Southeast Asia, Northern Australia and many smaller islands, especially in the Western Pacific Ocean. To people from elsewhere who have never lived close to mangrove forests and enjoyed the diverse benefits of doing so, their role in sustaining livelihoods is often difficult to comprehend. This explains why these complex ecosystems often became degraded and devalued during the colonial histories of these regions, exposing their inhabitants to unnecessary livelihood stress and weakening the resilience they had developed over millennia from sustainable interactions with mangrove forests. For food, for building material, for medicine, for shelter from storms both literal and societal.

And so, in an era of globalization, we bemoan the loss of mangroves. And while in places we may scramble to try and restore aspects of their original functionality, we also acknowledge the opportunities afforded by this loss and replace places where mangrove once grew with other food-producing systems or with other means of protecting exposed lands from erosion, the latter an issue of key importance as the pace of sea-level rise along many tropical coasts accelerates.

But for decades, our efforts to restore optimal function to such coasts have been hampered by the lack of a 'big picture'. We have worked locally, sometimes nationally, but often unaware of what is happening on the next island, in the adjoining country, and how this might affect our endeavours. We have also struggled to link climate change to our efforts to restore or repurpose mangrove forests, especially in understanding the game-changing effects of more acidic and deoxygenated ocean waters which are becoming warmer than those in which many mangrove ecosystems once thrived. And of course the effects of rising sea levels that shift the geographical baseline for mangrove restoration in many places, not least by mobilizing the shallow-water sediments in which propagules take root.

This superb collection of studies offers great hope for the future, providing information that allows for the effective and sustainable planning of mangrove-associated coastal management in an era of changing baselines, both human and non-human. It is easy to see its relevance, both in terms of large-scale even global planning, but also at a more localized level in places where the tools and data for understanding the opportunities and barriers to mangrove restoration and/or similar ecosystem-aligned interventions are now readily available. Yet perhaps the most important take-home message from this collection is that technology is evolving to meet the most profound of future challenges and that, thanks to the enterprise and commitment of scholars such as Sanjeev Kumar Srivastava and his co-editors, we can expect that the world's mangroves will be better understood and better managed decades from now than they have been for a long time.

University of the Sunshine Coast
Sippy Downs, QLD, Australia
University of Melbourne
Melbourne, VIC, Australia
September 2025

Patrick D. Nunn

Preface

Mangrove ecosystems play a crucial role in supporting coastal communities by providing a wide range of essential goods and services. They serve as natural barriers against shoreline erosion, offer protection from cyclone and storms, sequester carbon from atmosphere, and supply food, fuel, and construction materials to millions of people living near shorelines. Additionally, mangroves hold significant cultural and spiritual value for many communities. Currently, the primary driver of mangrove degradation is their conversion to other land uses, like aquaculture and infrastructure development. Climate-related factors such as rising temperatures, global warming, sea-level rise, and natural disasters further exacerbate the destruction of these ecosystems. Consequently, monitoring the distribution and dynamics of mangroves is vital for scientific research across both terrestrial and marine domains. Recent advancements in earth observation technology—including improvements in spectral, temporal, radiometric, and spatial resolution—alongside enhanced signal-to-noise ratios, increased data availability, machine learning techniques, sophisticated image-processing algorithms, high-performance computing, and information technology, have made it possible to monitor and manage mangroves at local to global scales in near real-time. Emerging remote sensing platforms such as unmanned aerial vehicles (UAVs), LiDAR, and other innovative sensors are now being effectively utilized for comprehensive mangrove monitoring.

The primary objective of this book is to explore the advancement of GIS and remote sensing technologies in mangrove ecosystem monitoring and management. It presents a comprehensive overview of both traditional and cutting-edge techniques for assessing mangrove types, biophysical and biochemical parameter retrieval, species identification, mapping and classification, machine learning applications, physics-based approaches, and the associated challenges and future directions. This book integrates the latest technological innovations, including the use of Synthetic Aperture Radar (SAR), hyperspectral data, and multi-source data fusion, to improve the accuracy and reliability of mangrove monitoring outcomes. It also highlights the role of geospatial technologies—such as multispectral, hyperspectral, radar, microwave, and LiDAR data—in enhancing the precision and scope of mangrove assessments. Designed to benefit academics and researchers, this book offers fresh insights, problem-solving strategies, and future perspectives in forestry and related disciplines, contributing to the evolving landscape of environmental monitoring and conservation.

This book explores advanced techniques and recent developments that support mangrove-related research and conservation efforts. Mangrove ecosystems are vital to coastal protection and development, yet they face increasing threats from climate change and human activities. Considering these challenges, there is an urgent need for comprehensive reference material to guide planners and policymakers in sustainable mangrove management. Beyond its relevance to decision-makers, the book is designed to engage a broad audience—including mangrove ecologists, climate scientists, academic institutions, government agencies, industry professionals, and developers of sensors and analytical tools. By showcasing practical applications of emerging technologies in mangrove monitoring and management, this publication aims to bridge the gap between research, innovation, and policy implementation.

This book becomes possible because of extensive and valuable contributions from interdisciplinary experts from all over the world working in the field of Mangrove Ecosystem and Earth observation. Based on contributions, this book has been divided into four parts: Part I contains introduction to mangrove ecosystem and earth observation. Part II details the conventional remote sensing and GIS techniques for mangrove ecosystems. Part III contains advanced remote sensing and GIS techniques for mangrove. Part IV provides the key challenges and future research need in mangrove ecosystems protection and conservation.

In Part I, Introduction to Mangrove Ecosystem and Earth Observation, the first chapter discusses various earth observation techniques and data sets, along with their relevance for monitoring and managing mangrove ecosystems, with a focus on the Sundarbans. The second chapter discusses the ecological role of the mangrove ecosystem, along with its socio-economic and cultural significance. The third chapter provides details of ground-based techniques for mangrove monitoring and management with an emphasis on a number of ecosystem services.

In Part II of this book, i.e. the Conventional Remote Sensing and GIS Techniques for Mangrove Ecosystems, the fourth chapter provided the important information on the role, status, and extent of the nursery developed in Indramayu Regency, West Java Province, Indonesia, and evaluates its similarities or differences with the Indian Sundarbans. The fifth chapter emphasizes the urgent need of accurate mapping of mangrove area due to changing climatic conditions and increasing anthropogenic threats. The sixth chapter provides a variety of data sources dating back to the mid-eighteenth century that can offer more in-depth insights into the dynamics of these Mangrove ecosystems. The seventh chapter in this section details about the different cyclone happened in the Bay of Bengal and its impact on the Indian Sundarbans.

Part III of the book provides the Advanced Remote Sensing and GIS Techniques for Mangroves Management and Monitoring. The eighth chapter of this section provides the role of Hyperspectral Remote Sensing (HRS) in mangrove species identification; methodologies involve in producing robust discrimination results and a few case studies to assess the real-time application of HRS in mangrove species identification. The ninth chapter provides a comparative assessment of a dedicated mangrove product, land cover datasets, spectral indices, and Machine Learning algorithm, with a particular focus on the identification of the mangrove class. The tenth chapter highlights the importance of EO in forest biomass estimation. Their study demonstrated that by integrating C-band SAR backscatter, textural parameters, and optical vegetation indices with machine learning models, a significant improvement in the accuracy of above-ground biomass (AGB) estimation of mangrove forests can be obtained. The eleventh chapter provides a cloud-based web application developed using Google Earth Engine (GEE) to monitor and analyse the Sundarbans' vegetation and hydrological dynamics. The developed web applications are equipped with time-series analysis, split-panel comparisons, and rainfall distribution mapping, providing researchers and policymakers with data-driven insights to support sustainable forest management and conservation planning. The twelfth chapter provides a detailed methodology for leaf chlorophyll content estimation by using the indices derived using AVIRIS-NG Hyperspectral images over Bhitarkanika mangrove ecosystem. The thirteenth chapter provides an in-depth investigation of the carbon dynamics in the mangroves of India by using carbon isotopes. The fourteenth chapter of this section provides a detailed spatiotemporal analysis of LULC across the Indian Sundarbans using MODIS decadal land cover data combined with Markov chain modelling to forecast future changes in mangrove cover changes.

Part IV of the book provides the different key challenges and future research needs in the field mangrove management and monitoring using different techniques. The fifteenth chapter provides a comprehensive overview of optical, SAR, LiDAR, hyperspectral, thermal infrared, and UAV-based platforms and sensors—and discusses their unique strengths, limitations, and suitability for monitoring mangrove ecosystems. In the same section, the last and final sixteenth chapter provides the different fields of EO applications in mangrove monitoring. Each platform's advantages and disadvantages are emphasized, along with the necessity of multi-sensor integration and consistent monitoring procedures. This chapter also emphasized on the obstacles that still exist that can prevent the full operationalization of EO for conservation planning, such as spectrum confusion, cloud cover, data fragmentation, a lack of ground truth data, and institutional limitations.

In conclusion, this volume has been carefully structured to enhance the scientific understanding, methodological development, and practical application of diverse Earth Observation (EO) techniques for the monitoring and management of mangrove ecosystems. By incorporating state-of-the-art approaches such as Geographic Information Systems (GIS), Light Detection and Ranging (LiDAR), SAR, and data from advanced satellite platforms including MODIS, hyperspectral remote sensing satellites, ground instruments, the book underscores the transformative potential of EO in supporting sustainable mangrove protection and conservation.

This work represents one of the first comprehensive efforts to promote synergistic and multidisciplinary collaboration among scientists, practitioners, and policy stakeholders engaged in the domain of mangrove research and EO applications. It is envisaged that the volume will serve as a reference for a diverse readership with shared interests in satellite remote sensing, ecological modelling, geospatial technologies, sustainable development, and ecosystem monitoring. In this regard, it will be of particular benefit to academicians, researchers, environmental scientists, consultants, meteorologists, and computational experts committed to advancing the field of mangrove science and management.

The Editors gratefully acknowledge the contributions of all authors and anonymous reviewers for their scholarly inputs, dedication, and adherence to strict timelines. Sincere appreciation is also extended to the staff of Springer, for their professionalism, patience, and unwavering support throughout the preparation of this volume.

Sippy Downs, QLD, Australia	Sanjeev Kumar Srivastava
Varanasi, India	Prashant K. Srivastava
Canberra, ACT, Australia	Tapas K. Biswas
Kharagpur, India	Mukunda Dev Behera
Dhaka, Bangladesh	Sharif A. Mukul

Acknowledgement

Studies conducted to prepare several chapters included in this book are partly funded by "Asia-Pacific Network for Global Change Research (CRRP2020-08MY; https://doi.org/10.30852/p.13609)". The University of the Sunshine Coast provided the necessary research infrastructure needed for this book. This included high-performance computing systems for analysing data sets and access to several software packages. The authors duly acknowledge the Bangladesh and Indian governments for their necessary permissions for all such studies.

Contents

Contributors

Md Shydul Amin Department of Forestry and Environmental Science, School of Agriculture and Mineral Sciences, Shahjalal University of Science and Technology, Sylhet, Bangladesh

Mohammed A. S. Arfin-Khan Department of Forestry and Environmental Science, School of Agriculture and Mineral Sciences, Shahjalal University of Science and Technology, Sylhet, Bangladesh

Mochamad Candra Wirawan Arief Center for Environment and Sustainability Science, Faculty of Fishery and Marine Science, Universitas Padjadjaran, Sumedang, West Java, Indonesia

Pintoo Bandopadhyay Department of Agronomy, Bidhan Chandra Krishi Viswavidyalaya, Mohanpur, West Bengal, India

Saon Banerjee Department of Agricultural Meteorology and Physics, Bidhan Chandra Krishi Viswavidyalaya, Mohanpur, West Bengal, India

Debangana Banik ICAR—Central Research Institute for Jute and Allied Fibers, Kolkata, West Bengal, India

Saroj K. Barik CSIR–National Botanical Research Institute, Lucknow, India

North-Eastern Hill University, Shillong, Meghalaya, India

Dhananjay Barman ICAR—Central Research Institute for Jute and Allied Fibers, Kolkata, West Bengal, India

Surbhi Barnwal Centre for Ocean, River, Atmosphere and Land Sciences (CORAL), Indian Institute of Technology Kharagpur, Kharagpur, India

Bulu Basak ICAR—Central Research Institute for Jute and Allied Fibers, Kolkata, West Bengal, India

Mukunda Dev Behera Centre for Ocean, River, Atmosphere and Land Sciences (CORAL), Indian Institute of Technology Kharagpur, Kharagpur, India

Soumit K. Behera CSIR–National Botanical Research Institute, Lucknow, India

Bimal K. Bhattyacharya Space Applications Centre, ISRO, Ahmedabad, India

Saroni Biswas Department of Agricultural Meteorology and Physics, Bidhan Chandra Krishi Viswavidyalaya, Mohanpur, West Bengal, India

Md. Sahinur Islam Fahim Department of Forestry and Environmental Science, School of Agriculture and Mineral Sciences, Shahjalal University of Science and Technology, Sylhet, Bangladesh

Amol Kumar Garg Department of Geography, School of Environment and Earth Sciences, Central University of Punjab, Bathinda, Punjab, India

Sujit M. Ghosh Centre for Ocean, River, Atmosphere and Land Sciences (CORAL), Indian Institute of Technology Kharagpur, Kharagpur, India

Reginawanti Hindersah Faculty of Agriculture, Universitas Padjadjaran, Sumedang, West Java, Indonesia

Center for Environment and Sustainability Science, Faculty of Fishery and Marine Science, Universitas Padjadjaran, Sumedang, West Java, Indonesia

Gouranga Kar ICAR—Central Research Institute for Jute and Allied Fibers, Kolkata, West Bengal, India

Md Rezaul Karim Institute of Forestry and Conservation, John H. Daniels Faculty of Architecture, Landscape, and Design, University of Toronto, Toronto, ON, Canada

Sadra Karimzadeh Department of Architecture and Building Engineering, Institute of Science Tokyo, Tokyo, Japan

Remote Sensing Laboratory, Department of Remote Sensing and GIS, University of Tabriz, Tabriz, Iran

Shariq Nawaz Khan Amity School of Applied Sciences, Amity University, Lucknow, India

Pavan Kumar Geospatial Laboratory, College of Horticulture and Forestry, Rani Lakshmi Bai Central Agricultural University, Jhansi, India

Rina Kumari School of Environmental Sciences and Sustainable Development, Central University of Gujarat, Gandhinagar, India

Sondi Kuswaryan Faculty of Animal Husbandry, Universitas Padjadjaran, Sumedang, West Java, Indonesia

Mostafa Mahdavifard Remote Sensing Laboratory, Department of Remote Sensing and GIS, University of Tabriz, Tabriz, Iran

Department of Geographical Sciences, University of Hormozgan, Bandar Abbas, Iran

Parul Maurya School of Environmental Sciences and Sustainable Development, Central University of Gujarat, Gandhinagar, India

Smrutisikha Mohanty Department of Life Sciences, School of Natural Sciences, Shiv Nadar Institution of Eminence (Deemed to be University), Tehsil Dadri, India

Pallab Mozumder Department of Earth and Environment, Florida International University, Miami, FL, USA

Sharif A. Mukul Department of Environment and Development Studies, United International University, Dhaka, Bangladesh

Geospatial Analytics for Conservation and Management, School of Science Technology and Engineering (SSTE), University of the Sunshine Coast, Sippy Downs, QLD, Australia

Department of Earth and Environment, Florida International University, Miami, FL, USA

Sk. Mustak Department of Geography, School of Environment and Earth Sciences, Central University of Punjab, Bathinda, Punjab, India

Somnath Paramanik Centre for Ocean, River, Atmosphere and Land Sciences (CORAL), Indian Institute of Technology Kharagpur, Kharagpur, India

Megha Paul Geospatial Laboratory, College of Horticulture and Forestry, Rani Lakshmi Bai Central Agricultural University, Jhansi, India

Sarathi Saha Department of Agricultural Meteorology and Physics, Bidhan Chandra Krishi Viswavidyalaya, Mohanpur, West Bengal, India

Md. Shamim Reza Saimun Department of Forestry and Environmental Science, School of Agriculture and Mineral Sciences, Shahjalal University of Science and Technology, Sylhet, Bangladesh

Piyali Sarkar Department of Agricultural Meteorology and Physics, Bidhan Chandra Krishi Viswavidyalaya, Mohanpur, West Bengal, India

Bhawana Sharma Remote Sensing Laboratory, Institute of Environment and Sustainable Development, Banaras Hindu University, Varanasi, Uttar Pradesh, India

Harikesh Singh Geospatial Analytics for Conservation and Management, School of Science Technology and Engineering (SSTE), University of the Sunshine Coast, Sippy Downs, QLD, Australia

SmartSat Cooperative Research Centre, Adelaide, SA, Australia

Pashupati Nath Singh Remote Sensing Laboratory, Institute of Environment and Sustainable Development, Banaras Hindu University, Varanasi, Uttar Pradesh, India

Prashant K. Srivastava Geospatial Analytics for Conservation and Management, School of Science Technology and Engineering (SSTE), University of the Sunshine Coast, Sippy Downs, QLD, Australia

Remote Sensing Laboratory, Institute of Environment and Sustainable Development, Banaras Hindu University, Varanasi, Uttar Pradesh, India

Sanjeev Kumar Srivastava Geospatial Analytics for Conservation and Management, School of Science Technology and Engineering (SSTE), University of the Sunshine Coast, Sippy Downs, QLD, Australia

Fahmida Sultana Department of Forestry and Environmental Science, School of Agriculture and Mineral Sciences, Shahjalal University of Science and Technology, Sylhet, Bangladesh

Department of Biosciences, Faculty of Science and Engineering, Swansea University, Swansea, UK

Akancha Tripathi School of Environmental Sciences and Sustainable Development, Central University of Gujarat, Gandhinagar, India

Upasana Yadav Amity School of Applied Sciences, Amity University, Lucknow, India

Abbreviations

AGB	Above ground biomass
AHP	Analytic hierarchy process
AI	Artificial intelligence
APN-GCR	Asia-Pacific Network for Global Change Research
ARVI	Atmospherically resistant vegetation index
AVIRIS	Airborne visible/infrared imaging spectrometer
BoB	Bay of Bengal
BSI	Bare soil index
CA	Cellular automata
CART	Classification and regression tree
CBM	Community-based monitoring
CDM	Clean development mechanisms
CHMs	Canopy height models
CI	Carbon isotopes
CNN	Convolutional neural network
CVA	Change vector analysis
DL	Deep learning
dNBR	differenced NBR
DSMs	Digital surface models
DT	Decision tree
DTMs	Digital terrain models
ECOSTRESS	ECOsystem Spaceborne Thermal Radiometer Experiment on Space Station
EM	Electromagnetic
EMVI	Enhanced mangrove vegetation index
EnMAP	Environmental mapping and analysis program
EO-1	Earth Observation-1
ESA	European Space Agency
EVI	Enhanced vegetation index
FAO	Food and Agriculture Organization of the United Nations
FRA	Forest resource assessment

GBM	Gradient boosted model
GEDI	Global ecosystem dynamics investigation
GEE	Google earth engine
GIS	Geographic information systems
GLAS	Geoscience laser altimeter system
GNDVI	Green normalized difference vegetation index
GPS	Global positioning system
HRS	Hyperspectral remote sensing
HyMap,	Hyperspectral mapper
HySpex	Hyperspectral explorer
IGBP	International geosphere-biosphere programme
IPCC	Intergovernmental Panel on Climate Change
IRCEI	Inverted red-edge chlorophyll index
ISODATA	Iterative self-organizing data analysis technique algorithm
IUCN	International Union for Conservation of Nature
LAI	Leaf area index
LiDAR	Light detection and ranging
LST	Land surface temperature
LSU	Linear spectral unmixing
LULC	Land use/land cover
MCA	Markov chain analysis
MCE	Multi-criteria evaluation
MD	Minimum distance
Mes	Mangrove ecosystems
MFs	Mangrove forests
MGDI	MODIS global disturbance index
ML	Machine learning
MLC	Maximum likelihood classifier
MODIS	Moderate resolution imaging spectroradiometer
MPVI	Mangrove probability vegetation index
MVI	Mangrove vegetation index
MWTP	Marginal willingness to pay
NASA	National Aeronautics and Space Administration
NBR	Normalized burn ratio
NDBaI	Normalized difference bareness index
NDBI	Normalized difference built-up index
NDFI	Normalized difference flood index
NDII	Normalized difference infrared index
NDSI	Normalized difference salinity index
NDTI	Normalized difference turbidity index
NDVI	Normalized difference vegetation index
NDWI	Normalized difference water index
NDWVI	Normalized difference wetland vegetation index
NIR	Near infrared
NOAA	National Oceanic and Atmospheric Administration

OBIA	Object-based image classifier
OCEANSAT-2	Ocean Satellite-2
OCM-2	Ocean Colour Monitor 2
PALSAR	Phased array type L-band synthetic aperture radar
PES	Payment for Ecosystem Services
PLSR	Partial least squares regression
PRISMA	PRecursore IperSpettrale della Missione Applicativa
QGIS	Quantum geographic information system
RaDAR	Radio detecting and ranging
REDD	Reducing emissions from deforestation and forest degradation
RF	Random forest
RS	Remote sensing
SAM	Spectral angle mapper
SAR	Synthetic aperture radar
SAVI	Soil adjusted vegetation index
SDGs	Sustainable development goals
SFF	Spectral feature fitting
SIAD	Shortwave infrared absorption depth
SID	Spectral information diversion
SOC	Soil organic carbon
SPAD	Soil plant analysis development
SVM	Support vector machines
SWIR	Short-wave infrared
TC	Tasseled cap
TIRS	Thermal infrared sensor
TNDVI	Transformed normalized difference vegetation index
UAV	Unmanned airborne vehicle
UNEP	United Nations Environment Programme
UNESCO	United Nations Educational, Scientific and Cultural Organization
UNFCCC	United Nations Framework Convention on Climate Change
USAID	United States Agency for International Development
VNIR	Visible and near-infrared
XGB	Extreme gradient boosting

Introduction to Mangrove Ecosystem and Earth Observation

Introduction to Remote Sensing and GIS in Mangrove Monitoring and Management

Prashant K. Srivastava, Bhawana Sharma, Harikesh Singh, and Sanjeev Kumar Srivastava

Abstract This chapter provides an overview of the role of hyperspectral, multispectral, and Light Detection and Ranging (LiDAR) data in mangrove species identification, discusses methodological frameworks for achieving reliable classification results, and presents case studies demonstrating their practical application in real-world mangrove conservation and restoration efforts. This chapter will also provide the state-of-the-art techniques in mangrove management through sensors installed on various platforms such as ground, airborne to spaceborne and geospatial technology. The chapter further incorporates case studies that illustrate the practical application of these methods in conservation and restoration contexts, thereby demonstrating their relevance to real-world management challenges.

Keywords Geospatial technology · Earth Observation · LiDAR · SAR · Optical

P. K. Srivastava (✉)
Remote Sensing Laboratory, Institute of Environment and Sustainable Development, Banaras Hindu University, Varanasi, India

Geospatial Analytics for Conservation and Management, School of Science Technology and Engineering (SSTE), University of the Sunshine Coast, Sippy Downs, QLD, Australia
e-mail: prashant.iesd@bhu.ac.in

B. Sharma
Remote Sensing Laboratory, Institute of Environment and Sustainable Development, Banaras Hindu University, Varanasi, India

H. Singh · S. K. Srivastava
Geospatial Analytics for Conservation and Management, School of Science Technology and Engineering (SSTE), University of the Sunshine Coast, Sippy Downs, QLD, Australia
e-mail: Sanjeev.Srivastava@usc.edu.au

S. K. Srivastava et al. (eds.), *Earth Observation for Monitoring Mangrove Ecosystems*, https://doi.org/10.1007/978-3-032-16668-5_1

1 Introduction

Mangrove forests are among the most biologically diverse coastal ecosystems worldwide, supporting an exceptionally rich variety of plant and animal species (Ewel et al., 1998). Their high ecological value stems from their roles in coastal protection, carbon sequestration, nursery habitats for marine life, and overall biodiversity maintenance (Barbier et al., 2011). Economically, they contribute to fisheries, tourism, and shoreline stability (Corral & Schling, 2017). Accurate identification of species composition within mangrove ecosystems (ME) is critical for informed decision-making in conservation, sustainable management, and the design of effective restoration programs (Sahana et al., 2022).

Multispectral remote sensing, particularly from high-resolution satellites and uncrewed aerial vehicle (UAV) platforms, offers broader spatial coverage and cost-effective monitoring, making it suitable for large-scale mapping and time-series analysis (Singh et al., 2023). With the advanced hyperspectral remote sensing (HRS), it has demonstrated superior capability for mangrove species discrimination compared to many other remote sensing techniques—owing to its fine spectral resolution (Pandey et al., 2019). Light Detection and Ranging (LiDAR) adds a complementary three-dimensional perspective by capturing canopy height, vertical structure, and terrain information, which enhances species classification when integrated with spectral data (Salum et al., 2020; Srivastava et al., 2022b; Penglase et al., 2023).

Recent advancements have seen the integration of traditional classification methods with machine learning algorithms—such as Random Forest (RF), support vector machines, and deep learning networks—applied to combined hyperspectral, multispectral, and LiDAR datasets for MEs monitoring and management (Li et al., 2021; Chen et al., 2025). The fusion of multisensory datasets improves the robustness and accuracy of species discrimination in mangrove forests by leveraging both spectral and structural information (Ilamathi & Chidambaram, 2025). Findings emphasized that vegetation indices derived from high spectral resolution data, when combined with ground measurements, are valuable tools for reliable mangrove biomass estimation, supporting forest monitoring and sustainable management (Heumann, 2011; Tran et al., 2022).

In view of the above, this chapter provides a comprehensive overview of advanced remote sensing approaches for mangrove monitoring and species identification. It examines the use of hyperspectral and multispectral imagery, LiDAR, and synthetic aperture radar (SAR), with attention to their individual strengths and complementary roles in capturing structural, spectral, and biophysical attributes of MEs. Methodological frameworks for ensuring accurate and reliable classification results are discussed, ranging from feature extraction and machine learning to data fusion techniques. The chapter also draws on case studies that demonstrate how these methods are applied in practical conservation and restoration initiatives, offering insights into challenges and lessons learned. In addition, it reviews state-of-the-art tools for mangrove management, focusing on the integration of geospatial technologies and sensors deployed across multiple platforms—including ground-based systems, drones and other airborne instruments, as well as satellite missions—illustrating how multi-platform observations can support effective monitoring, sustainable management, and long-term resilience of mangrove forests.

2 Geospatial Technology in Mangrove Monitoring and Management

Geographic Information Systems (GIS) are essential for mapping, monitoring, and managing mangrove ecosystems by integrating diverse Earth Observation datasets from satellites and field surveys (Krause et al., 2004). GIS is a computer-based framework for capturing, storing, managing, analyzing, and visualizing spatial (location-based) data (Chang, 2012). It integrates geographic coordinates with attribute information, allowing users to examine relationships, patterns, and trends in space and time (Mitchell, 1999). GIS combines data from maps, satellite imagery, surveys, sensors, and other sources to produce layered information that can be queried, modeled, and displayed in various formats, such as thematic maps and 3D visualizations (Srivastava et al., 2022a). This integration of spatial and nonspatial data supports informed decision-making in diverse fields, including environmental management, urban planning, disaster response, agriculture, and resource monitoring (Srivastava et al., 2013, 2019a).

In mangrove mapping, GIS enables precise boundary delineation, species distribution mapping, and change detection over time, supporting assessments of degradation, regeneration, and overall ecosystem health (Gnanappazham et al., 2021). It also facilitates species distribution modeling using environmental variables and helps identify priority areas for conservation or restoration (Wilson et al., 2011). By producing thematic maps, spatial statistics, and decision-support tools, GIS provides a comprehensive framework for evidence-based coastal management and climate adaptation strategies in mangrove regions (Johnson & Gillingham, 2005).

With GIS, often Analytic Hierarchy Process (AHP)-Multi-Criteria Evaluation (MCE) is combined for suitability analysis and mangrove ecosystem management. MCE is quite an effective technique for forest management, initially developed for business but now widely used in a variety of environmental applications. AHP-MCE is generally used for evaluating the effects of forestry interventions on ecosystem service provision. The Analytic Hierarchy Process (AHP), is a widely used multi-criteria evaluation (MCE) method that assigns weights to decision-making criteria through pairwise comparisons (Saaty & Wind, 1980). In this approach, the relative priorities of criteria are derived from the principal eigenvector (e) of the comparison matrix (M), corresponding to its maximum eigenvalue (λ_{max}) (Eq. 1). The normalized eigenvector provides the weight vector, representing the relative importance of each criterion in the decision process (Saaty, 2003).

$$Me = \lambda_{max} e \tag{1}$$

To ensure reliability, the consistency of the judgments is measured using the Consistency Index (CI) (Eq. 2) and consistency ratio (Fernandez Diaz et al., 2016) (Eq. 3). The CI is calculated from λ_{max} and the number of criteria (n), while the Consistency Ratio (CR) is obtained by comparing CI with the Random Inconsistency Index (RI):

$$\text{CR} = \text{CI} / \text{RI} \tag{2}$$

$$\text{CI} = (\lambda_{max} - n) / (n - 1) \tag{3}$$

A CR value below 0.1 indicates acceptable consistency, whereas values above this threshold suggest inconsistency in the pairwise comparisons, requiring reassessment (Srivastava et al., 2012). This framework ensures both precision and transparency in deriving weights for multi-criteria decision-making.

Multi-Criteria Decision Analysis (MCDA), a ranking-based, has been effectively applied to mangrove forest management and planning. Sahraei et al. (2024) used a hybrid Fuzzy-DEMATEL-ANP (FDANP) model based on several remote sensing and GIS datasets to identify suitable locations for mangrove plantations in Iran. Srivastava et al. (2019b) applied the Multi-Criteria Decision Making (MCDM) approach by determining the relative importance of different criteria and providing the ranking to the important parameters. Abd-El Monsef et al. (2017) used AHP-MCE-based site suitability analysis for delineating existing mangroves stands and alternate sites for mangroves plantation. Mohammadi and Limaei (2018) published a study at Iranian Caspian Forest and used multiple criteria decision-making approaches, Analytic Hierarchy Process (AHP) and Network Analysis (ANP) to prioritize the indicators of sustainable forest management. Forest fragmentation and prediction of land cover changes was done using hybrid combination of Fuzzy-AHP-MCE and the Markov Cellular Automata method (Ramachandra et al., 2017; Banerjee et al., 2018).

3 LiDAR-Based Technology in Mangrove Monitoring

3.1 LiDAR Concept

Light Detection and Ranging (LiDAR) is an active remote sensing technology that uses laser pulses to measure distances between a sensor and the Earth's surface (Fernandez Diaz et al., 2016). Airborne LiDAR systems, typically mounted on aircraft or drones, emit thousands of laser pulses per second, recording the time taken for each pulse to return after reflecting off the ground or vegetation (Srivastava et al., 2022b). The distance R is computed using the time-of-flight algorithm:

$$R = \frac{c\Delta t}{2}$$

where c is the speed of light and Δt is the two-way travel time of the pulse. This process generates a dense cloud of three dimensional points (point cloud) $\{x_i, y_i, z_i, I_i\}$, which can be processed to derive high-resolution Digital Terrain Models (DTMs), Digital Surface Models (DSMs), and Canopy Height Models (CHMs) (Olive et al., 2020). A common algorithm to compute canopy height is:

$$\text{CHM}(x,y) = \text{DSM}(x,y) - \text{DTM}(x,y)$$

Unlike passive optical sensors, LiDAR can penetrate vegetation canopy precise ground elevation data, even in complex and dense environments. Algorithms such as Progressive Morphological Filtering (PMF) or Cloth Simulation Filtering (CSF) are widely used to separate ground points from vegetation points, ensuring accurate modeling of mangrove structure and biomass (Cai et al., 2023).

3.2 Application in Mangrove Ecosystems

In mangrove ecosystems, LiDAR offers a powerful tool for mapping and monitoring structural characteristics such as canopy height, biomass, and species zonation (Wang et al., 2019). The ability of LiDAR to capture fine-scale vertical and horizontal vegetation structure is particularly valuable in mangroves, where dense canopy cover and tidal inundation often limit the effectiveness of optical imagery. LiDAR-derived CHMs can be used to assess mangrove forest health, detect degradation, and quantify aboveground carbon stocks (Salum et al., 2021).

For biomass estimation, allometric models are commonly applied using LiDAR-derived metrics (Fatoyinbo et al., 2018). A simplified equation is:

$$\text{AGB} = a \cdot \left(H^b \cdot D^c\right)$$

where AGB is aboveground biomass (AGB), *H* is canopy height derived from the CHM, *D* is tree diameter (which can be statistically related to LiDAR canopy metrics), and *a*, *b*, *c* are species-specific coefficients.

To separate ground and canopy points in tidal environments, filtering algorithms such as Progressive Morphological Filtering (PMF) or Cloth Simulation Filtering (CSF) are applied (Cai et al., 2023). Once clean ground and canopy surfaces are generated, canopy volume can be estimated as:

$$V = \Sigma\left(\text{CHM}(x,y) \cdot A_{\text{cell}}\right)$$

where CHM (x,y) is the canopy height at each grid cell and A_{cell} is the area of the cell. This volume is then linked with biomass models to estimate carbon storage and sequestration potential.

Furthermore, integration of LiDAR with multispectral or hyperspectral data enhances species classification and supports habitat suitability modeling (Li et al., 2021). These applications are critical for conservation planning, restoration monitoring, and understanding the role of mangroves in coastal protections and blue carbon sequestration.

3.3 LiDAR Studies in Mangrove

LiDAR has been extensively applied in mangrove ecosystems for structural mapping estimation and conservation monitoring. For instance, Salum et al. (2021) demonstrated that LiDAR-derived canopy height models provided highly accurate estimates of aboveground biomass (AGB) across mangroves in Darwin Harbour, Northern Australia, with strong correlations (R^2 up to 0.94) and low root mean square error (RMSE) values, enabling reliable extrapolation to other sites. In Guanabara Bay, Brazil, Pereira and colleagues applied discrete-return LiDAR metrics in combination with Random Forest and Auto-PLS (Partial Least Square) regression to map AGB over an area of ~58 km^2 with high accuracy. Recently, Zhu et al. (2024) integrated UAV LiDAR with WorldView-2 imagery using a GA-ANN wrapper model, which improved biomass estimation accuracy by selecting the most predictive variables (e.g., leaf area index (LAI), spectral indices). Beyond biomass estimation, structural mapping using combined Terrestrial LiDAR Scanning (TLS) and UAV LiDAR in Okinawa, Japan, helped resolve fine-scale microtopography and tree attributes, demonstrating the effectiveness of multi-platform LiDAR integration for detailed 3D characterization of mangrove forests (Niwa et al., 2023).

4 Microwave Remote Sensing-Based Techniques in Mangrove Monitoring

Since the 1990s, publicly available SAR systems such as JERS-1 SAR, SIR-C, and NASA's AIRSAR have offered valuable data for vegetation studies (Lucas et al., 2007). While these missions (Table 1) demonstrated strong potential for mangrove characterization and monitoring, their application remained limited due to funding constraints, in situ data collection challenges, and the lack of long-term archives for consistent change detection. The subsequent generation of SAR missions, including ALOS PALSAR and PALSAR-2, provided improved temporal continuity and global coverage, enabling more systematic assessments of forest and mangrove ecosystems. More recently, new sensors such as NovaSAR, RISAT satellite series, and

Table 1 Key SAR systems relevant for vegetation studies with mission timeline and frequency

SAR mission	Sensor type	Frequenc/ band	Operational period	Application to mangroves
JERS-1 SAR	Satellite	L-band (1.27 GHz)	1992–1998	Regional mangrove mapping, biomass estimation
SIR-C	Space shuttle	L/C/X-band	1994	Experimental mangrove mapping and structural studies
AIRSAR	Airborne	L/C/X-band	1991–2001	High-resolution canopy structure and biomass assessment
ALOS PALSAR	Satellite	L-band	2006–2011	Mangrove mapping, change detection, biomass studies
ALOS-2 PALSAR-2	Satellite	L-band	2014–present	Improved temporal coverage for mangrove monitoring
NovaSAR	Satellite	S-band	2018–present	Coastal vegetation monitoring, mangrove detection
Sentinel-1	Satellite	C-band	2014–present	Operational mangrove mapping, change detection
NISAR	Satellite	L/S-band	2025–present	Operational mangrove mapping, biomass estimation, long-term monitoring

the recently launched NISAR mission are expected to advance operational monitoring through enhanced spatial, temporal, and polarimetric capabilities, offering unprecedented opportunities for large-scale, long-term mangrove monitoring.

For operational applications, accurate information on the status and dynamics of mangroves is vital for shaping national and international policies on conservation, climate change mitigation, and sustainable coastal management. To address this need, the Japan Aerospace Exploration Agency (JAXA), under its Kyoto and Carbon (K&C) initiative, developed the Global Mangrove Watch (GMW) program (Lucas et al., 2014). The GMW provides a systematic and repeatable framework for monitoring mangrove extent and changes at local to global scales (Bunting et al., 2022). This is primarily achieved using L-band synthetic aperture radar (SAR) data, which offers the advantage of consistent observations under all weather conditions, including dense cloud cover common in tropical coastal regions (Ottinger & Kuenzer, 2020). By leveraging SAR's sensitivity to vegetation structure and biomass, these tools support evidence-based policy development, implementation, and evaluation, while also contributing to global initiatives such as REDD+, biodiversity conservation, and climate resilience planning (Ahmed & Glaser, 2016; Aziz et al., 2016).

A study explored the application of phased array-type L-band synthetic aperture radar (PALSAR) data for distinguishing different physiographic types of mangroves with varying forest structural characteristics in the subtropical mangrove region of Cananéia, located on the southern coast of São Paulo, Brazil (de Souza Pereira et al., 2016). Another study carried out in Malaysia to assess the (ALOS-PALSAR estimated backscattering for the aboveground biomass (AGB) estimation of a managed mangrove forest (Hamdan et al., 2014). Their findings showed the potential of PALSAR in capturing structural differences within mangrove ecosystems, offering valuable insights for ecological characterization (Pham et al., 2019b; Lucas et al., 2020). This is particularly important given the growing availability of freely accessible SAR datasets, which can significantly enhance large-scale analysis, long-term monitoring, and conservation planning for mangrove forests (Lu & Wang, 2022).

5 Optical Remote Sensing

5.1 Multispectral Sensing for MEs Monitoring

Multispectral imaging captures the optical wavelength regions in more or less than 10 broad bands. Most multispectral payloads (MSPs) acquire data in 3–4 bands covering a range of 0.3–0.9 μm (Lillesand et al., 2008). MSPs are prevalent across all types of platforms—airborne and spaceborne. These passive sensors possess electronic detectors that record, store, and transmit the radiance data of diverse earth features. However, the quality of their output imagery largely depends upon atmospheric elements—aerosols, water vapor, clouds, and ever-dynamic weather conditions (Pham et al., 2019b). Also, the broad bandwidth and low spectral resolution limit the users to distinguish unique earth features. Nevertheless, the fine spatial and temporal resolution of space-based MSPs facilitate high-quality and frequently available datasets, a very crucial input for time-series and change detection analysis. This explains the vast contribution of multispectral remote sensing in multitude of disciplines.

5.2 Application in Mangrove Monitoring

High temporal and spatial resolution in recently developed advanced multispectral sensors (Fig. 1) is effectively contributing to a vast scope of Earth Observation studies. Output of these sensors are rigorously used in mapping, monitoring, estimating, and predicting multiple attributes of mangrove species around the world. Multispectral imageries are used in mangrove forest classification (Purwanto & Asriningrum, 2019), species identification, chlorophyll content estimation, canopy closure and canopy density measurement (Jensen et al., 1991), and aboveground biomass (Baloloy et al., 2018) prediction in the mangrove ecosystem. Collaborative approach using multispectral images with hyperspectral, microwave, or lidar techniques along with machine learning algorithms is observed in many mangrove vegetation assessments (Gholizadeh et al., 2015; Pham & Brabyn, 2017; Li et al., 2021; Wang et al., 2022; Erandi et al., 2025) nowadays. The resultant accuracy of these collaborations shows pronounced enhancement and offers a cutting-edge strategy in mangrove ecosystem monitoring.

Fig. 1 Operational high-resolution multispectral satellite sensors

5.3 Multispectral Approach in Mangrove Studies

Understanding the species distribution and canopy structure of mangrove forests is critical for the conservation of both flora and fauna within these ecosystems. Li et al. (2021) conducted a detailed study on multilayered mangrove mapping that incorporated overstory and understory detection, along with species classification, using multispectral WorldView-3 data, airborne hyperspectral imagery, and LiDAR point clouds. The results revealed a strong correlation between ground and modeled understory canopy heights, showing that the LiDAR integration can help in better stratifying the multilayered canopies. Importantly, this work generated the first understory map for this ecologically significant and sensitive mangrove habitat, offering valuable insights for conservation and management (Pandey et al., 2019).

Following a similar intent, health monitoring in Mui Ca Mau mangroves (Hai et al., 2022) was implemented using fine-resolution time-series data of SPOT 4 and SPOT 5 satellites. Factors such as diversity, density, canopy width, and fragmentation that are having significant impact on mangrove health were adopted to map the mangrove health in this study. For each factor, spatial metrics—contiguity index, percentage of like adjacencies, patch density and Shannon's evenness index—were calculated and ranked using Analytic Hierarchy Process (AHP). Mangrove health map generated using these variables showed the degrading condition of mangroves due to persistent fragmentation as the mangrove cover keeps expanding in successive years. The study demonstrated the efficacy of the multispectral technique with spatial metrics in mangrove health mapping. Consequently, these results will play crucial role in further decision-making regarding the use of remote approach in mangrove monitoring.

The profound ecological significance of the mangrove ecosystem makes it one of the important habitats to protect, preserve, and conserve. It is important to assess the prevailing conditions of mangrove forests for future implications. Paramanik et al. (2022) have conducted a study of a similar nature. Specific biochemical and biophysical attributes of mangrove forests were explored to understand the dynamics of mangrove forest productivity. Temporal analysis was conducted for leaf area index (LAI), solar-induced chlorophyll fluorescence (SIF), and gross primary productivity (GPP) over Bhitarkanika Wildlife Sanctuary using MODIS data to assess the seasonal pattern of these attributes. In this study, normalized difference vegetation index (NDVI), Normalized Difference Water Index (NDWI), Ratio Vegetation Index (RVI), and Soil Adjusted Vegetation Index (SAVI) were calculated to predict LAI and chlorophyll content on Landsat-8 data products. These prediction models were further enhanced using the Random Forest algorithm. The overall results showed $R = 0.68$ and RMSE = 0.74 for LAI and RMSE = 0.54 mg/g for predicted leaf chlorophyll content (Paramanik et al., 2022).

5.4 Hyperspectral Remote Sensing in ME Monitoring

Hyperspectral imaging refers to the image acquisition in more than hundreds of bands across the optical region of the electromagnetic spectrum. These bands are narrower and contiguous when compared with discrete and broader bands of multispectral sensors (Lillesand et al., 2008). Higher spectral resolution of hyperspectral data contributes to detect unique spectral signature of disparate earth features. Notably, these sensors can accurately discriminate between multiple species using subtle spectral variations. These spectral variations mainly depend upon the physical and chemical constituents of the target object. In the case of vegetation, the biophysical and biochemical components greatly influence the spectral information (Srivastava et al., 2020). Besides the limitations of atmospheric interference, low

Signal to Noise Ratio (SNR), spectral redundancy, and calibration or processing complexities (Bioucas-Dias et al., 2013), data availability for time-series analysis and high spatial resolution will certainly be addressed in the near future by recently launched hyperspectral satellites. Missions like Global Hyperspectral Observation Satellite (GHOSt) constellation, launched by Orbital Sidekick Inc. in 2024, is providing 8 m spatial resolution, 512 bands, and 24 h revisit time, while the Firefly constellation launched in 2025 by Pixxel, an Indian origin satellite, will provide spatial resolution of 5 m in 250 bands with a global coverage and revisit period of 24 h (Pixxel, hyperspectral imaging for Earth Observation). These advancements will broaden the scope of hyperspectral remote sensing which served as a pivotal technology for more than 40 years to conduct comprehensive analysis in multiple domains.

5.5 Application in Mangrove Monitoring

Hyperspectral imaging is employed in assessing diverse attributes of the mangrove ecosystem. Spectral signatures of mangrove vegetation are frequently utilized to identify, discriminate, and map individuals at the species level with comparatively higher accuracy (Wang & Sousa, 2009; Cao et al., 2021; Maurya et al., 2021). Forest type and diversity index (Wang et al., 2022) assessments in mangroves also yielded reliable results in some research papers. Hyperspectral-derived reflectance data, textural features, and VIs often incorporated to detect pest or diseased infestation (Jiang et al., 2021). These attributes are greatly enhanced using an integrated machine learning approach to monitor overall health conditions (Vidhya et al., 2014) of mangrove vegetation. Many researchers used hyperspectral datasets in ever-dynamic pigment-based analysis (Axelsson et al., 2013; George et al., 2019; Zhao et al., 2019) at leaf and canopy-level, primarily investigating chlorophyll (Flores-de-Santiago et al., 2013), nitrogen (Fauzi et al., 2013), phosphorus, and leaf water contents. Biomass and carbon stock estimation (Pandey et al., 2019; Pham et al., 2019a; Anand et al., 2020) was also carried out across the mangrove stands for large-scale and robust predictions.

5.6 Hyperspectral Approach in Mangrove Studies

Pandey et al. (2019) utilized Earth Observation-1 (EO-1) Hyperion hyperspectral data to map species distribution and species-specific biomass in the Bhitarkanika Mangrove Forest. The study aimed at accurately estimating forest biomass and validated the results against field-based measurements. In another study by Behera et al. (2021), the composition of mangrove species within the Bhitarkanika Wildlife Sanctuary, India, was assessed using AVIRIS-NG and Sentinel-2 datasets through their red-edge spectral bands and chlorophyll absorption features. They concluded that the Random Forest (RF) algorithm was best for classifying the three major mangrove species, i.e., *Heritiera fomes*, *Excoecaria agallocha*, and *Avicennia officinalis*. Several allometric models were proposed, demonstrating strong positive correlations between estimated and observed biomass (Van Vinh et al., 2019; Rahman et al., 2021). Through the models, the study generated spatial distribution maps of mangrove species and their associated biomass, highlighting the effectiveness of integrating geospatial analysis with field inventory for biomass assessment in tropical MEs (Kamal & Phinn, 2011; Pirasteh et al., 2024).

Pigment analysis of mangrove vegetation for productivity and stress-related assessment was implemented by George et al. (2019) as he predicted the most effective chlorophyll index for the mangroves of Andaman Island using satellite-based hyperspectral data. Reflectance data of EO-1 Hyperion were used to compute Simple Ratio (SR), modified simple ratio (MSR), normalized difference vegetation index (NDVI), and nonlinear index (NLI) in the study. These hyperspectral indices are generated using multiple combinations of narrow bands lying in the range of specific wavelengths. Sensitivity analysis on predicted indices showed that 549 nm and 559 nm of the green wavelength region and 702 nm, 722 nm, 742 nm, and 763 nm of the red-edge wavelength region are most suited to predict chlorophyll in the target mangrove cover. Findings of the study indicate that the SR index showed the highest correlation to the manually calculated chlorophyll values with $R^2 = 0.75$, RMSE = 0.60 at $p < 0.05$. The study signifies the superiority of the hyperspectral technique to generate best performing novel indices.

In the series of pigment analysis, an integrated approach of multiple attributes for increased performance is more preferred in the present time. Wang et al. (2020) adopted this approach to study the leaf biochemistry of a mangrove forest. With the aid of a handheld hyperspectral spectroradiometer, leaf chlorophyll index, leaf water content, nitrogen concentration, and phosphorus concentration of mangrove vegetation were derived using reflectance data of 135 samples. These biochemical parameters of mangrove forests were estimated using the continuous wavelet transform (CWT) tool, Random Forest Regressor (RFR) and three data-splitting methods. The study showed that RFR model with sample set partitioning based on the joint x-y distance (SPXY) method provides higher accuracy than other data splitting methods. Estimated chlorophyll index and nitrogen concentration showed better accuracy than the phosphorous and water content. It was concluded that species information and soil properties merged with CWT and SPXY method significantly improve the biochemical attributes predicted in mangrove stands.

6 Airborne Techniques

Airborne remote sensing in MEs monitoring dates back to the early 1980s. Terchunian et al. (1986) used aerial photographic interpretation techniques to monitor the decreasing extent of mangroves from 1966 to 1982 due to continuously increasing shrimp pond developments. The sensor captured wavelengths ranging from 1.2 to 2.4 μm in 128 bands. He reported 16% decline in mangrove stretch with a subsequent 27% rise in mangrove areas being transformed to shrimp pond development. The study highlighted the significant loss of mangrove biodiversity which instigated country-level mangrove mapping initiative for future management of MEs.

From RGB images, captured using analog and digital cameras, to image acquisition using multispectral, hyperspectral, thermal, and LiDAR sensors, the airborne remote sensing techniques are remarkably developed in the past four decades. This is substantiated by Chen et al. (2025), who conducted a study using a DJI Phantom 4RTK multispectral drone, which integrates a visible band with five monochromatic bands, to acquire baseline data in October 2023 over the mangrove area of Jinniu Island. The study compared the performance of the proposed MangroveNet model against other deep learning architectures, including SegNet, UNet, and DeepUNet. Results indicated that MangroveNet achieved a stronger generalization capability and delivered more accurate mangrove extraction than the other models.

A recent study by de Oliveira et al. (2024) quantifies individual trees in a dense mangrove cover using three different individual tree detection (ITD) methods on fine-resolution multispectral data retrieved by an uncrewed aerial vehicle (UAV). The two of the ITD methods used multispectral image segmentation, while the third one relied on variable surface height (Gaussian) model. Validation of predicted results was performed using three rules—comparison of plot-level tree counts, spatial relationship of predicted canopy to referenced polygon canopy, and referenced point canopy. In conclusion, both segmentation and Gaussian method attain an acceptable quality limit (AQL) of 15% while considering the plot size in the flight area. The study emphasized that spatially calibrated plots will increase the model accuracy which consequently contributes to biomass and carbon prediction of MEs.

Use of hyperspectral sensors for airborne remote sensing commenced in 1982 when the first flight of Airborne Imaging Spectrometer (AIS), mounted on an aircraft, was executed by Jet Propulsion Laboratory (JPL). Now, Airborne Visible-Infrared Imaging Spectrometer-Next Generation (AVIRIS-NG), developed by JPL, is rigorously used in mangrove monitoring using hyperspectral technique. Hati et al. (2021) conducted health monitoring in estuarine mangrove forests of Lothian Island using hyperspectral imageries acquired by AVIRIS-NG. Nine VIs corresponding to almost all health aspects were computed. Weighted overlay analysis of derived VIs contributed to stress map generation for the years 2016 and 2018. Sentinel-2 derived spectral metric—Discriminant Normalized Vegetation Index (DNVI)—of the same years, used in this study to validate the previously generated stress maps. The results showed increasing stress conditions in the whole mangrove stretch. The authors described that the plantation of new species and species-specific stress tolerance were the major causes of the existing mangrove cover.

7 Conclusion

Mangrove forests are vital yet vulnerable ecosystems, and their effective monitoring is essential for conservation. Earth observation technologies provide consistent, large-scale insights into mangrove extent, condition, and change, enabling informed management and restoration efforts. This chapter discusses the roles of various sensors including hyperspectral and multispectral, LiDAR, and synthetic aperture radar (SAR) for monitoring and management of MEs. With the advent of advanced sensors onboard AVIRIS-NG, Sentinel series, Landsat series, and microwave missions, it is now possible to provide a more accurate distribution of MEs and their management. Recently, L and S bands NASA-ISRO SAR mission is also launched to provide more detailed observations of the MEs through high penetration wavelengths. Several case studies are also provided in this chapter to provide a clearer understanding of the processes and approaches that can be applied for improved monitoring of the MEs.

References

Abd-El Monsef, H., Hassan, M. A., & Shata, S. (2017). Using spatial data analysis for delineating existing mangroves stands and siting suitable locations for mangroves plantation. *Computers and Electronics in Agriculture, 141*, 310–326.

Ahmed, N., & Glaser, M. (2016). Coastal aquaculture, mangrove deforestation and blue carbon emissions: Is REDD+ a solution? *Marine Policy, 66*, 58–66.

Anand, A., Pandey, P. C., Petropoulos, G. P., Pavlides, A., Srivastava, P. K., Sharma, J. K., & Malhi, R. K. M. (2020). Use of hyperion for mangrove forest carbon stock assessment in Bhitarkanika forest reserve: A contribution towards blue carbon initiative. *Remote Sensing, 12*, 597.

Axelsson, C., Skidmore, A. K., Schlerf, M., Fauzi, A., & Verhoef, W. (2013). Hyperspectral analysis of mangrove foliar chemistry using PLSR and support vector regression. *International Journal of Remote Sensing, 34*, 1724–1743.

Aziz, A. A., Thomas, S., Dargusch, P., & Phinn, S. (2016). Assessing the potential of REDD+ in a production mangrove forest in Malaysia using stakeholder analysis and ecosystem services mapping. *Marine Policy, 74*, 6–17.

Baloloy, A. B., Blanco, A. C., Candido, C. G., Argamosa, R. J. L., Dumalag, J. B. L. C., Dimapilis, L. L. C., & Paringit, E. C. (2018). Estimation of mangrove forest aboveground biomass using multispectral bands, vegetation indices and biophysical variables derived from optical satellite imageries: Rapideye, planetscope and sentinel-2. *ISPRS Annals of the Photogrammetry, Remote Sensing and Spatial Information Sciences, 4*, 29–36.

Banerjee, R., Srivastava, P. K., Pike, A., & Petropoulos, G. P. (2018). Identification of painted rock-shelter sites using GIS integrated with a decision support system and fuzzy logic. *ISPRS International Journal of Geo-Information, 7*, 326.

Barbier, E. B., Hacker, S. D., Kennedy, C., Koch, E. W., Stier, A. C., & Silliman, B. R. (2011). The value of estuarine and coastal ecosystem services. *Ecological Monographs, 81*, 169–193.

Behera, M. D., Barnwal, S., Paramanik, S., Das, P., Bhattyacharya, B. K., Jagadish, B., Roy, P. S., Ghosh, S. M., & Behera, S. K. (2021). Species-level classification and mapping of a mangrove forest using random forest—Utilisation of AVIRIS-NG and sentinel data. *Remote Sensing, 13*, 2027.

Bioucas-Dias, J. M., Plaza, A., Camps-Valls, G., Scheunders, P., Nasrabadi, N., & Chanussot, J. (2013). Hyperspectral remote sensing data analysis and future challenges. *IEEE Geoscience and Remote Sensing Magazine, 1*, 6–36.

Bunting, P., Rosenqvist, A., Hilarides, L., Lucas, R. M., Thomas, N., Tadono, T., Worthington, T. A., Spalding, M., Murray, N. J., & Rebelo, L.-M. (2022). Global mangrove extent change 1996–2020: Global mangrove watch version 3.0. *Remote Sensing, 14*, 3657.

Cai, S., Yu, S., Hui, Z., & Tang, Z. (2023). ICSF: An improved cloth simulation filtering algorithm for airborne LiDAR data based on morphological operations. *Forests, 14*, 1520.

Cao, J., Liu, K., Zhuo, L., Liu, L., Zhu, Y., & Peng, L. (2021). Combining UAV-based hyperspectral and LiDAR data for mangrove species classification using the rotation forest algorithm. *International Journal of Applied Earth Observation and Geoinformation, 102*, 102414.

Chang, K.-T. (2012). *Introduction to geographic information systems*. McGraw-Hill.

Chen, X., Zhang, X., Zhuang, C., Dai, X., Kong, L., Xie, Z., & Hu, X. (2025). An approach for detecting mangrove areas and mapping species using multispectral drone imagery and deep learning. *Sensors (Basel), 25*, 2540.

Corral, L. R., & Schling, M. (2017). The impact of shoreline stabilization on economic growth in small Island developing states. *Journal of Environmental Economics and Management, 86*, 210–228.

de Oliveira, P. A., Conti, L. A., Neto, F. C. N., Barcellos, R. L., & Cunha-Lignon, M. (2024). Mangrove individual tree detection based on the uncrewed aerial vehicle multispectral imagery. *Remote Sensing Applications: Society and Environment, 33*, 101100.

de Souza Pereira, F. R., Kampel, M., & Cunha-Lignon, M. (2016). Mangrove vegetation structure in Southeast Brazil from phased array L-band synthetic aperture radar data. *Journal of Applied Remote Sensing, 10*, 036021.

Erandi, M.-M., Rubicel, T.-A., Jose, T.-V., & Patricia, M.-C. (2025). An approach for accurate identification and monitoring of species in mangrove forests based on multi-source spectral data and deep learning. *Ecological Informatics, 85*, 102961.

Ewel, K., Twilley, R., & Ong, J. (1998). Different kinds of mangrove forests provide different goods and services. *Global Ecology & Biogeography Letters, 7*, 83–94.

Fatoyinbo, T., Feliciano, E. A., Lagomasino, D., Lee, S. K., & Trettin, C. (2018). Estimating mangrove aboveground biomass from airborne LiDAR data: A case study from the Zambezi River delta. *Environmental Research Letters, 13*, 025012.

Fauzi, A., Skidmore, A. K., van Gils, H., Schlerf, M., & Heitkönig, I. M. (2013). Shrimp pond effluent dominates foliar nitrogen in disturbed mangroves as mapped using hyperspectral imagery. *Marine Pollution Bulletin, 76*, 42–51.

Fernandez Diaz, J. C., Carter, W. E., Shrestha, R. L., & Glennie, C. L. (2016). Lidar remote sensing. In *Handbook of satellite applications* (pp. 1–52). Springer.

Flores-de-Santiago, F., Kovacs, J. M., & Flores-Verdugo, F. (2013). The influence of seasonality in estimating mangrove leaf chlorophyll-a content from hyperspectral data. *Wetlands Ecology and Management, 21*, 193–207.

George, R., Padalia, H., Sinha, S., & Kumar, A. S. (2019). Evaluating sensitivity of hyperspectral indices for estimating mangrove chlorophyll in Middle Andaman Island, India. *Environmental Monitoring and Assessment, 191*, 785.

Gholizadeh, H., Robeson, S. M., & Rahman, A. F. (2015). Comparing the performance of multispectral vegetation indices and machine-learning algorithms for remote estimation of chlorophyll content: A case study in the Sundarbans mangrove forest. *International Journal of Remote Sensing, 36*, 3114–3133.

Gnanappazham, L., Prasad, K. A., & Dadhwal, V. (2021). Geospatial tools for mapping and monitoring coastal mangroves. In *Mangroves: Ecology, biodiversity and management* (pp. 475–551). Springer.

Hai, P. M., Tinh, P. H., Son, N. P., Thuy, T. V., Hong Hanh, N. T., Sharma, S., Hoai, D. T., & Duy, V. C. (2022). Mangrove health assessment using spatial metrics and multi-temporal remote sensing data. *PLoS One, 17*, e0275928.

Hamdan, O., Aziz, H. K., & Hasmadi, I. M. (2014). L-band ALOS PALSAR for biomass estimation of Matang Mangroves, Malaysia. *Remote Sensing of Environment, 155*, 69–78.
Hati, J. P., Goswami, S., Samanta, S., Pramanick, N., Majumdar, S. D., Chaube, N. R., Misra, A., & Hazra, S. (2021). Estimation of vegetation stress in the mangrove forest using AVIRIS-NG airborne hyperspectral data. *Modeling Earth Systems and Environment, 7*, 1877–1889.
Heumann, B. W. (2011). Satellite remote sensing of mangrove forests: Recent advances and future opportunities. *Progress in Physical Geography, 35*, 87–108.
Ilamathi, P., & Chidambaram, S. (2025). Integration of hyperspectral imaging and deep learning for sustainable mangrove management and sustainable development goals assessment. *Wetlands, 45*, 9.
Jensen, J. R., Lin, H., Yang, X., Ramsey, E., III, Davis, B. A., & Thoemke, C. W. (1991). The measurement of mangrove characteristics in southwest Florida using SPOT multispectral data. *Geocarto International, 6*, 13–21.
Jiang, X., Zhen, J., Miao, J., Zhao, D., Wang, J., & Jia, S. (2021). Assessing mangrove leaf traits under different pest and disease severity with hyperspectral imaging spectroscopy. *Ecological Indicators, 129*, 107901.
Johnson, C. J., & Gillingham, M. P. (2005). An evaluation of mapped species distribution models used for conservation planning. *Environmental Conservation, 32*, 117–128.
Kamal, M., & Phinn, S. (2011). Hyperspectral data for mangrove species mapping: A comparison of pixel-based and object-based approach. *Remote Sensing, 3*, 2222–2242.
Krause, G., Bock, M., Weiers, S., & Braun, G. (2004). Mapping land-cover and mangrove structures with remote sensing techniques: A contribution to a synoptic GIS in support of coastal management in North Brazil. *Environmental Management, 34*, 429–440.
Li, Q., Wong, F. K. K., & Fung, T. (2021). Mapping multi-layered mangroves from multispectral, hyperspectral, and LiDAR data. *Remote Sensing of Environment, 258*, 112403.
Lillesand, T., Kiefer, R., & Chipman, J. (2008). Multispectral, thermal, and hyperspectral sensing. In *Remote sensing and image interpretation* (pp. 325–391).
Lu, Y., & Wang, L. (2022). The current status, potential and challenges of remote sensing for large-scale mangrove studies. *International Journal of Remote Sensing, 43*, 6824–6855.
Lucas, R. M., Mitchell, A. L., Rosenqvist, A., Proisy, C., Melius, A., & Ticehurst, C. (2007). The potential of L-band SAR for quantifying mangrove characteristics and change: Case studies from the tropics. *Aquatic Conservation: Marine and Freshwater Ecosystems, 17*, 245–264.
Lucas, R., Rebelo, L.-M., Fatoyinbo, L., Rosenqvist, A., Itoh, T., Shimada, M., Simard, M., Souza-Filho, P. W., Thomas, N., & Trettin, C. (2014). Contribution of L-band SAR to systematic global mangrove monitoring. *Marine and Freshwater Research, 65*, 589–603.
Lucas, R., Van De Kerchove, R., Otero, V., Lagomasino, D., Fatoyinbo, L., Omar, H., Satyanarayana, B., & Dahdouh-Guebas, F. (2020). Structural characterisation of mangrove forests achieved through combining multiple sources of remote sensing data. *Remote Sensing of Environment, 237*, 111543.
Maurya, K., Mahajan, S., & Chaube, N. (2021). Remote sensing techniques: Mapping and monitoring of mangrove ecosystem—A review. *Complex & Intelligent Systems, 7*, 2797–2818.
Mitchell, A. (1999). *The ESRI guide to GIS analysis: Geographic patterns & relationships*. ESRI, Inc..
Mohammadi, Z., & Limaei, S. (2018). Multiple criteria decision-making approaches for forest sustainability (Case study: Iranian Caspian forests). *Forest Research, 7*, 1–7.
Niwa, H., Ise, H., & Kamada, M. (2023). Suitable LiDAR platform for measuring the 3D structure of mangrove forests. *Remote Sensing, 15*, 1033.
Olive, K., Lewis, T., Ghaffariyan, M. R., & Srivastava, S. K. (2020). Comparing canopy height estimates from satellite-based photogrammetry, airborne laser scanning and field measurements across Australian production and conservation eucalypt forests. *Journal of Forest Research, 25*, 108–112.
Ottinger, M., & Kuenzer, C. (2020). Spaceborne L-band synthetic aperture radar data for geoscientific analyses in coastal land applications: A review. *Remote Sensing, 12*, 2228.

Pandey, P. C., Anand, A., & Srivastava, P. K. (2019). Spatial distribution of mangrove forest species and biomass assessment using field inventory and earth observation hyperspectral data. *Biodiversity and Conservation, 28*, 2143–2162.

Paramanik, S., Varghese, R., Behera, M. D., Barnwal, S., Behera, S. K., & Bhattyacharya, B. K. (2022). Photosynthetic variables estimation in a mangrove forest. In *Advances in remote sensing for forest monitoring* (pp. 126–149).

Penglase, K., Lewis, T., & Srivastava, S. K. (2023). A new approach to estimate fuel budget and wildfire hazard assessment in commercial plantations using drone-based photogrammetry and image analysis. *Remote Sensing, 15*, 2621.

Pham, L. T., & Brabyn, L. (2017). Monitoring mangrove biomass change in Vietnam using SPOT images and an object-based approach combined with machine learning algorithms. *ISPRS Journal of Photogrammetry and Remote Sensing, 128*, 86–97.

Pham, T. D., Xia, J., Ha, N. T., Bui, D. T., Le, N. N., & Takeuchi, W. (2019a). A review of remote sensing approaches for monitoring blue carbon ecosystems: Mangroves, seagrasses and salt marshes during 2010–2018. *Sensors, 19*, 1933.

Pham, T. D., Yokoya, N., Bui, D. T., Yoshino, K., & Friess, D. A. (2019b). Remote sensing approaches for monitoring mangrove species, structure, and biomass: Opportunities and challenges. *Remote Sensing, 11*, 230.

Pirasteh, S., Mafi-Gholami, D., Li, H., Fang, Z., Nouri-Kamari, A., & Khorrami, B. (2024). Precision in mapping and assessing mangrove biomass: Insights from the Persian Gulf coasts. *International Journal of Applied Earth Observation and Geoinformation, 128*, 103769.

Purwanto, A. D., & Asriningrum, W. (2019). Identification of mangrove forests using multispectral satellite imageries. *International Journal of Remote Sensing and Earth Sciences, 16*, 63–86.

Rahman, M. S., Donoghue, D. N., Bracken, L. J., & Mahmood, H. (2021). Biomass estimation in mangrove forests: A comparison of allometric models incorporating species and structural information. *Environmental Research Letters, 16*, 124002.

Ramachandra, T., Setturu, B., Rajan, K., & Subash Chandran, M. (2017). Modelling the forest transition in central Western Ghats, India. *Spatial Information Research, 25*, 117–130.

Saaty, T. L. (2003). Decision-making with the AHP: Why is the principal eigenvector necessary. *European Journal of Operational Research, 145*, 85–91.

Saaty, T. L., & Wind, Y. (1980). Marketing applications of the analytic hierarchy process. *Management Science, 26*, 641–658.

Sahana, M., Areendran, G., & Sajjad, H. (2022). Assessment of suitable habitat of mangrove species for prioritizing restoration in coastal ecosystem of Sundarban Biosphere Reserve, India. *Scientific Reports, 12*, 20997.

Sahraei, R., Ghorbanian, A., Kanani-Sadat, Y., Jamali, S., & Homayouni, S. (2024). Mangrove plantation suitability mapping by integrating multi criteria decision making geospatial approach and remote sensing data. *Geo-spatial Information Science, 27*, 1290–1308.

Salum, R. B., Souza-Filho, P. W. M., Simard, M., Silva, C. A., Fernandes, M. E., Cougo, M. F., do Nascimento Junior, W., & Rogers, K. (2020). Improving mangrove above-ground biomass estimates using LiDAR. *Estuarine, Coastal and Shelf Science, 236*, 106585.

Salum, R. B., Robinson, S. A., & Rogers, K. (2021). A validated and accurate method for quantifying and extrapolating mangrove above-ground biomass using LiDAR data. *Remote Sensing, 13*, 2763.

Singh, P., Srivastava, P. K., Verrelst, J., Mall, R., Rivera, J. P., Dugesar, V., & Prasad, R. (2023). High resolution retrieval of leaf chlorophyll content over Himalayan pine forest using Visible/IR sensors mounted on UAV and radiative transfer model. *Ecological Informatics, 75*, 102099.

Srivastava, P. K., Han, D., Gupta, M., & Mukherjee, S. (2012). Integrated framework for monitoring groundwater pollution using a geographical information system and multivariate analysis. *Hydrological Sciences Journal, 57*, 1453–1472.

Srivastava, P. K., Singh, S. K., Gupta, M., Thakur, J. K., & Mukherjee, S. (2013). Modeling impact of land use change trajectories on groundwater quality using remote sensing and GIS. *Environmental Engineering and Management Journal, 12*, 2343.

Srivastava, P. K., Pandey, P. C., Petropoulos, G. P., Kourgialas, N. N., Pandey, V., & Singh, U. (2019a). GIS and remote sensing aided information for soil moisture estimation: A comparative study of interpolation techniques. *Resources, 8*, 70.
Srivastava, P. K., Petropoulos, G. P., Gupta, M., Singh, S. K., Islam, T., & Loka, D. (2019b). Deriving forest fire probability maps from the fusion of visible/infrared satellite data and geospatial data mining. *Modeling Earth Systems and Environment, 5*, 627–643.
Srivastava, P. K., Malhi, R. K. M., Pandey, P. C., Anand, A., Singh, P., Pandey, M. K., & Gupta, A. (2020). Revisiting hyperspectral remote sensing: Origin, processing, applications and way forward. In *Hyperspectral remote sensing* (pp. 3–21). Elsevier.
Srivastava, S. K., Scott, G., & Rosier, J. (2022a). Use of geodesign tools for visualisation of scenarios for an ecologically sensitive area at a local scale. *Environment and Planning B: Urban Analytics and City Science, 49*, 23–40.
Srivastava, S. K., Seng, K. P., Ang, L. M., Pachas, A. N. A., & Lewis, T. (2022b). Drone-based environmental monitoring and image processing approaches for resource estimates of private native forest. *Sensors, 22*, 7872.
Terchunian, A., Klemas, V., Segovia, A., Alvarez, A., Vasconez, B., & Guerrero, L. (1986). Mangrove mapping in Ecuador: The impact of shrimp pond construction. *Environmental Management, 10*, 345–350.
Tran, T. V., Reef, R., & Zhu, X. (2022). A review of spectral indices for mangrove remote sensing. *Remote Sensing, 14*, 4868.
Van Vinh, T., Marchand, C., Linh, T. V. K., Vinh, D. D., & Allenbach, M. (2019). Allometric models to estimate above-ground biomass and carbon stocks in Rhizophora apiculata tropical managed mangrove forests (Southern Viet Nam). *Forest Ecology and Management, 434*, 131–141.
Vidhya, R., Vijayasekaran, D., Ahamed Farook, M., Jai, S., Rohini, M., & Sinduja, A. (2014). Improved classification of mangroves health status using hyperspectral remote sensing data. *The International Archives of the Photogrammetry, Remote Sensing and Spatial Information, 40*, 667–670.
Wang, L., & Sousa, W. P. (2009). Distinguishing mangrove species with laboratory measurements of hyperspectral leaf reflectance. *International Journal of Remote Sensing, 30*, 1267–1281.
Wang, D., Wan, B., Qiu, P., Zuo, Z., Wang, R., & Wu, X. (2019). Mapping height and aboveground biomass of mangrove forests on Hainan Island using UAV-LiDAR sampling. *Remote Sensing, 11*, 2156.
Wang, J., Xu, Y., & Wu, G. (2020). The integration of species information and soil properties for hyperspectral estimation of leaf biochemical parameters in mangrove forest. *Ecological Indicators, 115*, 106467.
Wang, D., Qiu, P., Wan, B., Cao, Z., & Zhang, Q. (2022). Mapping α-and β-diversity of mangrove forests with multispectral and hyperspectral images. *Remote Sensing of Environment, 275*, 113021.
Wilson, C. D., Roberts, D., & Reid, N. (2011). Applying species distribution modelling to identify areas of high conservation value for endangered species: A case study using Margaritifera margaritifera (L.). *Biological Conservation, 144*, 821–829.
Zhao, Y., Yan, C., Lu, S., Wang, P., Qiu, G. Y., & Li, R. (2019). Estimation of chlorophyll content in intertidal mangrove leaves with different thicknesses using hyperspectral data. *Ecological Indicators, 106*, 105511.
Zhu, Y., Myint, S. W., Liu, K., Liu, L., & Cao, J. (2024). Integration of UAV LiDAR and WorldView-2 images for modeling mangrove aboveground biomass with GA-ANN wrapper. *Ecological Processes, 13*, 85.

Different Roles of Mangrove Ecosystems and Their Significance

Md Rezaul Karim, Md. Shamim Reza Saimun, Md Shydul Amin, Md. Sahinur Islam Fahim, Fahmida Sultana, Dhananjay Barman, Sanjeev Kumar Srivastava , Sharif A. Mukul, and Mohammed A. S. Arfin-Khan

Abstract Mangrove forests, covering ~167,000–181,000 km^2 across 123 countries, are among the most carbon-dense and functionally diverse ecosystems on Earth. Their capacity to sequester 5.5–6.5 Mg C ha^{-1} year^{-1} in deltaic settings and store up to 2200 Mg C ha^{-1} in carbonate soils underscores their disproportionate role in global climate mitigation. Beyond carbon regulation, mangroves attenuate wave energy by as much as 99% over 500 m of forest, preventing an estimated US$65 billion in annual flood damages and safeguarding more than 15 million people worldwide. Yet, despite these services, ~35% of global mangrove cover has been lost in recent

M. R. Karim
Institute of Forestry and Conservation, John H. Daniels Faculty of Architecture, Landscape, and Design, University of Toronto, Toronto, ON, Canada

M. S. R. Saimun · M. S. Amin · M. S. I. Fahim · M. A. S. Arfin-Khan (✉)
Department of Forestry and Environmental Science, School of Agriculture and Mineral Sciences, Shahjalal University of Science and Technology, Sylhet, Bangladesh
e-mail: khan-for@sust.edu

F. Sultana
Department of Forestry and Environmental Science, School of Agriculture and Mineral Sciences, Shahjalal University of Science and Technology, Sylhet, Bangladesh

Department of Biosciences, Faculty of Science and Engineering, Swansea University, Swansea, UK

D. Barman
ICAR—Central Research Institute for Jute and Allied Fibers, Kolkata, West Bengal, India

S. K. Srivastava
Geospatial Analytics for Conservation and Management, School of Science Technology and Engineering (SSTE), University of the Sunshine Coast, Sippy Downs, QLD, Australia

S. A. Mukul
Geospatial Analytics for Conservation and Management, School of Science Technology and Engineering (SSTE), University of the Sunshine Coast, Sippy Downs, QLD, Australia

Department of Environment and Development Studies, United International University, Dhaka, Bangladesh

S. K. Srivastava et al. (eds.), *Earth Observation for Monitoring Mangrove Ecosystems*, https://doi.org/10.1007/978-3-032-16668-5_2

decades, largely driven by aquaculture expansion, coastal development, and deforestation for timber and fuelwood. This chapter argues that mangrove conservation must be reframed not simply as an ecological priority but as an urgent global imperative for climate adaptation and socio-economic stability. By synthesizing ecological functions, socio-economic benefits, and valuation frameworks, we demonstrate that the erosion of mangrove resilience directly undermines fisheries that support millions of households, coastal infrastructure protection, and cultural heritage embedded in local communities. At the same time, fragmented research and policy silos have limited effective interventions, particularly in Asia where 36% of global losses have occurred. We contend that sustaining mangroves requires integrated strategies that align ecological restoration with local livelihood security and international climate commitments. Incorporating advances in remote sensing, economic valuation, and community-led conservation, this synthesis highlights pathways for embedding mangrove ecosystem services into adaptive governance and global sustainability agendas. Given their dual role in climate mitigation and disaster risk reduction, safeguarding mangroves represents a decisive opportunity to couple.

Keywords Carbon regulation · Mangrove conservation · Mangrove resilience · Adaptive governance

1 Introduction

Mangrove ecosystems represent one of the most ecologically significant and functionally diverse coastal biomes, uniquely positioned at the interface of terrestrial and marine systems (Giri et al., 2011). These intertidal forests have been dominated by salt-tolerant tree species, serve as critical buffers against coastal erosion, mitigate the impacts of storm surges and sea-level rise, and regulate global biogeochemical cycles (Blankespoor et al., 2017; Sunkur et al., 2023). Their ability to sequester substantial amounts of carbon has positioned them as key components of blue carbon ecosystems, reinforcing their role in climate change mitigation. Beyond their ecological significance, mangroves sustain millions of coastal inhabitants by providing essential resources such as timber, fuelwood, fisheries, and medicinal products, while also supporting traditional livelihoods and cultural practices (Giri et al., 2008). However, despite their invaluable services, mangrove forests are experiencing significant declines due to deforestation, habitat degradation, and climate change-induced stressors. The urgency of their conservation necessitates a comprehensive understanding of their ecological, socio-economic, and adaptive functions.

Mangroves are distributed across tropical and subtropical coastlines in over 123 countries (Cummings & Shah, 2018), covering an estimated 167,000 to 181,000 km^2 globally (Fig. 1) (Giri et al., 2011). Their distribution can be broadly categorized into two major biogeographic zones: the western zone, encompassing the Atlantic

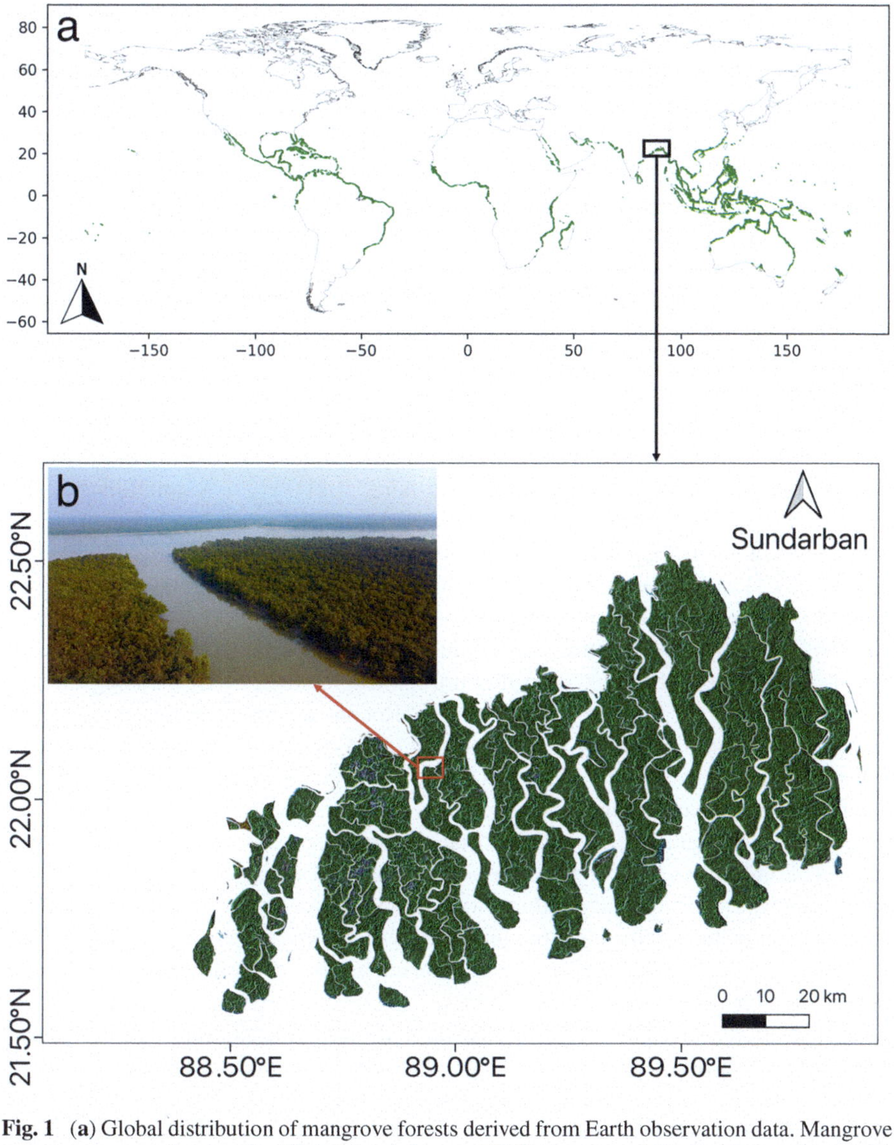

Fig. 1 (**a**) Global distribution of mangrove forests derived from Earth observation data. Mangrove extent was mapped using the Global Land Survey (GLS) dataset and the Landsat archive, with approximately 1000 Landsat scenes classified through a hybrid supervised–unsupervised digital image classification approach. Mangrove forests are shown in green, and country boundaries are delineated in black. (**b**) The Sundarbans, the world's largest contiguous mangrove forest, illustrated in high resolution using ESRI Sentinel-2 land cover data. The inset shows a drone image of a major river confluence with tidal channels within the Sundarbans, captured on 25 December 2021

coasts of the Americas and Africa, and the eastern zone, which includes the Indian Ocean and Indo-Pacific regions. The eastern zone harbors the highest species diversity, particularly in Southeast Asia, which accounts for nearly one-third of the world's mangrove cover (Gerona-Daga & Salmo III, 2022). In contrast, mangroves in the Americas, though extensive, exhibit lower species richness, dominated by a few genera such as *Rhizophora, Avicennia*, and *Laguncularia* (Moreira, 2020). The Sundarbans, spanning Bangladesh and India, represent the world's largest contiguous mangrove forest, providing a habitat for numerous endangered species, including the Bengal tiger (*Panthera tigris*) (Sarker et al., 2016). Despite their vast expanse, mangrove forests continue to decline, with approximately 35% of global mangrove cover lost in the past few decades due to anthropogenic pressures, including land reclamation for aquaculture, urban expansion, and pollution (Richards et al., 2020).

The resilience of mangrove ecosystems is attributed to their remarkable physiological and structural adaptations, which enable them to thrive in saline, waterlogged, and anoxic conditions (Kumari & Rathore, 2021). These adaptations include aerial root systems, salt-excreting mechanisms, and viviparous propagules, all of which contribute to their survival in fluctuating coastal environments (Naskar & Palit, 2015). However, regional variations in environmental conditions and human-induced pressures result in differing patterns of degradation and restoration potential. While extensive research has been conducted on mangrove ecology and conservation, much of the existing literature is fragmented, often focusing on isolated ecological or socio-economic aspects. Recent advances in remote sensing have provided complementary global and regional assessments of mangrove extent, dynamics, and degradation trajectories (e.g., Lu & Wang, 2021; Tran et al., 2022), which offer important spatial baselines. This chapter seeks to bridge these gaps by presenting an integrative perspective that encompasses the ecological functions, socio-economic roles, and conservation imperatives of mangroves. Unlike conventional reviews that emphasize either biodiversity conservation or economic valuation, this chapter highlights the interconnectedness of these dimensions, emphasizing how mangrove resilience is intrinsically linked to human well-being and climate adaptation strategies.

By synthesizing contemporary research with emerging findings on mangrove adaptation to climate change, this chapter provides a comprehensive discussion on the different roles of mangrove ecosystems and their significance. It explores the mechanisms through which mangroves persist in dynamic environments and the sociopolitical frameworks that govern their conservation. In this chapter, case studies and infographics are incorporated, providing tangible insights into successful restoration initiatives and policy interventions. Moreover, the urgent need for sustainable mangrove management to safeguard these critical ecosystems for future generations is emphasized in this chapter, integrating scientific evidence with real-world conservation strategies.

To provide a coherent framework for this chapter, a conceptual flowchart (Fig. 2) is presented, illustrating the progression from mangrove ecological functions, through anthropogenic and climate-related threats, to socio-economic implications,

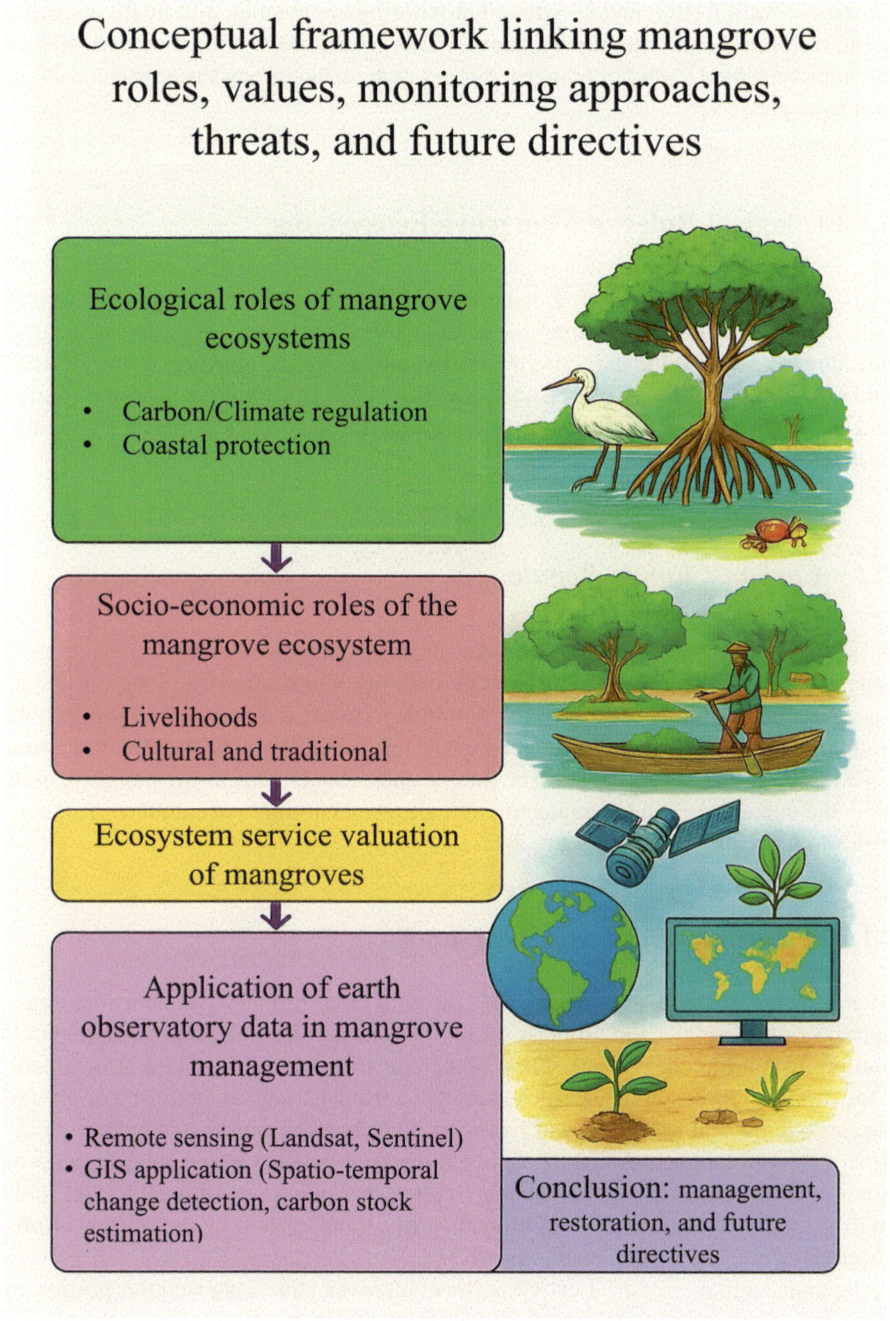

Fig. 2 Conceptual framework illustrating the ecological roles, socio-economic values, monitoring approaches, threats, and future directives of mangrove ecosystems

Earth observation (EO) and GIS-based monitoring approaches, and finally management, restoration, and future perspectives. This sequential organization underscores the interconnected nature of mangrove ecosystem services, pressures, and conservation strategies.

2 Ecological Roles of Mangrove Ecosystems

Mangrove ecosystems provide a diverse array of ecological services that are vital to both the environment and human societies. These roles can be broadly categorized into three main aspects: carbon/climate regulation, coastal protection and disaster mitigation, and biodiversity hotspots. Each of these roles underscores the unique and indispensable value of mangroves in maintaining ecological balance and supporting human livelihoods.

2.1 Carbon/Climate Regulation

Among the most effective carbon sinks on Earth, mangrove forests are playing a crucial role in climate regulation through carbon sequestration and long-term storage. As blue carbon ecosystems, they store significant amounts of carbon in both biomass and sediments, with their capacity far exceeding that of many terrestrial forests (Choudhary et al., 2024). The unique anaerobic conditions in mangrove soils slow the decomposition of organic matter, allowing carbon to accumulate over centuries to millennia (Dasat & Sam, 2022).

2.1.1 Mechanisms of Carbon Sequestration

Mangrove ecosystems regulate climate through three primary mechanisms: atmospheric carbon uptake and biomass accumulation, sediment carbon burial, and outwelling to marine ecosystems. First, mangrove trees actively capture atmospheric CO_2 through photosynthesis, subsequently storing carbon in both above-ground (leaves, stems, and branches) and below-ground (roots) biomass (Choudhary et al., 2024). Research indicates that mangrove forests exhibit an average net primary production of 226 g C m^2 $year^{-1}$, a rate that exceeds that of many terrestrial forests. This high productivity contributes significantly to global carbon sequestration efforts (Mcleod et al., 2011; Tang et al., 2018).

Second, unlike terrestrial ecosystems, mangroves store a substantial portion of their carbon stock in waterlogged, anoxic soils, which create conditions that markedly slow organic matter decomposition (Choudhary et al., 2024). As a result, 50–90% of total carbon in mangrove ecosystems is retained within sediments for extended periods. Moreover, due to the influence of regional hydrological dynamics, species composition, and sedimentation rates, carbon burial rates in these ecosystems vary from 0.17 to 4.3 Mg C ha^{-1} $year^{-1}$ (Kelleway et al., 2017a, 2017b; Lamont et al., 2020). This long-term carbon storage in mangrove sediments plays a dynamic role in mitigating atmospheric CO_2 concentrations and buffering the effects of climate change.

Lastly, through the outwelling process, organic carbon is exported to adjacent marine ecosystems, resulting in significant contribution to the broader carbon cycle. This mechanism facilitates the transfer of mangrove-derived organic material to seagrass meadows and offshore sediments, where it is further sequestered (Hemminga et al., 1994; Chen et al., 2017). Empirical studies have reported that up to 83% of the carbon buried in seagrass beds in proximity to mangroves originates from these coastal forests (Potouroglou et al., 2017). This relationship highlights the role of mangroves as critical components of coastal carbon dynamics, emphasizing their importance in sustaining global blue carbon stocks.

2.1.2 Regional Variability in Carbon Sequestration

Environmental conditions such as tidal hydrodynamics, soil characteristics, and species composition drive the carbon dynamics, resulting in different sequestration rates in different mangrove ecosystems (Table 1). The following table presents a comparison of global mangrove ecosystems in terms of carbon sequestration efficiency:

2.2 *Coastal Protection and Disaster Mitigation*

Mangrove ecosystems function as natural coastal defenses, reducing the impact of extreme weather events, mitigating erosion, and stabilizing shorelines. Their dense root networks dissipate wave energy, trap sediments, and prevent shoreline retreat, making them essential for coastal resilience in the face of climate change (Marois & Mitsch, 2015; Gijsman et al., 2021). Previous literature has shown that mangrove forests can mitigate wave energy by up to 99% across a 500-m-wide forest, resulting significant decrease in coastal vulnerability to storm surges and flooding (McIvor et al., 2012). One of the primary mechanisms is named wave energy dissipation, through which mangroves reduce disaster risk. Mangrove forests act as wave breakers; the velocity of waves passing through them is gradually reduced by friction from trunks, aerial roots, and pneumatophores. Research indicates that mangroves lower wave height by 13–66% over a 100-m stretch, with larger and denser forests providing greater protection (McIvor et al., 2012). In addition to wave attenuation, mangroves act as storm surge buffers, reducing both the extent and severity of coastal flooding (Montgomery et al., 2019). For example, regions

Table 1 Global variation in mangrove soil organic carbon (SOC) storage across different coastal environmental settings (CES)

Mangrove ecosystem	Carbon sequestration (Mg/ha/year)	Total carbon stock (Mg/ha)	Role in climate mitigation
Deltaic (Amazon, Caravelas, etc.)	5.5–6.5	400–800	High sediment input supports rapid SOC accumulation, but past estimates overestimated SOC by 86%
Estuarine (Florida Coastal Everglades, Costa Rica)	3.5–5.0	900–1500	Tidal influence enhances soil aeration, reducing decomposition, leading to high SOC stability
Carbonate (Karstic, Bahamas, Pacific Atolls)	4.0–5.5	1200–2200	Peat-dominated soils store 50% more SOC than previously estimated, making these settings key blue carbon reservoirs
Bedrock (Brazil, Caribbean, SE Asia)	2.5–3.5	600–1000	Limited sediment supply reduces sequestration rates, but long-term carbon storage potential remains significant
Lagoonal (West Africa, SE Asia, Central America)	3.0–4.5	700–1100	Moderate SOC retention with a mix of marine and riverine carbon sources
Composite (Mixed River & Wave, Indo-Pacific, Gulf of Mexico)	4.0–6.0	1000–1800	Dynamic sediment processes enhance carbon burial efficiency, with stable SOC pools

The table summarizes estimated carbon sequestration rates, total carbon stocks, and their role in climate mitigation. Data highlight the influence of geomorphological and hydrological processes on SOC variability, with carbonate settings storing more carbon than previously estimated and deltaic settings showing past overestimations. Estimates are based on a global compilation of field measurements integrated into an eco-geomorphological modeling framework (Rovai et al., 2018)

with mangrove forests wider than 1 km experienced 24% lower economic losses from hurricanes compared to areas with degraded mangroves in Central America (Del Valle et al., 2020). Mangroves contribute to socio-economic resilience by safeguarding infrastructure, agricultural land, and fisheries, besides their biophysical functions. By ensuring sustainable livelihoods through resources such as timber, fish, and other marine products while simultaneously serving as crucial carbon sinks, mangroves play critical roles in the coastal zone of several regions. Economically, their protective functions are substantial, with estimates suggesting that mangroves help prevent approximately $65 billion in annual flood-related damages and shield over 15 million people globally from coastal hazards (Menéndez et al., 2020).

Table 2 Key species associated with mangrove ecosystems, highlighting their ecological roles, economic importance, and dependence on mangrove habitats

Species	Ecological role	Economic importance	Dependence on mangroves	Source
Bengal tiger (*Panthera tigris*)	Apex predator; maintains ecological balance	Tourism, cultural value, conservation importance	Habitat and prey availability within mangrove ecosystems	Ghosh et al. (2015)
Mud crab (*Scylla* spp.)	Keystone species; ecosystem engineer, food source	Fishing industry, culinary value, local livelihoods	Spawning and nursery grounds in mangrove areas	Lee (1998); Pati et al. (2023)
Mangrove snapper (*Lutjanus argentimaculatus*)	Predatory fish; maintains marine food web	Fisheries, important for local economies	Habitat in mangrove roots and estuarine areas	Aburto-Oropeza et al. (2009); Sandilyan and Kathiresan (2012)
Black mangrove (*Avicennia germinans*)	Critical to ecosystem structure; primary producer	Timber, fuelwood, medicinal uses, ecological services	Fundamental role in mangrove forest structure and regeneration	Satyanarayana et al. (2012); Friis and Killilea (2024)
Palaemonid shrimp (*Palaemon* spp.)	Important detritivore; contributes to nutrient cycling	Aquaculture, shrimp farming	Breeding and juvenile habitats in mangrove swamps	Solari et al. (2017)
Saltwater crocodile (*Crocodylus porosus*)	Apex predator, maintains balance in ecosystems	Tourism, cultural significance	Nesting and feeding habitats in mangrove forests	Semeniuk et al. (2011); Hanson et al. (2015)
Mangrove oyster (*Crassostrea* spp.)	Filter feeder; water quality improvement	Fisheries, shellfish industry	Thrives on mangrove roots and substrates	Chacin et al. (2025)
Herons and egrets (*Ardeidae* spp.)	Birds that control invertebrate and fish populations	Ecotourism, biodiversity indicator species	Nesting, roosting, and foraging within mangrove habitats	Etezadifar et al. (2010); Gopi and Pandav (2011); Ghasemi et al. (2012)

Mangroves support a diverse range of organisms, from apex predators to keystone species, contributing to ecosystem stability, fisheries, and ecotourism

2.3 *Biodiversity Hotspot*

Mangrove ecosystems are globally recognized as biodiversity hotspots (Table 2), providing critical habitats for over 1400 mangrove-affiliated species, including fish, crustaceans, mollusks, and birds (Rodrigues et al., 2004). These forests serve as nurseries and breeding grounds for 30–50% of commercially important fish species, playing an essential role in sustaining global fisheries (Cau et al., 2020). Their structural complexity, including aerial roots and dense canopies, offers refuge and feeding grounds for a wide array of marine organisms, enhancing coastal productivity and ecosystem stability.

Mangroves not only sustain aquatic ecology and marine biodiversity but also make a substantial contribution to terrestrial biodiversity, especially in areas where they coexist alongside tropical forests. Specialized species have evolved in response to the distinct environmental conditions of mangrove ecosystems, characterized by high salinity, waterlogged soils, and anoxic conditions (Alvarenga et al., 2015; Perri et al., 2023). Mangrove forests like the Sundarbans provide habitat and prey for apex predators like the Bengal tiger (*Panthera tigris*) (Loucks et al., 2010; Ghosh et al., 2015). The biological link between these coastal forests and neighburing ecosystems is further highlighted by the fact that migratory bird species, such as herons and egrets, rely on mangroves for both nesting and foraging (Zakaria & Rajpar, 2015; Kelleway et al., 2017b).

In addition to its economic significance, mangrove biodiversity supports a variety of businesses, including tourism, aquaculture, fisheries, and the harvesting of non-timber forest products. Both domestic and international seafood markets greatly benefit from commercially valuable species, such as mud crabs (*Scylla* spp.) and mangrove snappers (*Lutjanus argentimaculatus*). Furthermore, filter-feeding species like mangrove oysters (*Crassostrea* spp.) improve water quality while sustaining shellfish industries, and saltwater crocodiles (*Crocodylus porosus*) attract ecotourism revenue (Dame, 1993; Layman et al., 2014). The safeguarding of mangrove biodiversity is therefore crucial for both the long-term viability of these economic benefits as well as for preserving the integrity of the ecosystem.

3 Socio-economic Roles of Mangrove Ecosystem

Beyond their ecological significance, mangrove forests play a crucial role in sustaining livelihoods and preserving cultural heritage in coastal regions. These dynamic ecosystems support diverse economic activities, from fisheries and aquaculture to ecotourism and small-scale industries. At the same time, they hold deep cultural and spiritual value for many indigenous and local communities. However, the extent to which people benefit from these forests varies, shaped by economic dependency, regional traditions, and community awareness.

3.1 Livelihood

For millions of coastal residents, mangrove resources provide a primary means of sustenance and economic stability. Fisheries, aquaculture, and the harvesting of crabs, mollusks, and honey contribute significantly to household incomes, particularly in rural settlements where alternative employment opportunities are scarce. A study conducted in Myanmar involving over 185 households showed that 43% of households generated their income through selling forest products (Aye et al., 2019). Small-scale enterprises centered around mud crab (*Scylla* spp.) farming, shrimp aquaculture, and beekeeping have emerged as profitable ventures, enhancing financial resilience for coastal communities (Mirera et al., 2014; Hua et al., 2025). Additionally, livestock grazing in mangrove areas provides a supplementary source of income, particularly in regions where agricultural land is limited (Ahouangan et al., 2022). However, despite these benefits, economic disparities exist. In some areas, a negative correlation between ecosystem services and community well-being has been observed, largely due to unsustainable resource extraction, economic instability, and the lack of diversified income sources.

A shift toward sustainable economic models can help balance conservation and economic needs. Encouraging responsible aquaculture, adding value to mangrove-derived products, and integrating eco-friendly enterprises into local economies can enhance both community well-being and the long-term sustainability of these ecosystems.

3.2 Cultural/Traditional

Across many coastal societies, mangroves hold deep cultural and historical importance. From traditional fishing techniques and folklore to religious practices, these forests are woven into the identity of local communities. In certain regions, they are even regarded as protective natural barriers against storms and tidal surges, reinforcing their spiritual significance. Beyond their role in tradition, these coastal forests provide recreational and aesthetic benefits, fostering ecotourism and nature-based cultural activities. A study in the Hara Biosphere Reserve which is a key mangrove area, shows that 81.2% of visitors are willing to pay for access, indicating people's attraction to this ecosystem including birdwatchers, researchers, and visitors seeking immersive experiences in nature (Dehghani et al., 2010). The rise of boat tours, wildlife excursions, and eco-cultural festivals has not only increased environmental awareness but also generated revenue for local economies. However, the degree to which mangroves remain embedded in cultural practices varies. Urbanization and industrial development have led to a decline in traditional knowledge and reduced direct engagement with these forests in some communities (Rangel et al., 2024). Preserving this cultural heritage requires proactive efforts, such as community-led conservation programs, integration of traditional ecological knowledge in resource management, and promotion of sustainable ecotourism initiatives.

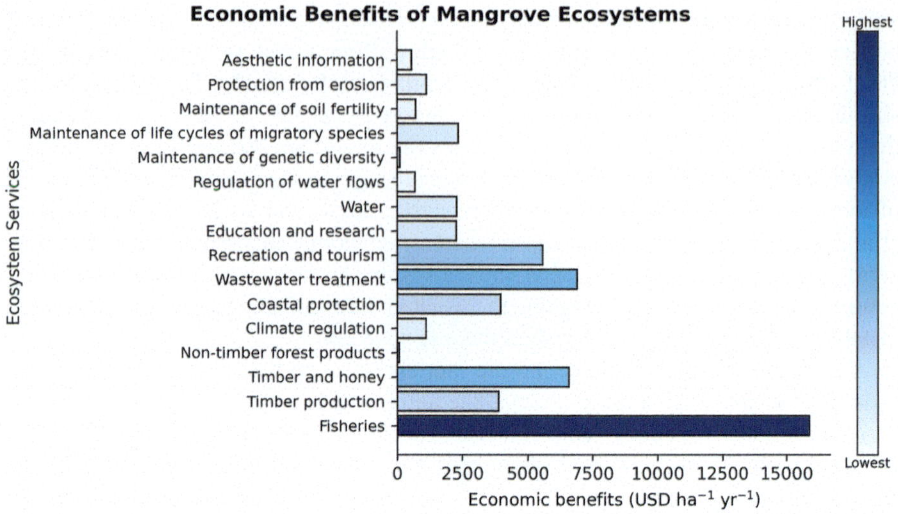

Fig. 3 Economic benefits of mangrove ecosystems, comparing restored and natural mangrove ecosystems across various ecosystem services. The bar length represents the total economic value (USD ha^{-1} $year^{-1}$) associated with each ecosystem service. The color gradient reflects the range of values from lowest to highest. Data source: (Su et al., 2021)

4 Ecosystem Service Valuation of Mangroves

Ecosystem service valuation is a fundamental tool for quantifying the ecological, economic, and social benefits provided by mangroves (Fig. 3). These intertidal forests deliver a wide array of services, categorized into provisioning, regulating, and cultural services. Among provisioning services, fisheries rank as the most valuable, with an average estimated value of 17,090 USD per hectare per year, reflecting their critical role as nurseries for commercially important fish and crustacean species (Mukherjee et al., 2014). Wood and timber resources, essential for fuel and construction in many developing countries, are another key provisioning service, though often undervalued at approximately 247 USD per hectare per year (Mukherjee et al., 2014). Mangroves offer significant regulating services, particularly in coastal protection by buffering storm surges and mitigating erosion, valued at 8459 USD per hectare per year. They are also highly effective in carbon sequestration, storing 1023 Mg C per hectare, surpassing other tropical forest ecosystems (Donato et al., 2011).

Despite their vital ecological function, services like bioremediation of pollutants and protection from salt intrusion are rarely represented in economic assessments, leaving significant gaps in understanding their full value. Cultural services such as ecotourism and aesthetic appreciation provide additional socio-economic benefits, particularly in regions where mangroves attract visitors. However, these services remain underrepresented in valuation exercises, especially in developing countries where community livelihoods heavily depend on mangrove ecosystems. The disparity between expert-based assessments and monetary valuations highlights the need for inclusive valuation frameworks that recognize the plurality of benefits mangroves provide. It is also important to recognize the limitations of economic valuation itself. Many non-market services, particularly cultural, spiritual, and relational values, are inherently difficult to monetize and risk being undervalued or overlooked in purely monetary frameworks (Hejnowicz & Rudd, 2017). Over-reliance on market-based estimates can therefore bias decision-making toward easily quantified services, potentially marginalizing benefits that are equally critical for local communities and long-term sustainability. Such comprehensive approaches can better inform policy decisions, improve conservation incentives, and promote sustainable management of mangrove ecosystems.

5 Application of Remote Sensing and GIS in Mangrove Management

Advances in Earth observation (EO) and geographic information systems (GIS) have primarily transformed mangrove monitoring, providing robust tools to quantify ecosystem extent, structure, dynamics, and vulnerability at multiple spatial and temporal scales. Global-scale datasets, such as the Global Mangrove Forest Cover (Hamilton & Casey, 2016) and the Hansen Global Forest Change (GFC) dataset (Hansen et al., 2013), have enabled consistent mapping of mangrove distribution and long-term loss trends. More recently, the Global Mangrove Watch (GMW) integrates multi-sensor satellite data to provide near-real-time monitoring of mangrove extent, allowing for high-resolution assessments of deforestation, degradation, and restoration trajectories (Bunting et al., 2022). These datasets provide essential baselines for assessing ecosystem services and informing climate mitigation strategies, particularly in the context of blue carbon accounting.

Remote sensing platforms offer complementary capabilities for detailed mangrove characterization. Optical sensors such as Landsat (Hemati et al., 2021) and Sentinel-2 (Misra et al., 2020) facilitate multidecadal change detection, canopy phenology monitoring, and biomass estimation, while MODIS data supports large-scale temporal analyses of productivity and greenness of mangroves (Ishtiaque et al., 2016). High-resolution commercial satellites, including PlanetScope, enable localized mapping for mangrove management interventions with carbon stock estimation (Purnamasari et al., 2021), whereas LiDAR and synthetic aperture radar (SAR) provide three-dimensional structural information, canopy height, and aboveground biomass estimates even under cloud-prone conditions (Kaasalainen et al., 2015). Hyperspectral sensors further allow species-level discrimination, supporting biodiversity assessments and ecological niche mapping.

GIS-based analyses extend the utility of EO data by integrating spatial layers to model mangrove vulnerability, carbon stock distribution, and ecosystem service provisioning (Kuenzer & Tuan, 2013). Change detection algorithms, spatial overlay analyses, and hydrodynamic modeling have been employed to identify high-risk areas for erosion, sea-level rise, and anthropogenic encroachment (Youssef et al., 2021). Carbon stock estimation using EO-derived biomass models informs carbon credit and payment-for-ecosystem-service programs, while spatial prioritization tools support restoration planning, protected area design, and policy enforcement (Le et al., 2021).

Collectively, these Earth observation and GIS approaches provide a quantitative, spatially explicit framework that bridges ecological assessment with management action. By integrating multi-source satellite data and geospatial analyses, practitioners can identify degradation hotspots, monitor recovery trajectories, evaluate ecosystem service flows, and implement evidence-based conservation strategies. The convergence of remote sensing, GIS, and ecological modeling thus underpins contemporary mangrove management, offering scalable solutions to safeguard these critical coastal ecosystems under accelerating anthropogenic and climate pressures.

6 Major Threats and Consequences

Mangrove ecosystems face escalating threats driven by both anthropogenic activities and environmental changes (Table 3). Over the past five decades, approximately 35% of global mangrove coverage has been lost (Gouvêa et al., 2022), with Asia contributing 36% of these losses (Yousefi & Naderloo, 2022). The primary drivers of mangrove degradation include coastal development, aquaculture expansion, and deforestation for timber and fuelwood. Coastal infrastructure projects such as resorts, ports, and dams disrupt natural hydrological flows, resulting in increased soil erosion, altered salinity, and habitat fragmentation. Aquaculture, particularly large-scale shrimp farming, has converted up to 52% of mangrove forests globally in recent decades, with Indonesia, Vietnam, and Bangladesh among the most affected regions (Bhowmik et al., 2022).

Deforestation for timber and fuelwood extraction accounts for 26% of mangrove loss, driven by local demand for construction materials and energy sources (Ferreira et al., 2022). Beyond human-induced pressures, climate change exacerbates mangrove vulnerability through rising sea levels, increased storm intensity, and shifts in precipitation patterns, which alter species composition and productivity. The rise in sea level, projected to reach 1.5–2.5 m by 2099, poses a significant threat to low-lying mangrove forests, particularly in deltaic regions (Akram et al., 2023).

Pollution from industrial discharge, agricultural runoff, and oil spills further degrades mangrove habitats, impairing their water filtration capacity and leading to biodiversity loss. Additionally, eutrophication—driven by excessive nutrient input—stimulates harmful algal blooms that suffocate mangrove roots and reduce photosynthetic efficiency (Sarkar & Sarkar, 2018; Tsikoti & Genitsaris, 2021). The combined impacts of these threats not only compromise the ecological integrity of mangrove forests but also diminish their capacity to deliver vital ecosystem services, including coastal protection, carbon sequestration, and fishery support.

Table 3 Major anthropogenic and natural threats to mangrove ecosystems and their associated ecological consequences

Threat category	Major threats	Consequences	Reference
Anthropogenic	Coastal development	Habitat fragmentation, erosion, altered hydrology	Arévalo-Mejía et al. (2020); Vousdoukas et al. (2020)
	Deforestation for timber	Reduction in carbon storage, habitat degradation	Lee et al. (2019); Arias-Ortiz et al. (2021)
	Pollution (industrial and agricultural runoff)	Water quality degradation, eutrophication, biodiversity loss	Tsikoti and Genitsaris (2021); Tekman et al. (2022); Bindiya et al. (2023)
	Biological invasion	Altered species composition, competition with native species	Mandal et al. (2022)
Natural	Climate change and sea-level rise	Coastal erosion, habitat loss, salt intrusion, altered species composition	Godoy et al. (2018); De Lacerda et al. (2019); Li et al. (2022)
	Storm surges and cyclones	Physical destruction, increased soil salinity, vegetation mortality	Hoppe-Speer et al. (2011); Asbridge et al. (2015)
	Tidal inundation	Waterlogging, sedimentation changes, reduced seedling recruitment	Xie et al. (2020)
	Temperature fluctuations	Range shifts, altered growth patterns, species mortality	Li et al. (2022)
	Coastal development	Habitat fragmentation, erosion, altered hydrology	Akram et al. (2023)

Effective mitigation strategies require integrated management approaches that address both direct human pressures and the broader impacts of climate change.

7 Conclusion

Mangrove ecosystems are among the most functionally diverse and ecologically vital coastal biomes, providing essential services such as carbon sequestration, coastal protection, biodiversity support, and socio-economic benefits. Their role as blue carbon reservoirs underscores their significance in climate mitigation, while their ability to attenuate wave energy and buffer storm surges highlights their importance as natural coastal defenses. Additionally, mangroves serve as nurseries for commercially valuable marine species and support millions of coastal livelihoods, reinforcing the link between ecosystem integrity and human well-being.

Despite their critical ecological and economic roles, mangrove forests are increasingly threatened by deforestation, habitat conversion, pollution, and climate change. Addressing these challenges requires integrated conservation strategies, including habitat restoration, sustainable resource management, and policy interventions informed by ecological and socio-economic research. Given their resilience and capacity for regeneration, targeted conservation initiatives, coupled with community engagement and adaptive governance, can enhance their long-term sustainability.

Moving forward, bridging knowledge gaps and implementing science-based policies that recognize the full value of mangrove ecosystem services will be essential. Collaboration among ecologists, policymakers, and local communities will ensure conservation efforts align with socio-economic priorities, balancing environmental protection with human development. By safeguarding mangroves, we not only preserve critical coastal ecosystems but also contribute to global climate stability and long-term resilience.

References

Aburto-Oropeza, O., et al. (2009). Recruitment and ontogenetic habitat shifts of the yellow snapper (*Lutjanus argentiventris*) in the Gulf of California. *Marine Biology, 156*(12), 2461–2472. https://doi.org/10.1007/s00227-009-1271-5

Ahouangan, B. S. C. M., et al. (2022). Ruminant keeping around mangrove forests in Benin (West Africa): Herders' perceptions of threats and opportunities for conservation of mangroves. *Discover Sustainability, 3*(1), 13. https://doi.org/10.1007/s43621-022-00082-x

Akram, H., et al. (2023). Mangrove health: A review of functions, threats, and challenges associated with mangrove management practices. *Forests, 14*(9). https://doi.org/10.3390/f14091698

Alvarenga, D. O., et al. (2015). Cyanobacteria in mangrove ecosystems. *Biodiversity and Conservation, 24*(4), 799–817. https://doi.org/10.1007/s10531-015-0871-2

Arévalo-Mejía, R., et al. (2020). A baseline assessment of hydrologic alteration degree for the Mexican catchments at gauged rivers (2016). *Science of the Total Environment, 729*, 139041. https://doi.org/10.1016/j.scitotenv.2020.139041

Arias-Ortiz, A., et al. (2021). Losses of soil organic carbon with deforestation in mangroves of Madagascar. *Ecosystems, 24*(1), 1–19. https://doi.org/10.1007/s10021-020-00500-z

Asbridge, E., et al. (2015). Mangrove response to environmental changes predicted under varying climates: Case studies from Australia. *Current Forestry Reports, 1*, 178–194.

Aye, W. N., et al. (2019). Contribution of mangrove forest to the livelihood of local communities in Ayeyarwaddy region, Myanmar. *Forests*. https://doi.org/10.3390/f10050414

Bhowmik, A. K., et al. (2022). Global mangrove deforestation and its interacting social-ecological drivers: A systematic review and synthesis. *Sustainability*. https://doi.org/10.3390/su14084433

Bindiya, E. S., Sreekanth, P. M., & Bhat, S. G. (2023). Conservation and management of mangrove ecosystem in diverse perspectives. In S. T. Sukumaran & T. R. Keerthi (Eds.), *Conservation and sustainable utilization of bioresources* (pp. 323–352). Springer Nature Singapore. https://doi.org/10.1007/978-981-19-5841-0_13

Blankespoor, B., Dasgupta, S., & Lange, G.-M. (2017). Mangroves as a protection from storm surges in a changing climate. *Ambio, 46*(4), 478–491.

Bunting, P., et al. (2022). Global mangrove extent change 1996–2020: Global mangrove watch version 3.0. *Remote Sensing, 14*(15). https://doi.org/10.3390/rs14153657

Cau, A., et al. (2020). The nursery role of marine animal forests. In S. Rossi & L. Bramanti (Eds.), *Perspectives on the marine animal forests of the world* (pp. 309–331). Springer International Publishing. https://doi.org/10.1007/978-3-030-57054-5_10

Chacin, D. H., Bell, S. S., & Stallings, C. D. (2025). Red mangrove (Rhizophora mangle) roots serve as additional valuable habitat for the eastern oyster (*Crassostrea virginica*) in a subtropical estuary. *Journal of Experimental Marine Biology and Ecology, 583*, 152083. https://doi.org/10.1016/j.jembe.2025.152083

Chen, G., et al. (2017). Mangroves as a major source of soil carbon storage in adjacent seagrass meadows. *Scientific Reports, 7*(1), 42406.

Choudhary, B., Dhar, V., & Pawase, A. S. (2024). Blue carbon and the role of mangroves in carbon sequestration: Its mechanisms, estimation, human impacts and conservation strategies for economic incentives. *Journal of Sea Research, 199*, 102504. https://doi.org/10.1016/j.seares.2024.102504

Cummings, A. R., & Shah, M. (2018). Mangroves in the global climate and environmental mix. *Geography Compass, 12*(1), e12353.

Dame, R. (1993). Bivalve filter feeders and estuarine and coastal ecosystem processes: Conclusions. In R. F. Dame (Ed.), *Bivalve filter feeders* (pp. 565–569). Springer Berlin Heidelberg.

Dasat, G. S., & Sam, C. E. (2022). Carbon sequestration and the enzymic latch mechanism in red, black and white mangrove soils of Florida USA. *World Journal of Advanced Research and Reviews, 13*(3), 421–430. https://doi.org/10.30574/wjarr.2022.13.3.0235

De Lacerda, L. D., Borges, R., & Ferreira, A. C. (2019). Neotropical mangroves: Conservation and sustainable use in a scenario of global climate change. *Aquatic Conservation: Marine and Freshwater Ecosystems, 29*(8), 1347–1364.

Dehghani, M., et al. (2010). Recreation value of hara biosphere reserve using willingness-to-pay method. *International Journal of Environmental Research, 4*(2), 271–280. https://doi.org/10.22059/ijer.2010.19

Del Valle, A., et al. (2020). Mangroves protect coastal economic activity from hurricanes. *Proceedings of the National Academy of Sciences, 117*(1), 265–270.

Donato, D. C., et al. (2011). Mangroves among the most carbon-rich forests in the tropics. *Nature Geoscience, 4*(5), 293–297.

Etezadifar, F., et al. (2010). Breeding success of western reef heron in Hara Biosphere Reserve, Persian Gulf. *Waterbirds, 33*(4), 527–533.

Ferreira, A. C., Borges, R., & de Lacerda, L. D. (2022). Can sustainable development save mangroves? *Sustainability, 14*(3), 1263.

Friis, G., & Killilea, M. E. (2024). Mangrove ecosystems of the United Arab Emirates. In J. A. Burt (Ed.), *A natural history of the Emirates* (pp. 217–240). Springer Nature Switzerland. https://doi.org/10.1007/978-3-031-37397-8_7

Gerona-Daga, M. E. B., & Salmo, S. G., III. (2022). A systematic review of mangrove restoration studies in Southeast Asia: Challenges and opportunities for the United Nation's Decade on Ecosystem Restoration. *Frontiers in Marine Science, 9*, 987737.

Ghasemi, S., et al. (2012). Relative abundance and diversity of waterbirds in a Persian Gulf mangrove forest, Iran. *Tropical Zoology, 25*(1), 39–53.

Ghosh, A., et al. (2015). The Indian Sundarban mangrove forests: History, utilization, conservation strategies and local perception. *Diversity, 7*, 149–169. https://doi.org/10.3390/d7020149

Gijsman, R., et al. (2021). Nature-based engineering: A review on reducing coastal flood risk with mangroves. *Frontiers in Marine Science, 8*, 702412.

Giri, C., et al. (2008). Mangrove forest distributions and dynamics (1975–2005) of the tsunami-affected region of Asia. *Journal of Biogeography, 35*(3), 519–528.

Giri, C., et al. (2011). Status and distribution of mangrove forests of the world using earth observation satellite data. *Global Ecology and Biogeography, 20*(1), 154–159. https://doi.org/10.1111/j.1466-8238.2010.00584.x

Godoy, M. D. P., de Andrade Meireles, A. J., & de Lacerda, L. D. (2018). Mangrove response to land use change in estuaries along the semiarid coast of Ceará, Brazil. *Journal of Coastal Research, 34*(3), 524–533.

Gopi, G. V., & Pandav, B. (2011). Nest space partitioning among colonial nesting waterbirds at Bhitarkanika Mangroves, India. *World Journal of Zoology, 6*(1), 61–72.

Gouvêa, L. P., et al. (2022). Global impacts of projected climate changes on the extent and aboveground biomass of mangrove forests. *Diversity and Distributions, 28*(11), 2349–2360.

Hamilton, S. E., & Casey, D. (2016). Creation of a high spatio-temporal resolution global database of continuous mangrove forest cover for the 21st century (CGMFC-21). *Global Ecology and Biogeography, 25*(6), 729–738. https://doi.org/10.1111/geb.12449

Hansen, M. C., et al. (2013). High-resolution global maps of 21st-century forest cover change. *Science, 342*(6160), 850–853. https://doi.org/10.1126/science.1244693

Hanson, J. O., et al. (2015). Feeding across the food web: The interaction between diet, movement and body size in estuarine crocodiles (*C rocodylus porosus*). *Austral Ecology, 40*(3), 275–286.

Hejnowicz, A. P., & Rudd, M. A. (2017). The value landscape in ecosystem services: Value, value wherefore art thou value? *Sustainability, 9*(5). https://doi.org/10.3390/su9050850

Hemati, M., et al. (2021). A systematic review of Landsat data for change detection applications: 50 years of monitoring the earth. *Remote Sensing, 13*(15). https://doi.org/10.3390/rs13152869

Hemminga, M. A., et al. (1994). Carbon outwelling from a mangrove forest with adjacent seagrass beds and coral reefs (Gazi Bay, Kenya). *Marine Ecology-Progress Series, 106*, 291.

Hoppe-Speer, S. C. L., et al. (2011). The response of the red mangrove Rhizophora mucronata Lam. to salinity and inundation in South Africa. *Aquatic Botany, 95*(2), 71–76.

Hua, H. H., et al. (2025). A comparative analysis of livelihood sustainability among farmers in Ca Mau Province, Vietnam: A case study of organic mangrove–shrimp and rice–shrimp farming systems. *Cogent Social Sciences, 11*(1), 2451736. https://doi.org/10.1080/23311886.2025.2451736

Ishtiaque, A., Myint, S. W., & Wang, C. (2016). Examining the ecosystem health and sustainability of the world's largest mangrove forest using multi-temporal MODIS products. *Science of the Total Environment, 569–570*, 1241–1254. https://doi.org/10.1016/j.scitotenv.2016.06.200

Kaasalainen, S., et al. (2015). Combining lidar and synthetic aperture radar data to estimate forest biomass: Status and prospects. *Forests, 6*(1), 252–270. https://doi.org/10.3390/f6010252

Kelleway, J. J., et al. (2017a). Geochemical analyses reveal the importance of environmental history for blue carbon sequestration. *Journal of Geophysical Research: Biogeosciences, 122*(7), 1789–1805.

Kelleway, J. J., et al. (2017b). Review of the ecosystem service implications of mangrove encroachment into salt marshes. *Global Change Biology, 23*(10), 3967–3983.

Kuenzer, C., & Tuan, V. Q. (2013). Assessing the ecosystem services value of Can Gio Mangrove Biosphere Reserve: Combining earth-observation- and household-survey-based analyses. *Applied Geography, 45*, 167–184. https://doi.org/10.1016/j.apgeog.2013.08.012

Kumari, A., & Rathore, M. S. (2021). Roles of mangroves in combating the climate change. In *Mangroves: Ecology, biodiversity and management* (pp. 225–255).

Lamont, K., et al. (2020). Thirty-year repeat measures of mangrove above-and below-ground biomass reveals unexpectedly high carbon sequestration. *Ecosystems, 23*, 370–382.

Layman, C. A., et al. (2014). Provision of ecosystem services by human-made structures in a highly impacted estuary. *Environmental Research Letters, 9*(4), 44009. https://doi.org/10.1088/1748-9326/9/4/044009

Le, N. N., et al. (2021). Learning from multimodal and multisensor earth observation dataset for improving estimates of mangrove soil organic carbon in Vietnam. *International Journal of Remote Sensing, 42*(18), 6866–6890. https://doi.org/10.1080/01431161.2021.1945158

Lee, S. Y. (1998). Ecological role of grapsid crabs in mangrove ecosystems: A review. *Marine and Freshwater Research, 49*(4), 335–343.

Lee, S. Y., et al. (2019). Better restoration policies are needed to conserve mangrove ecosystems. *Nature Ecology & Evolution, 3*(6), 870–872. https://doi.org/10.1038/s41559-019-0861-y

Li, X., et al. (2022). Correlations between photosynthetic heat tolerance and leaf anatomy and climatic niche in Asian mangrove trees. *Plant Biology, 24*(6), 960–966.

Loucks, C., et al. (2010). Sea level rise and tigers: Predicted impacts to Bangladesh's Sundarbans mangroves. *Climatic Change, 98*(1), 291–298. https://doi.org/10.1007/s10584-009-9761-5

Lu, Y., & Wang, L. (2021). How to automate timely large-scale mangrove mapping with remote sensing. *Remote Sensing of Environment, 264*, 112584. https://doi.org/10.1016/j.rse.2021.112584

Mandal, B., Ganguly, A., & Mukherjee, A. (2022). A review for understanding the reasons of vanishing Sundari tree Heritiera fomes Buchanan-Hamilton from Sundarban mangroves. *Environment and Ecology, 39*(4), 813–817. ISSN 0970-0420.

Marois, D. E., & Mitsch, W. J. (2015). Coastal protection from tsunamis and cyclones provided by mangrove wetlands—A review. *International Journal of Biodiversity Science, Ecosystem Services & Management, 11*(1), 71–83.

McIvor, A. L., et al. (2012). *Storm surge reduction by mangroves* (2) (p. 35). The Nature Conservancy and Wetlands International. Retrieved from http://www.naturalcoastalprotection.org/documents/storm-surge-reduction-by-mangroves

Mcleod, E., et al. (2011). A blueprint for blue carbon: Toward an improved understanding of the role of vegetated coastal habitats in sequestering CO2. *Frontiers in Ecology and the Environment, 9*(10), 552–560.

Menéndez, P., et al. (2020). The global flood protection benefits of mangroves. *Scientific Reports, 10*(1), 4404.

Mirera, D. O., Ochiewo, J., & Munyi, F. (2014). Social and economic implications of small-scale mud crab (Scylla serrata) aquaculture: The case of organized community groups. *Aquaculture International, 22*(4), 1499–1514. https://doi.org/10.1007/s10499-014-9762-x

Misra, G., Cawkwell, F., & Wingler, A. (2020). Status of phenological research using Sentinel-2 data: A review. *Remote Sensing, 12*(17). https://doi.org/10.3390/rs12172760

Montgomery, J. M., et al. (2019). Attenuation of storm surges by coastal mangroves. *Geophysical Research Letters, 46*(5), 2680–2689.

Moreira, N. M. (2020). *Mangroves of Latin America*. Grupo Compas. p-211. ISBN 9789978251591.

Mukherjee, N., et al. (2014). Ecosystem service valuations of mangrove ecosystems to inform decision making and future valuation exercises. *PLoS One, 9*(9), e107706.

Naskar, S., & Palit, P. K. (2015). Anatomical and physiological adaptations of mangroves. *Wetlands Ecology and Management, 23*(3), 357–370.

Pati, S. G., et al. (2023). Impacts of habitat quality on the physiology, ecology, and economical value of mud crab Scylla sp.: A comprehensive review. *Water*. https://doi.org/10.3390/w15112029

Perri, S., et al. (2023). Salinity-induced limits to mangrove canopy height. *Global Ecology and Biogeography, 32*(9), 1561–1574. https://doi.org/10.1111/geb.13720

Potouroglou, M., et al. (2017). Measuring the role of seagrasses in regulating sediment surface elevation. *Scientific Reports, 7*(1), 11917.

Purnamasari, E., Kamal, M., & Wicaksono, P. (2021). Comparison of vegetation indices for estimating above-ground mangrove carbon stocks using PlanetScope image. *Regional Studies in Marine Science, 44*, 101730. https://doi.org/10.1016/j.rsma.2021.101730

Rangel, J. M. L., do Nascimento, A. L. B., & Ramos, M. A. (2024). The influence of urbanization on local ecological knowledge: A systematic review. *Journal of Ethnobiology and Ethnomedicine, 20*(1), 106. https://doi.org/10.1186/s13002-024-00747-z

Richards, D. R., Thompson, B. S., & Wijedasa, L. (2020). Quantifying net loss of global mangrove carbon stocks from 20 years of land cover change. *Nature Communications, 11*(1), 4260. https://doi.org/10.1038/s41467-020-18118-z

Rodrigues, A. S. L., et al. (2004). Global gap analysis: Priority regions for expanding the global protected-area network. *BioScience, 54*(12), 1092–1100. https://doi.org/10.1641/0006-3568(2004)054[1092:GGAPRF]2.0.CO;2

Rovai, A. S., et al. (2018). Global controls on carbon storage in mangrove soils. *Nature Climate Change, 8*(6), 534–538. https://doi.org/10.1038/s41558-018-0162-5

Sandilyan, S., & Kathiresan, K. (2012). Mangrove conservation: A global perspective. *Biodiversity and Conservation, 21*(14), 3523–3542. https://doi.org/10.1007/s10531-012-0388-x

Sarkar, S. K., & Sarkar, S. K. (2018). Algal blooms: Potential drivers, occurrences and impact. In *Marine algal bloom: Characteristics, causes and climate change impacts* (pp. 53–109).

Sarker, S. K., et al. (2016). Are we failing to protect threatened mangroves in the Sundarbans world heritage ecosystem? *Scientific Reports, 6*(1), 21234.

Satyanarayana, B., et al. (2012). A socio-ecological assessment aiming at improved forest resource management and sustainable ecotourism development in the mangroves of Tanbi Wetland National Park, The Gambia, West Africa. *Ambio, 41*(5), 513–526. https://doi.org/10.1007/s13280-012-0248-7

Semeniuk, V., et al. (2011). The saltwater crocodile, *Crocodylus porosus* Schneider, 1801, in the Kimberley coastal region. *Journal of the Royal Society of Western Australia, 94*(2), 407.

Solari, P., et al. (2017). Antennular morphology and contribution of aesthetascs in the detection of food-related compounds in the shrimp *Palaemon adspersus* Rathke, 1837 (Decapoda: Palaemonidae). *The Biological Bulletin, 232*(2), 110–122.

Su, J., Friess, D. A., & Gasparatos, A. (2021). A meta-analysis of the ecological and economic outcomes of mangrove restoration. *Nature Communications, 12*(1), 5050. https://doi.org/10.1038/s41467-021-25349-1

Sunkur, R., et al. (2023). Mangroves' role in supporting ecosystem-based techniques to reduce disaster risk and adapt to climate change: A review. *Journal of Sea Research, 196*, 102449.

Tang, J., et al. (2018). Coastal blue carbon: Concept, study method, and the application to ecological restoration. *Science China Earth Sciences, 61*, 637–646.

Tekman, M., et al. (2022). *Impacts of plastic pollution in the Oeans on marine species, biodiversity and ecosystems*. WWF Germany. https://doi.org/10.5281/zenodo.5898684

Tran, T. V., Reef, R., & Zhu, X. (2022). A review of spectral indices for mangrove remote sensing. *Remote Sensing, 14*(19). https://doi.org/10.3390/rs14194868

Tsikoti, C., & Genitsaris, S. (2021). Review of harmful algal blooms in the coastal Mediterranean Sea, with a focus on Greek waters. *Diversity*. https://doi.org/10.3390/d13080396

Vousdoukas, M. I., et al. (2020). Sandy coastlines under threat of erosion. *Nature Climate Change, 10*(3), 260–263. https://doi.org/10.1038/s41558-020-0697-0

Xie, D., et al. (2020). Mangrove diversity loss under sea-level rise triggered by bio-morphodynamic feedback and anthropogenic pressures. *Environmental Research Letters, 15*(11), 114033.

Yousefi, M., & Naderloo, R. (2022). Global habitat suitability modeling reveals insufficient habitat protection for mangrove crabs. *Scientific Reports, 12*(1), 21713.

Youssef, Y. M., et al. (2021). Natural and anthropogenic coastal environmental hazards: An integrated remote sensing, GIS, and geophysical-based approach. *Surveys in Geophysics, 42*(5), 1109–1141. https://doi.org/10.1007/s10712-021-09660-6

Zakaria, M., & Rajpar, M. N. (2015). Assessing the Fauna diversity of Marudu Bay mangrove Forest, Sabah, Malaysia, for future conservation. *Diversity, 7*, 137–148. https://doi.org/10.3390/d7020137

Ground-Based Techniques in Mangrove Monitoring and Management

Sharif A. Mukul, Mohammed A. S. Arfin-Khan, Saon Banerjee, Dhananjay Barman, Gouranga Kar, Prashant K. Srivastava, Pallab Mozumder, and Sanjeev Kumar Srivastava

Abstract Mangroves provide a wide range of ecosystem services, including coastal protection, biodiversity conservation, and fisheries support. While offering cultural and recreational benefits to local communities, they also play a crucial role in climate regulation and nutrient cycling. Despite their immense importance, mangrove ecosystems around the world are disappearing at an alarming rate. To safeguard this

---

S. A. Mukul (✉)
Department of Environment and Development Studies, United International University, Dhaka, Bangladesh

Geospatial Analytics for Conservation and Management, School of Science Technology and Engineering (SSTE), University of the Sunshine Coast, Sippy Downs, QLD, Australia

Department of Earth and Environment, Florida International University, Miami, FL, USA
e-mail: mukul@eds.uiu.ac.bd

M. A. S. Arfin-Khan
Department of Forestry and Environmental Science, School of Agriculture and Mineral Sciences, Shahjalal University of Science and Technology, Sylhet, Bangladesh

S. Banerjee
Department of Agricultural Meteorology and Physics, Bidhan Chandra Krishi Viswavidyalaya, Mohanpur, West Bengal, India

D. Barman · G. Kar
ICAR—Central Research Institute for Jute and Allied Fibers, Kolkata, West Bengal, India

P. K. Srivastava
Geospatial Analytics for Conservation and Management, School of Science Technology and Engineering (SSTE), University of the Sunshine Coast, Sippy Downs, QLD, Australia

Institute of Environment and Sustainable Development, Banaras Hindu University, Varanasi, Uttar Pradesh, India

P. Mozumder
Department of Earth and Environment, Florida International University, Miami, FL, USA

S. K. Srivastava
Geospatial Analytics for Conservation and Management, School of Science Technology and Engineering (SSTE), University of the Sunshine Coast, Sippy Downs, QLD, Australia

S. K. Srivastava et al. (eds.), *Earth Observation for Monitoring Mangrove Ecosystems*, https://doi.org/10.1007/978-3-032-16668-5_3

critical coastal ecosystem, effective management and continuous monitoring are essential. Ground-based monitoring techniques play a crucial role by providing high-resolution, site-specific data that complement remote sensing technologies, offering detailed insights into mangrove health, dynamics, and resilience. This chapter presents an overview of ground-based approaches to mangrove management and monitoring, emphasizing their significance; the key biological, physical, chemical, and social parameters involved; important methodological considerations; and the challenges associated with their application. While ground-based monitoring faces challenges such as limited accessibility, high resource demands, and environmental risks, it remains indispensable for the long-term conservation and management of mangroves. By combining these techniques with community engagement and technological innovation, stakeholders can develop robust strategies to strengthen mangrove resilience and ensure the sustainable management and protection of this imperiled ecosystem.

Keywords Biogeochemical process · Climate change · Ecosystem services · Livelihood dependence · Mangrove forest

1 Introduction

Mangroves are coastal ecosystems that occur in the intertidal zones of tropical and subtropical regions worldwide (Fig. 1; Spalding et al., 2010). Globally, they cover about 15 million hectares across 118 countries, accounting for nearly 15% of the world's coastlines (IUCN, 2024; Giri et al., 2011; Table 1). Mangrove ecosystems are characterized by the presence of "true mangroves," i.e., plant species that are specially adapted to saline environments, featuring unique traits such as

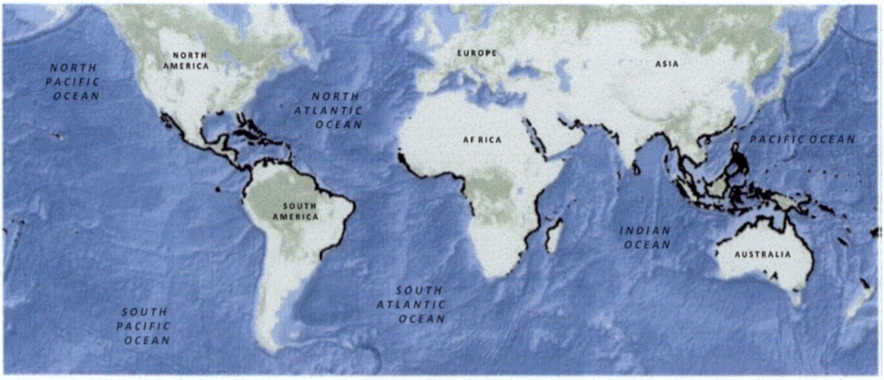

Fig. 1 Map showing the global distribution of mangroves (in black). Source: UNEP-WCMC (https://data.unep-wcmc.org; accessed on 15 June 2024)

Table 1 Mangrove distribution by region

Region	Share of global mangroves (%)	Major countries
Asia	42	Indonesia, Malaysia, Thailand
Africa	21	Mozambique, Kenya, Tanzania
North and Central America	15	Mexico, Belize, USA (Florida)
Australia and Oceania	12	Australia, Papua New Guinea
South America	10	Brazil, Colombia, Venezuela

pneumatophores (aerial roots), prop roots, and salt-excreting leaves (Biswas & Biswas, 2019; Lugo & Snedaker, 1974, Table 2).

Mangrove ecosystems provide a wide range of critical ecosystem services (Ho & Mukul, 2021). They act as natural barriers, protecting coastlines from cyclones, tsunamis, and storm surges (Mohammed et al., 2024; Costanza et al., 2008). They also play a significant role in carbon storage and sequestration (Hamilton & Friess, 2018; Donato et al., 2011), contribute to water quality regulation (Romañach et al., 2018), and serve as critical breeding and nursery habitats for numerous fish and wildlife species (zu Ermgassen et al., 2025). Beyond their ecological functions, mangroves also support local livelihoods by providing resources such as food, fuel, timber, construction materials, and opportunities for ecotourism (Friess et al., 2019).

Despite their immense ecological and socio-economic importance, mangroves are disappearing at a rate faster than that of tropical rainforests (Thomas et al., 2017). According to the first global assessment of mangroves under the IUCN Red List of Ecosystems, more than half of the world's mangrove ecosystems are currently at risk of collapse (Fig. 2; IUCN, 2024). The primary drivers of mangrove loss include coastal infrastructure development, aquaculture, conversion of land for agriculture, climate change, and sea-level rise (Leal & Spalding, 2024; Mukul et al., 2019, 2020a). Between 2000 and 2016, over half of global mangrove loss was attributed to land-use change, mainly the conversion of mangrove areas into aquaculture and agricultural lands, with the majority of these losses concentrated in six Southeast Asian countries (FAO, 2023; Goldberg et al., 2020).

Given their critical role in carbon sequestration, coastal protection, biodiversity conservation, and community livelihoods, the sustainable management and effective monitoring of mangroves are imperative to safeguard these vital ecosystems from ongoing and future threats (Friess et al., 2024; Schmitt & Duke, 2015). Ground-based techniques are essential for providing detailed, localized, and highly accurate data for the assessment and management of mangroves and for complementing remote sensing and aerial data (Misiukas et al., 2021).

This chapter provides an overview of the rationale for mangrove management and monitoring, emphasizing the importance of ground-based approaches. It discusses the key biological, physical, and chemical processes that are essential for effective mangrove monitoring and management, including human dimensions, main methodological considerations, and the challenges associated with ground-based monitoring and management of mangroves.

Table 2 List of true mangrove species

Family	Species
Acanthaceae	*Avicennia alba*
	Avicennia germinans (black mangrove)
	Avicennia marina
	Avicennia officinalis
Combretaceae	*Laguncularia racemosa* (white mangrove)
	Lumnitzera littorea
	Lumnitzera racemosa
Lythraceae	*Sonneratia alba*
	Sonneratia apetala
	Sonneratia caseolaris
	Sonneratia griffithii
	Sonneratia ovata
Rhizophoraceae	*Bruguiera cylindrica*
	Bruguiera exaristata
	Bruguiera gymnorhiza
	Bruguiera hainesii
	Bruguiera parviflora
	Bruguiera sexangula
	Ceriops australis
	Ceriops decandra
	Ceriops tagal
	Kandelia candel
	Kandelia obovata
	Rhizophora apiculata
	Rhizophora harrisonii
	Rhizophora mangle (red mangrove)
	Rhizophora mucronata
	Rhizophora racemosa
	Rhizophora stylosa
Arecaceae (Palms)	*Nypa fruticans* (special mangrove palm)
Malvaceae	*Heritiera fomes*
	Heritiera littoralis
	Hibiscus tiliaceus
Rubiaceae	*Scyphiphora hydrophylacea*
Euphorbiaceae	*Excoecaria agallocha*
	Excoecaria indica
Plumbaginaceae	*Aegialitis annulata*
	Aegialitis rotundifolia

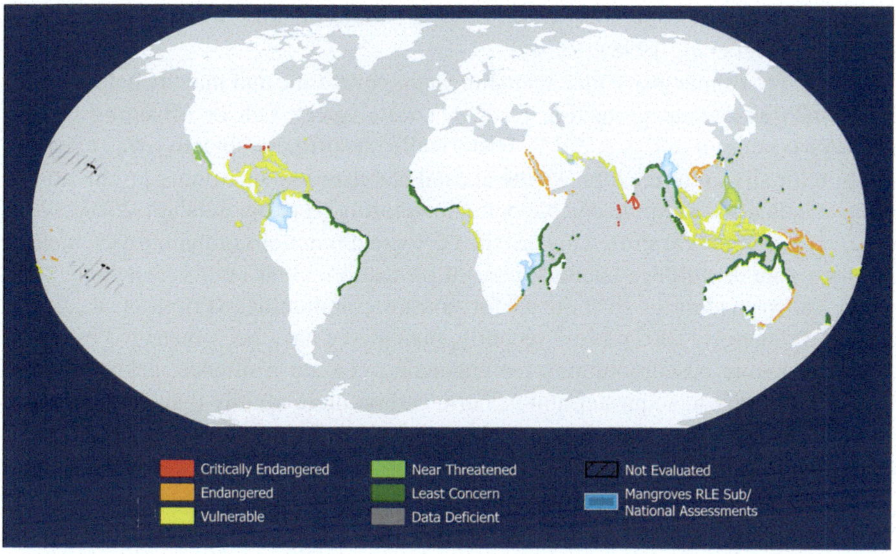

Fig. 2 Global red list of mangrove ecosystems. Source: Modified after IUCN (2024)

Fig. 3 Vertical (left) and horizontal (right) representation of a mangrove ecosystem. Photo credits: Sharif A. Mukul

2 The Rationale for Mangrove Management and Monitoring

The objective of mangrove management typically falls into several key categories: sustainable timber production, conservation of forests and wildlife habitats, coastal and shoreline protection, ecological restoration, and support of essential ecological functions such as soil stabilization, site protection, and carbon sequestration (Akram et al., 2023). The rationale for managing mangroves may also vary depending on the position or type of mangrove forest (Schmitt & Duke, 2015). For instance, as shown in Fig. 3, fringe mangrove forests play a vital role in protecting shorelines, and riverine forests are especially important for supporting plant and animal productivity, while basin or interior mangroves serve as key nutrient sinks and sources of wood products (Ewel et al., 1998).

Objectives of mangrove management change over time and may include specific goals such as coastal protection through coastal green belt or offsetting carbon emissions (Schmitt & Duke, 2015). Historically, the primary objective was on afforestation for silviculture, emphasizing coastal stabilization and timber production to support industries like pulp and paper manufacturing (Iftekhar & Islam, 2004). Over time, greater attention shifted toward the ecological functions of mangroves, including their role as wildlife habitats, coastal protection, contributors to pelagic food webs, and providers of livelihoods for coastal communities (Friess et al., 2019; Stubbs & Saenger, 2002). More recently, mangroves have been managed as cost-effective, nature-based solutions for enhancing climate resilience, acknowledged for their capacity to sequester atmospheric carbon more rapidly than other tropical ecosystems and their capacity to endure rising sea levels (Mukul et al., 2020b; Krauss et al., 2014).

Mangroves grow in different hydrogeomorphic settings, such as river-, tide-, wave-dominated, or interior mangrove forests, and rainfall, temperature, and freshwater supply also affect mangrove growth (Biswas & Biswas, 2019). Figure 4 presents a generalized decision-making flowchart that can guide mangrove management processes. The primary objective of any mangrove management strategy should be to protect and conserve existing mangrove ecosystems while minimizing the disturbances driving their loss. This approach is often more efficient and more cost-effective than efforts focused solely on planting new mangroves (Leal & Spalding, 2024). Alongside legal protection, local communities play a vital role in safeguarding mangrove forests from degradation and loss (Datta et al., 2012). Raising awareness and actively involving local stakeholders in mangrove management can further strengthen conservation outcomes (Walters et al., 2008).

In areas where mangroves have been present in the past, it is important to distinguish between sites that have been degraded since the loss of mangrove cover and those that remain relatively intact. Based on this distinction and having determined that planting is necessary, three different types of mangrove planting can be employed: reforestation, rehabilitation, and restoration (Schmitt & Duke, 2015). Additionally, afforestation is used in areas where mangroves did not previously grow (Siddiqi, 2001).

Because species distribution is closely tied to specific site conditions, careful attention must be given first to selecting a suitable planting site, followed by choosing the appropriate species and the most effective planting technique for that location (Lewis & Brown, 2014). Planting can be carried out using seeds, propagules, or seedlings, and site selection should be guided by an analysis of historical changes and natural processes (Schmitt & Duke, 2015). Historical data helps clarify coastal dynamics, such as patterns of accretion and erosion, and supports the selection of species that naturally thrived in the area prior to human disturbance (Lewis & Brown, 2014). This process should be complemented by observing natural regeneration, which signals that a site is suitable for mangrove growth and provides valuable insights into appropriate species choices and planting methods (Siddiqi, 2001).

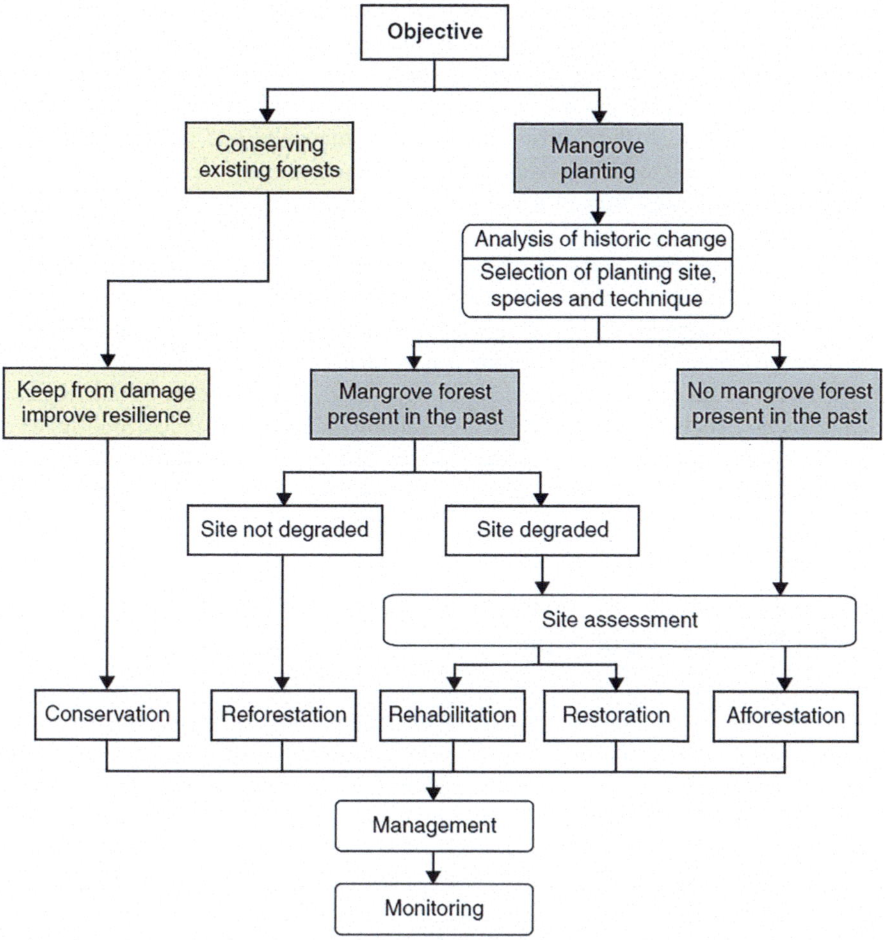

Fig. 4 Decision-making flow chart for mangrove management. Source: Modified after Schmitt and Duke (2015)

Mangrove forests have a strong capacity for natural regeneration following disturbances (Siddiqi, 2001). In areas where mangroves have been destroyed or degraded, a thorough site assessment is required to determine whether interventions such as soil or hydrological restoration are necessary before planting (Lewis & Brown, 2014). Whenever possible, rehabilitation approaches that promote natural regeneration should be prioritized (Schmitt & Duke, 2015). However, in cases where habitat loss or degradation is so severe that natural processes are insufficient for recovery, tailored, site-specific, and cost-effective rehabilitation or restoration methods must be implemented (Lewis & Brown, 2014).

Large-scale afforestation is typically undertaken to establish mangrove forests on treeless mudflats or newly accreted lands (Siddiqi, 2001). In such areas, afforestation requires specialized and carefully designed planting techniques. After planting, seedlings must be protected from human disturbances, such as destructive fishing practices and grazing by cattle or sheep (Schmitt & Duke, 2015). In some locations, additional protection from wave action may also be necessary. Once mangroves are established, they must be effectively managed and safeguarded from human activities such as felling or conversion to other land uses. Traditionally, these management and protection efforts have been overseen by government forestry departments.

Mangrove monitoring refers to the systematic collection of data and the processing of this data into information about the condition, health, and status of mangrove forests (Schmitt & Duke, 2015). It also helps to understand why changes are occurring. Ground-based monitoring of mangroves involves direct, on-site assessments of this ecosystem using appropriate approaches. While recent technological advancements, such as the availability of satellite imagery, artificial intelligence, remote sensing, and machine learning, have provided new opportunities for monitoring mangroves, ground-based data remains essential (Wang et al., 2025).

3 Considerations for Ground-Based Management and Monitoring of Mangroves

Mangrove ecosystems are shaped by a complex interplay of biological, physical, and chemical processes that sustain their structure, function, and resilience (Lugo & Snedaker, 1974). The ecological integrity of these ecosystems is maintained through the synergistic interaction of these processes, which operate across various spatial and temporal scales (Fig. 5). In addition to these natural dynamics, human dimensions—including anthropogenic pressures such as disturbances and local dependence on mangrove resources—play a significant role in influencing mangrove health and stability (Biswas et al., 2009). Disruption of any of these interconnected components, whether ecological or human-driven, can compromise the functionality of mangroves and the vital ecosystem services they provide (Schmitt & Duke, 2015). Therefore, a comprehensive understanding of these processes is essential for the effective monitoring and management of mangrove ecosystems, as outlined in the sections that follow.

3.1 Biological Processes

Biological processes in mangrove ecosystems are essential for sustaining productivity, maintaining biodiversity, and ensuring resilience to environmental change (Lugo & Snedaker, 1974). Primary production, driven by mangrove trees, algae, and epiphytic organisms, supports both local consumers and the export of energy and

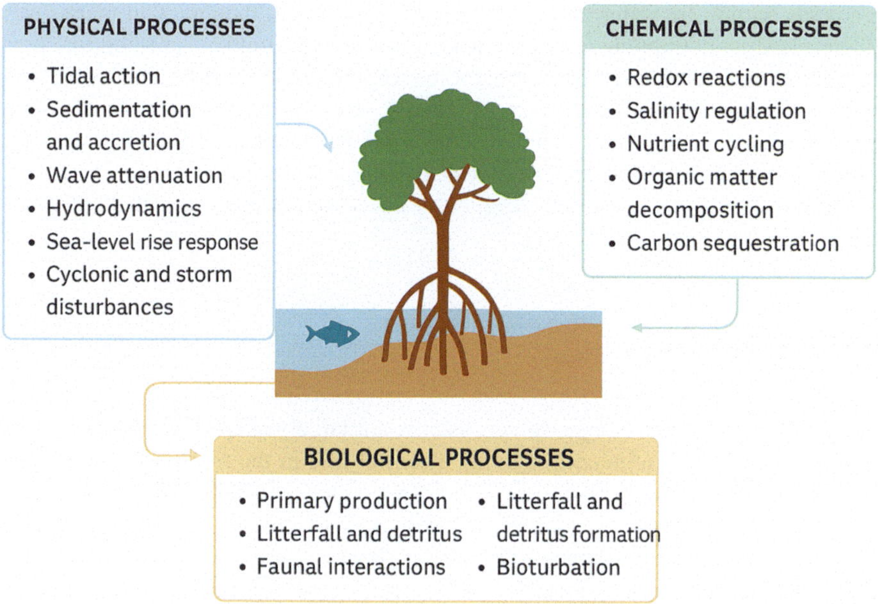

Fig. 5 The interplay of physical, chemical, and biological processes in the mangrove ecosystem

nutrients to adjacent ecosystems (Alongi, 2002). Litterfall and detrital pathways provide organic material that fuels microbial decomposition and detritivore food webs (Twilley et al., 1992). Faunal interactions and bioturbation, carried out by organisms such as crabs, mollusks, and polychaetes, play crucial roles in sediment aeration, nutrient cycling, and organic matter turnover (Kristensen, 2008). Their burrowing activities enhance oxygen penetration and stimulate microbial processes in otherwise anoxic sediments. Symbiotic relationships, such as nitrogen-fixing bacteria associated with mangrove roots, further contribute to nitrogen inputs in these typically nutrient-poor coastal environments (Kristensen et al., 2008). Finally, mangrove regeneration is shaped by recruitment, seedling establishment, and species interactions, which together drive successional dynamics and determine zonation patterns within mangrove forests (Siddiqi, 2001).

3.2 *Physical Processes*

Physical processes in mangroves shape their zonation patterns, habitat formation, and interactions with neighboring ecosystems such as seagrass beds and coral reefs. These processes are largely driven by hydrodynamics, geomorphology, and climatic forces. For example, tidal action regulates inundation patterns, sediment deposition, and nutrient fluxes across the intertidal zone (Alongi, 2009). Tidal exchange also plays a crucial role in oxygenating mangrove soils and maintaining salinity gradients within the water column. Meanwhile, sediment dynamics in mangroves enhance sediment trapping and vertical accretion, contributing to land-building processes that help buffer against sea-level rise (Krauss et al., 2014). The dense root networks of mangroves attenuate wave energy, stabilize shorelines, reduce coastal erosion, and promote further sediment deposition (Mazda et al., 1997). Additionally, storms and cyclonic disturbances can reshape mangrove stands by altering sediment profiles and clearing vegetation, triggering successional processes that influence ecosystem recovery (Duke, 2001).

3.3 Chemical Processes

Chemical processes in mangrove ecosystems are largely shaped by anaerobic soil conditions and tidal influences, playing a central role in the mangroves' contribution to global biogeochemical cycles (Kristensen et al., 2008). Mangroves help regulate salinity through specialized physiological adaptations, such as salt exclusion at the roots, salt secretion via specialized glands, or salt storage within vacuoles (Lugo & Snedaker, 1974). The sediments in mangrove systems are typically anoxic, fostering reduction reactions like sulfate reduction, methanogenesis, and denitrification, all of which influence nutrient dynamics and availability (Alongi, 1994). In addition, nitrogen cycling—including ammonification, nitrification, and denitrification—occurs within both the sediments and rhizosphere, further regulating nitrogen availability (Kristensen et al., 2008). Phosphorus cycling is primarily governed by adsorption-desorption processes involving iron and aluminum oxides in the sediments. Another key chemical process is the decomposition of organic matter, where mangrove leaf litter and woody debris break down through microbial activity, releasing dissolved organic carbon and nutrients that sustain detrital food webs (Alongi, 2002). Importantly, mangrove soils act as major carbon sinks, with the long-term burial of organic carbon under anaerobic conditions significantly contributing to "blue carbon" storage (Mukul et al., 2020b).

3.4 Human Dimensions

The human dimensions of mangroves encompass the complex social, cultural, and economic relationships between coastal communities and these vital ecosystems (Biswas et al., 2009). Mangroves provide a diverse array of ecosystem services, including fisheries, timber, fuelwood, non-timber forest products (NTFPs), coastal protection, and ecotourism opportunities, upon which millions of people worldwide depend for their livelihoods and well-being (Sarkar et al., 2024; Fig. 6). Therefore, the effective management and monitoring of mangroves must account for socio-economic factors and actively engage local communities in decision-making to ensure both ecological sustainability and community benefits (Datta et al., 2012).

Fig. 6 Mangroves provide key non-timber products like *Nypa* palm (left) and support ecotourism (right), as seen in the Sundarbans forest, where both generate important revenue and showcase the ecosystem's rich biodiversity. Photo credits: Sharif A. Mukul

4 Ground-Based Approaches for Monitoring and Managing Mangroves

Ground-based approaches play a crucial role in effective mangrove monitoring and management by providing detailed, site-specific information. These approaches include field-based surveys to assess forest structure, species composition, biomass, and regeneration status, as well as monitoring of soil and water parameters such as salinity, nutrient levels, and sedimentation (Bosire et al., 2008; Lugo & Snedaker, 1974). Participatory monitoring, involving local communities and stakeholders, enhances the accuracy and relevance of data collection by integrating traditional ecological knowledge and local observations (Walters et al., 2008). Ground-based methods also support adaptive management by enabling regular assessments of human pressures, such as harvesting intensity, land-use changes, and pollution, which are critical for designing locally appropriate conservation and restoration strategies (Datta et al., 2012). When combined with socio-economic data, these field-based approaches provide a holistic framework for managing mangroves in ways that balance ecological health with community needs (Biswas et al., 2009).

Ground-based techniques are also essential for validating and calibrating remote sensing data. Field measurements—such as vegetation density, species identification, and canopy structure—provide critical ground-truth information that improves the accuracy of satellite or aerial imagery analysis (Maurya et al., 2021). By integrating ground-based and remote sensing approaches, the overall effectiveness and precision of mangrove monitoring have significantly improved (Karsch et al., 2023). Recent technological advancements have further improved ground-based monitoring, with tools like handheld GPS units, drones, and mobile apps enabling more efficient data collection and real-time reporting. For instance, drones can capture high-resolution images of mangrove areas, while mobile apps allow field teams to record and upload data directly to centralized databases, streamlining both analysis and management efforts.

4.1 Field-Based Surveys and Sampling

Field surveys and sampling are the cornerstone of ground-based mangrove monitoring and management, providing essential, on-the-ground data. These activities involve direct observations and measurements of vegetation, fauna, soil, water parameters, and biogeochemical processes (Fig. 7). Key components include:

- **Biodiversity surveys:** Recording the presence and abundance of mangrove flora and fauna—including fish, crabs, birds, predators, and prey—to assess species composition, diversity, and ecological interactions (Mukul et al., 2019).
- **Vegetation assessment:** Measuring tree height, diameter at breast height (DBH), species composition, and canopy cover from representative plots of different size and shape to evaluate mangrove forest health, biomass, and carbon storage potential (Mukul et al., 2020b).
- **Seedling recruitment:** Monitoring the density and distribution of mangrove seedlings to gauge regeneration potential and forest recovery capacity (Siddiqi, 2001).
- **Phenology monitoring:** Tracking seasonal changes in flowering, fruiting, and leaf development to better understand mangrove life cycles and reproductive patterns in the face of climate or salinity change (Siddiqi, 2001).
- **Soil sampling:** Analyzing soil properties such as salinity, pH, organic matter content, and nutrient levels, which are critical for understanding growth conditions and detecting stressors like pollution or sediment imbalance (Twilley et al., 1992).
- **Sedimentation and erosion monitoring:** Using techniques such as sediment traps, erosion pins, and marker horizons to measure sediment deposition and

Fig. 7 Snapshots from field surveys conducted in the Sundarbans mangrove forest of Bangladesh. Photo credits: Sharif A. Mukul

erosion rates, helping assess the effects of coastal development, sea-level rise, and storm events on mangrove stability (Bhargava & Friess, 2022).
- **Water quality monitoring:** Measuring parameters such as salinity, temperature, dissolved oxygen, turbidity, and nutrient concentrations to understand water conditions influencing mangrove ecosystems (Kristensen et al., 2008). Regular monitoring of water also helps detect pollution sources, including agricultural runoff and industrial discharges, and their impacts on mangrove health.

4.2 Permanent Sample Plots

Establishing permanent plots is a common ground-based technique for long-term mangrove monitoring. These fixed plots are repeatedly surveyed over time to document changes in vegetation structure, species composition, and overall ecosystem dynamics. By tracking key indicators such as growth rates, regeneration, and mortality, permanent plots offer valuable insights into how mangrove forests respond to environmental changes and help predict their future under the influence of drivers such as climate variability, natural disturbances, and species interactions (Misiukas et al., 2021).

4.3 Socio-economic Surveys

Socio-economic surveys explore how mangroves support local livelihoods through traditional and commercial uses, such as harvesting NTFPs, fisheries, and honey production (Sarkar et al., 2024). These surveys help capture the diverse human benefits derived from mangroves and inform management strategies that are tailored to local contexts and community needs (Datta et al., 2012). Mangrove economic value, however, varies depending on valuation methods, the types of ecosystem services assessed, and local socio-economic contexts (Himes-Cornell et al., 2018). Some assessments even use composite indicators like "saved wealth" (avoided damage to property and income) and "saved health" (reduced risk of disease, injury, or death) to estimate the broader adaptation and resilience benefits provided by mangroves (Vo et al., 2012). For instance, coastal mangroves play a critical role in attenuating wave energy, reducing storm surge impacts, and preventing shoreline erosion—services that can be quantified using such approaches (Akber et al., 2018). In addition, mangroves generate substantial revenue through ecotourism, offering opportunities for local employment and environmental education. Socio-economic surveys can offer valuable insights for understanding not only the direct and indirect benefits of ecotourism but also visitors' willingness to pay for specific ecosystem services, helping guide sustainable tourism development and conservation efforts (Saha & Mukul, 2022).

4.4 Community-Based Monitoring (CBM)

Engaging local communities in mangrove monitoring is a powerful ground-based approach (Walters et al., 2008). Local knowledge and participation enhance data collection efforts and foster a sense of ownership and responsibility for mangrove conservation (Datta et al., 2012). CBM often includes activities like mangrove planting, clean-up drives, and awareness campaigns.

5 Challenges in Ground-Based Mangrove Management and Monitoring

Despite its many benefits, ground-based monitoring and management of mangroves face several significant challenges. One of the primary difficulties is accessibility, as mangrove ecosystems are often located in remote, isolated, or difficult-to-reach coastal zones, making fieldwork logistically complex and time-consuming (Halder et al., 2021). These activities are also highly resource-intensive, requiring substantial investments of time, labor, specialized skills, and financial resources to conduct surveys, sampling, and data collection effectively. Additionally, field teams working in mangrove environments are often exposed to environmental hazards, including extreme weather, unstable terrain, strong tidal currents, and encounters with potentially dangerous wildlife. Security concerns also arise, particularly in remote or less-monitored areas where the presence of armed groups, pirates, or ferocious animals may pose serious threats to field personnel, sometimes necessitating the use of armed guards or specialized protection measures to ensure safe operations.

6 Conclusion

Ground-based techniques are indispensable for effective monitoring and management of mangrove ecosystems. Despite the rapid advancements in remote sensing technologies, field-based observations provide irreplaceable, fine-scale data on species composition, forest structure, regeneration status, soil and hydrological conditions, and anthropogenic pressures. These techniques—ranging from simple plot surveys and transect measurements to more advanced methods such as soil sampling and sediment monitoring—offer critical ground-truthing for interpreting remotely sensed data and are essential for understanding ecological processes that cannot yet be fully captured from above. Furthermore, ground-based approaches play a crucial role in participatory monitoring of mangroves, engaging local communities and stakeholders, and integrating traditional ecological knowledge into management practices. This not only improves the accuracy and relevance of monitoring outcomes but also fosters local stewardship and enhances the sustainability of conservation efforts in mangroves. Integrating ground-based methods with modern geospatial tools in a complementary, hybrid monitoring framework also holds promise for delivering more comprehensive, multi-scale assessments of mangrove health and change. As mangroves continue to face mounting pressures from climate change and sea-level rise, robust ground-based monitoring will remain a cornerstone of efforts to protect this invaluable coastal ecosystem.

References

Akber, M. A., Patwary, M. M., Islam, M. A., & Rahman, M. R. (2018). Storm protection service of the Sundarbans mangrove forest, Bangladesh. *Natural Hazards, 94*, 405–418.

Akram, H., Hussain, S., Mazumdar, P., Chua, K. O., Butt, T. E., & Harikrishna, J. A. (2023). Mangrove health: A review of functions, threats, and challenges associated with mangrove management practices. *Forests, 14*, 1698.

Alongi, D. M. (1994). The role of bacteria in nutrient recycling in tropical mangrove and other coastal benthic ecosystems. *Hydrobiologia, 285*, 19–32.

Alongi, D. M. (2002). Present state and future of the world's mangrove forests. *Environmental Conservation, 29*, 331–349.

Alongi, D. M. (2009). The energetics of mangrove forests. Springer.

Bhargava, R., & Friess, D. A. (2022). Previous shoreline dynamics determine future susceptibility to cyclone impact in the Sundarban Mangrove Forest. *Frontiers in Marine Science, 9*, 814577.

Biswas, P. L., & Biswas, S. R. (2019). Mangrove forests: Ecology, management, and threats. In W. Leal Filho, A. M. Azul, L. Brandli, A. Lange Salvia, & T. Wall (Eds.), *Life on land. Encyclopedia of the UN Sustainable Development Goals*. Springer.

Biswas, S. R., Mallik, A. U., Choudhury, J. K., & Nishat, I. (2009). A unified framework for the restoration of Southeast Asian mangroves—Bridging ecology, society and economics. *Wetlands Ecology and Management, 17*, 365–383.

Bosire, J. O., Dahdouh-Guebas, F., Walton, M., Crona, B. I., Lewis, R. R., Field, C., Kairo, J. G., & Koedam, N. (2008). Functionality of restored mangroves: A review. *Aquatic Botany, 89*, 251–259.

Costanza, R., Pérez-Maqueo, O., Martinez, M. L., Sutton, P., Anderson, S. J., & Mulder, K. (2008). The value of coastal wetlands for hurricane protection. *Ambio, 37*, 241–248.

Datta, D., Chattopadhyay, R. N., & Guha, P. (2012). Community based mangrove management: A review on status and sustainability. *Journal of Environmental Management, 107*, 84–95.

Donato, D. C., Kauffman, J. B., Murdiyarso, D., Kurnianto, S., Stidham, M., & Kanninen, M. (2011). Mangroves among the most carbon-rich forests in the tropics. *Nature Geoscience, 4*, 293–297.

Duke, N. C. (2001). Gap creation and regenerative processes driving diversity and structure of mangrove ecosystems. *Wetlands Ecology and Management, 9*, 257–269.

Ewel, K. C., Twilley, R. R., & Ong, J. E. (1998). Different kinds of mangrove forests provide different goods and services. *Global Ecology and Biogeography Letters*, 7, 83–94.

FAO. (2023). *The world's mangroves 2000–2020.* Food and Agriculture Organization of the United Nations (FAO).

Friess, D. A., Rogers, K., Lovelock, C. E., Krauss, K. W., Hamilton, S. E., Lee, S. Y., Lucas, R., Primavera, J., Rajkaran, A., & Shi, S. (2019). The state of the world's mangrove forests: Past, present, and future. *Annual Review of Environment and Resources, 44*, 89–115.

Friess, D. A., Adams, J., Andradi-Brown, D. A., Bhargava, R., Carrasco, G., Dahdouh-Guebas, F., Heck, N., et al. (2024). Mangrove forests: Their status, threats, conservation and restoration. In *Treatise on estuarine and coastal science* (2nd ed.). Elsevier.

Giri, C., Ochieng, E., Tieszen, L. L., Zhu, Z., Singh, A., Loveland, T., Masek, J., & Duke, N. (2011). Status and distribution of mangrove forests of the world using earth observation satellite data. *Global Ecology and Biogeography, 20*, 154–159.
Goldberg, L., Lagomasino, D., Thomas, N., & Fatoyinbo, T. (2020). Global declines in human-driven mangrove loss. *Global Change Biology, 26*, 5844–5855.
Halder, N. K., Merchant, A., Misbahuzzaman, K., Wagner, S., & Mukul, S. A. (2021). Why some trees are more vulnerable during catastrophic cyclone events in the Sundarbans mangrove forest of Bangladesh? *Forest Ecology and Management, 490*, 119117.
Hamilton, S. E., & Friess, D. A. (2018). Global carbon stocks and potential emissions due to mangrove deforestation from 2000 to 2012. *Nature Climate Change, 8*, 240–244.
Himes-Cornell, A., Grose, S. O., & Pendleton, L. (2018). Mangrove ecosystem service values and methodological approaches to valuation: Where do we stand? *Frontiers in Marine Science, 5*, 376.
Ho, Y. S., & Mukul, S. A. (2021). Publication performance and trends in mangrove forest: A bibliometric analysis. *Sustainability, 13*, 12409.
Iftekhar, M. S., & Islam, M. R. (2004). Managing mangroves in Bangladesh: A strategy analysis. *Journal of Coastal Conservation, 10*, 139–146.
IUCN. (2024). *Red list of mangrove ecosystems*. The World Conservation Union (IUCN). Retrieved March 25, 2025, from https://iucn.org/resources/conservation-tool/iucn-red-list-ecosystems/red-list-mangrove-ecosystems
Karsch, G., Mukul, S. A., & Srivastava, S. K. (2023). Annual mangrove vegetation cover changes (2014–2020) in Indian Sundarbans National Park using Landsat-8 and Google Earth Engine. *Sustainability, 15*, 5592.
Krauss, K. W., McKee, K. L., Lovelock, C. E., Cahoon, D. R., Saintilan, N., Reef, R., & Chen, L. (2014). How mangrove forests adjust to rising sea level. *New Phytologist, 202*, 19–34.
Kristensen, E. (2008). Mangrove crabs as ecosystem engineers; with emphasis on sediment processes. *Journal of Sea Research, 59*, 30–43.
Kristensen, E., Bouillon, S., Dittmar, T., & Marchand, C. (2008). Organic carbon dynamics in mangrove ecosystems: A review. *Aquatic Botany, 89*, 201–219.
Leal, M., & Spalding, M. D. (Eds.). (2024). *The state of the world's mangroves 2024*. Global Mangrove Alliance.
Lewis, R. R., & Brown, B. (2014). *Ecological mangrove rehabilitation—A field manual for practitioners*. Mangrove Action Project, Canadian International Development Agency and OXFAM.
Lugo, A. E., & Snedaker, S. C. (1974). The ecology of mangroves. *Annual Review of Ecology and Systematics, 5*, 39–64.
Maurya, K., Mahajan, S., & Chaube, N. (2021). Remote sensing techniques: Mapping and monitoring of mangrove ecosystem—A review. *Complex & Intelligent Systems, 7*, 2797–2818.
Mazda, Y., Magi, M., Kogo, M., & Nguyen Hong, P. (1997). Mangroves as a coastal protection from waves in the Tong King delta, Vietnam. *Mangroves and Salt Marshes, 1*, 127–135.
Misiukas, J. M., Carter, S., & Herold, M. (2021). Tropical forest monitoring: Challenges and recent progress in research. *Remote Sensing, 13*, 2252.
Mohammed, S. F., Khan, A., Ahammed, S., Saimun, M. S. R., Bhuiyan, M. S., Srivastava, S. K., Mukul, S. A., & Arfin-Khan, M. A. S. (2024). Assessing vulnerability to cyclone hazards in the world's largest mangrove forest, The Sundarbans: A geospatial analysis. *Forests, 15*, 1722.
Mukul, S. A., Alamgir, M., Sohel, M. S. I., Pert, P. L., Turton, S. M., Herbohn, J., Khan, M. S. I., Ali Reza, A. H. M., Munim, S. A., & Laurance, W. F. (2019). Combined effects of climate change and sea level rise project dramatic habitat loss of critically endangered Bengal tiger in the Bangladesh Sundarbans. *Science of the Total Environment, 663*, 830–840.
Mukul, S. A., Huq, S., Herbohn, J., Seddon, N., & Laurance, W. F. (2020a). Saving the Sundarbans from development. *Science, 368*, 1198.
Mukul, S. A., Halim, M. A., & Herbohn, J. (2020b). Forest carbon stock and fluxes: Distribution, biogeochemical cycles, and measurement techniques. In W. Leal Filho, A. Azul, L. Brandli,

A. Lange Salvia, & T. Wall (Eds.), *Life on land. Encyclopedia of the UN Sustainable Development Goals*. Springer.
Romañach, S. S., DeAngelis, D. L., Koh, H. L., Li, Y., Teh, S. Y., Barizan, R. S. R., & Zhai, L. (2018). Conservation and restoration of mangroves: Global status, perspectives, and prognosis. *Ocean & Coastal Management, 154*, 72–82.
Saha, N., & Mukul, S. A. (2022). Visitor's willingness to pay for cultural ecosystem services in Bangladesh: An assessment for Lawachara National Park, a biodiversity hotspot. *Small-Scale Forestry, 21*, 185–201.
Sarkar, P., Banerjee, S., Biswas, S., Saha, S., Pal, D., Naskar, M. K., Srivastava, S. K., Barman, D., Kar, G., & Mukul, S. A. (2024). Contribution of mangrove ecosystem services to local livelihoods in the Indian Sundarbans. *Sustainability, 16*, 6804.
Schmitt, K., & Duke, N. C. (2015). Mangrove management, assessment and monitoring. In M. Köhl & L. Pancel (Eds.), *Tropical Forestry Handbook*. Springer.
Siddiqi, N. A. (2001). *Mangrove forestry in Bangladesh*. Institute of Forestry and Environmental Science, Chittagong University.
Spalding, M., Kainuma, M., & Collins, L. (2010). *Word Atlas of Mangroves*. Earthscan.
Stubbs, B. J., & Saenger, P. (2002). The application of forestry principles to the design, execution and evaluation of mangrove restoration projects. *Bois et Forêts des Tropiques, 273*, 5–21.
Thomas, N., Lucas, R., Bunting, P., Hardy, A., Rosenqvist, A., & Simard, M. (2017). Distribution and drivers of global mangrove forest change, 1996–2010. *PLoS One, 12*, e0179302.
Twilley, R. R., Chen, R. H., & Hargis, T. (1992). Carbon sinks in mangroves and their implications to the carbon budget of tropical coastal ecosystems. *Water, Air, and Soil Pollution, 64*, 265–288.
Vo, Q. T., Kuenzer, C., Vo, Q. M., Moder, F., & Oppelt, N. (2012). Review of valuation methods for mangrove ecosystem services. *Ecological Indicators, 23*, 431–446.
Walters, B. B., Rönnbäck, P., Kovacs, J. M., Crona, B., Hussain, S. A., Badola, R., et al. (2008). Ethnobiology, socio-economics and management of mangrove forests: A review. *Aquatic Botany, 89*, 220–236.
Wang, T., Zuo, Y., Manda, T., Hwarari, D., & Yang, L. (2025). Harnessing artificial intelligence, machine learning and deep learning for sustainable forestry management and conservation: Transformative potential and future perspectives. *Plants, 14*, 998.
zu Ermgassen, P. S. E., Worthington, T. A., Gair, J. R., et al. (2025). Mangroves support an estimated annual abundance of over 700 billion juvenile fish and invertebrates. *Communications Earth & Environment, 6*, 299.

Conventional Remote Sensing and GIS Techniques for Mangrove Ecosystems

Protecting Mangrove Ecosystems Through Proper Nursery Management in Karangsong and Pabean Ilir of Indramayu Regency: A Comparative Study with Indian Sundarbans with Remote Sensing Approaches

Reginawanti Hindersah, Mochamad Candra Wirawan Arief, Sondi Kuswaryan, Saon Banerjee, Saroni Biswas, and Piyali Sarkar

Abstract The mangrove ecosystem is highly impacted by climate change phenomena, especially the rise of sea levels and the severity of extreme weather events like cyclones. The overall coastal areas of Asia are mostly suffered by sea level rise and land degradation. During 1989 to 2007, around 70% of coastal land is deformed in West Bengal, India. Similarly, 7 mm of sea level rise was observed in Sundarbans during the said time period. Moreover, social pressure and deforestation are posing a threat to this ecosystem. This study appraises the role, status, and extent of the nursery developed in Indramayu Regency, West Java Province, Indonesia, and evaluates its similarities or differences with Indian Sundarbans. As mangrove naturally grows in the intertidal zone where sea water in the ocean meets the land, the nursery for mangrove seedlings also needs saline water to grow. Generally, seedlings are grown at moderate salinity, and those seedlings show better vegetative performance than seedlings grown under very high or low salt concentrations. The role of shading in developing mangrove nurseries was evaluated in this study. For example, *Rhizophora apiculata* seedlings prefer 75% shading. The key mangrove species grown in nursery in Indramayu are *Avicenna marina* (Forssk) Vierh, *Avicennia alba*

R. Hindersah
Faculty of Agriculture, Universitas Padjadjaran, Sumedang, West Java, Indonesia

M. C. W. Arief
Center for Environment and Sustainability Science, Faculty of Fishery and Marine Science, Universitas Padjadjaran, Sumedang, West Java, Indonesia

S. Kuswaryan
Faculty of Animal Husbandry, Universitas Padjadjaran, Sumedang, West Java, Indonesia

S. Banerjee (✉) · S. Biswas · P. Sarkar
Department of Agricultural Meteorology and Physics, Bidhan Chandra Krishi Viswavidyalaya, Mohanpur, West Bengal, India

S. K. Srivastava et al. (eds.), *Earth Observation for Monitoring Mangrove Ecosystems*, https://doi.org/10.1007/978-3-032-16668-5_4

Blume, *Bruguiera gymnorhiza*, *Rhizophora apiculata* Blume, *Rhizophora mucronata* Lam., and *Rhizophora stylosa* Griff. In contrast, in India the prominent mangrove species are *Heritiera fomes*, *Xylocarpus mekongensis*, *Xylocarpus granatum*, *Sonneratia apetala*, *Bruguiera gymnorhiza*, etc. In both countries, seedlings are grown in polybags with drainage holes filled with local soil. Mangrove restoration in the Indian Sundarbans is boosted by using grasses, especially in the erosion--riddled and nutrient-deficient patches. The intervention of local colleges and universities would also help the restoration process. In Indonesia, the Directorate of Conservation Areas and the Ministry of Forestry have disseminated guidelines for mangrove nurseries. Likewise, the Indian Government may involve local educational Institutions for this purpose. The demand for good-quality mangrove-seedling growers will increase in the future, and in turn, it will be very helpful to enhance the income of local inhabitants. The continuous involvement of mangrove experts, universities, and the government, as well as mangrove-related institutions, is needed to improve the quality of mangrove seedlings, which ensures the success of the rehabilitation of the Indramayu mangrove forest and Indian Sundarbans.

Keywords Mangroves · Nursery method · Seedling quality · Coastal livelihood · Ecosystem

1 Introduction

The impacts of climate change, population pressure, sea level rise, and land degradation on mangroves have received much attention at present. The annual rate of mangrove deforestation is occurring at a rate of 1–2%, which indicates that most forests will disappear within the current century. However, mangrove ecosystems exhibit a high level of their unique plant architecture, nutrient uptake pattern and self-biotic controls (Alongi, 2002). The importance of seedlings in mangrove forest rehabilitation is very much important. The growth and sustainability of a mangrove ecosystem depend on the survival and establishment of seedlings, accelerating restoration and the resistance of different plant species to environmental stress. The stress may be hydrodynamic, soil-related stress, and climatic stress (Biswas et al., 2017).

The survival of seedlings in the mangrove nursery and its performance after transplantation depend on many factors. The medium soil salinity is needed for preparation of mangrove seedling; hence the top soil management is very important (Rizki & Novi, 2018). The role of shading in developing mangrove nurseries is another important aspect. For example, *Rhizophora apiculata* seedlings prefer 75% shading. The dominant mangrove species are very much specific for a particular region. The *Avicenna marina* (Forssk) Vierh, *A. alba* Blume, *Bruguiera gymnorhiza*, *Rhizophora apiculata* Blume, *Rhizophora mucronata* Lam., etc. are grown in

Indramayu region of Indonesia. Hence, the seedlings of those species should be grown in nursery (Rahmattunisa et al., 2019).

Mangrove forests have declined dramatically, especially in Asia. Camacho et al. (2020) note that Asia holds the largest extent of the world's mangroves yet also suffers the fastest deforestation. According to FAO (2015), mangrove cover fell from about 6.03 million hectares in 2010 to 5.33 million hectares by 2015 in Asia. In response, many countries have undertaken mangrove rehabilitation programs. However, most past efforts have failed when they emphasized mass planting without scientific guidance. For example, initiatives in the Philippines (World Bank, 1984–1992) saw fewer than 20% of three million nursery seedlings survive beyond a few years, largely because planting methods ignored site conditions (Primavera & Esteban, 2008). Similarly, in Thailand, thousands of corporate-planted seedlings at Bang Kaew died when tidal flooding inundated an improperly chosen site (Lewis, 2005). These failures highlight that restoration requires more than sheer numbers of transplants—it needs high-quality seedlings, proper species choice, and site preparation (Ellison et al., 2020). Nurseries play a critical role by providing robust seedlings adapted to stresses. Seedlings raised in controlled nurseries benefit from careful tending (watering, weeding, hardening) that promotes vigour. Such conditioning can markedly improve outplant survival. For example, Toledo et al. (2001) found that 3-month-old nursery-grown *Avicennia germinans* seedlings had 100% survival in the nursery and 74% survival after 24 months in the field. Well-developed nursery plants often withstand hydrodynamic and soil stresses better than spindly propagules. In fact, meta-analysis guidelines suggest using "well-established (larger) saplings" of hardy species (such as *Avicennia marina*) can enhance restoration success in temperate and subtropical zones (Bayraktarov et al., 2016). Good nursery practice—growing seedlings near coastal nurseries, providing shade, watering, cleaning algae, and hardening off—ensures each plant can resist salinity, waves, and other stresses upon out planting. Properly reared seedlings also establish more quickly, accelerating forest recovery and extending ecosystem services sooner after planting (Sasmito et al., 2023).

Despite the advantages of nurseries, mangrove replanting still faces major challenges. High mortality is common when ecological conditions are unsuitable. Many projects simply "grow seedlings in greenhouses and transplant them" without addressing hydrology—a strategy proven ineffective (Lewis, 2005; Ellison et al., 2020). For instance, an area in Thailand saw 1000 logo-adorned seedlings planted in April 2023, but all but one perished because the low-lying plot flooded at high tide (Waters, 2023). Such examples underline that site selection is paramount: planting a healthy seedling in the wrong spot (too deep, too dry, polluted, etc.) dooms it. In general, restoration failures often stem from an "oversimplified view of mangroves" that focuses on tree count rather than ecology (Primavera & Esteban, 2008). Waters (2023) emphasizes that dynamic tidal and soil conditions must be managed—otherwise, even millions of planted seedlings will largely die. The lesson is clear: rehabilitation success depends on matching seedling stock to suitable habitat and providing ongoing care after planting. Modern geospatial tools are increasingly used to improve mangrove nursery and planting programs. Geographic Information

Systems (GIS) and satellite remote sensing can guide decisions on where and when to plant. For example, Syahid et al. (2019) developed an Analytic Hierarchy Process-based (AHP-based) land suitability model using satellite data and climate projections to identify nearly four million hectares of "very suitable" mangrove planting sites in Southeast Asia. Such GIS-based mapping helps allocate resources by highlighting areas with appropriate hydrodynamics, elevation, and protection needs. Likewise, satellite imagery (e.g. Landsat, Sentinel) can monitor mangrove cover change, biomass, and plant health indices over time. Kanniah et al. (2015) note that regular satellite observations provide up-to-date information on mangrove extent and condition, enabling managers to detect stress or damage early. In practice, remote sensing has been used to map species composition, chlorophyll content and stress signals in mangrove nurseries and replanted areas (Bhadra et al., 2023). By combining GIS site analyses and ongoing satellite monitoring, projects can be fine-tuned (e.g. adjusting irrigation or follow-up planting) and restoration planners can make precise, data-driven decisions.

Asia's restoration programs are pioneering innovative nursery techniques and technologies. In Indonesia, researchers at Universitas Diponegoro have introduced biodegradable "eco-polybags" for seedlings. These bags replace conventional plastic pots so each seedling can be planted directly in the ground without removing it from the bag, boosting root survival (Sodikin et al., 2023). This approach reduces transplant shock and prevents plastic waste, effectively increasing the survival rate of out planted seedlings. In Thailand, community-led nurseries emphasize best practices: for example, local practitioners often plant paired seedlings ("buddy trees"), recognizing that neighbouring stems support each other's root systems as they grow (Das, 2025). These community efforts have yielded results—for example, a small village's mangrove grove now hosts otters, kingfishers, and even dolphins, indicating a recovering ecosystem (Das, 2025).

High-tech approaches are also emerging. The Ericsson "Connected Mangroves" project in Malaysia equips saplings with IoT sensors that record soil moisture, temperature, water levels, and intrusion in real time. Data are transmitted to a cloud dashboard for local managers, allowing rapid interventions (e.g. irrigation or debris removal). Based on pilot results, this ICT system is anticipated to halve seedling mortality (Ericsson, 2021). Such sensor-driven nurseries demonstrate how Asian projects integrate technology and community participation to overcome traditional challenges. Other innovations include mobile drip-irrigation systems in Malaysian coastal villages and remote-monitoring apps for Bangladeshi mangrove nurseries (FAO, 2020).

Hence, successful mangrove rehabilitation in Asia hinges on high-quality nursery seedlings, careful planning, and adaptive management. Seedling-based approaches have accelerated restoration where nurseries produce resilient plants, whereas replanting without sufficient stock or scientific guidance has repeatedly failed. The latest integration of GIS/site modelling and satellite or sensor monitoring further enhances nursery programs, enabling precise site choice and early stress detection. Together, these practices, backed by both new research and local innovation, are raising survival rates and giving Asian mangrove nurseries better chances of restoring these critical coastal forests.

Keeping these points in mind, the present chapter evaluates the status and extent of the nursery developed in Indramayu Regency, West Java Province, Indonesia, and compares its similarities or differences with the Indian Sundarbans.

2 Study Region: Indramayu Regency

Indramayu Regency is located in the West Java Province of Indonesia (Fig. 1). The coastal area of Indramayu used to be a dense mangrove forest, but the government's policy in 1993 allowed the development of shrimp ponds in the region, leading to a significant reduction of mangrove areas and degradation. In 2012, Pabean Ilir village had 562.1 ha of fish and shrimp ponds, which were developed on mangrove forests. The shrimp pond area in 2012 covered 30.89% of the village area (Sodikin, 2012). Public awareness of coastal abrasion and saltwater intrusion resulted in a will to restore a mangrove movement aimed at rehabilitating and replanting the degraded mangrove area. Mangrove rehabilitation activities in Indramayu were initiated in 2008 by the Pantai Lestari (Sustainable Beach) Communities of Karangsong

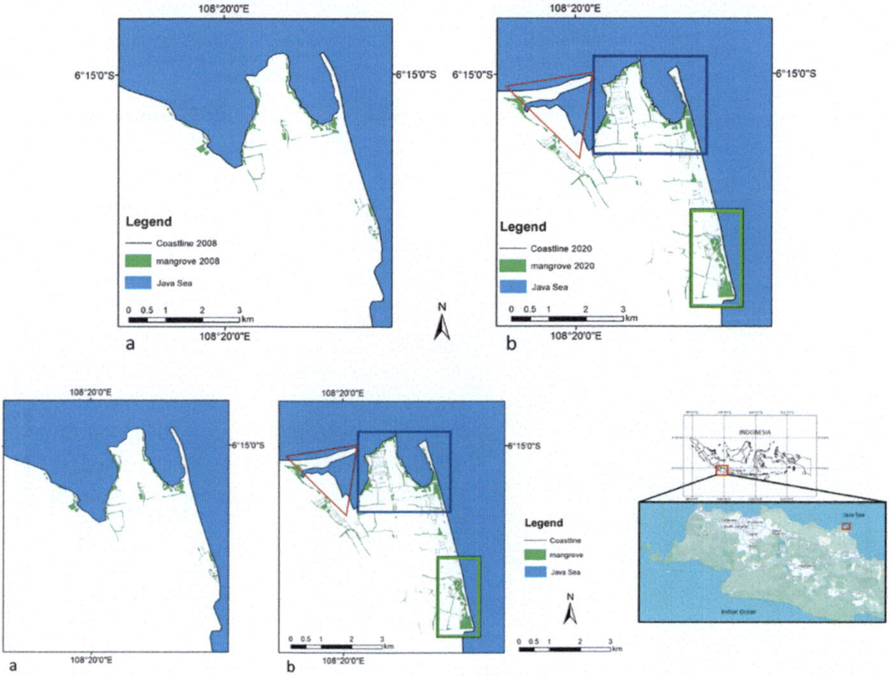

Fig. 1 Mangrove belt in the coastal area of Karangsong (in green) and Pabean Ilir (in blue) village of West Java, Indonesia in 2008 (**a**) and 2020 (**b**). A red triangle indicates the new emerging land

Village. This initiative leads to the recovery of the mangrove forest from 2008 onwards in Karangsong about 6.58 ha and Pabean Ilir about 51.72 ha. In 2017, rehabilitated mangrove forests in Karangsong and Pabean Ilir covered about 35 ha and 132.9 ha, respectively (Gunawan et al., 2018; Rahmattunisa et al., 2019). Due to severe sedimentation in the mainland, the new emerging land (deltaber) is developing in the northern part of Indramayu.

3 Study Region: Sundarbans

One of the world's largest mangrove forests is the Sundarbans, located in the Northern Bay of Bengal (Fig. 2). The mangrove forest in Sundarbans, covering about 10,000 km^2, is spread across India and Bangladesh, sharing approximately 40% and 60% of the total forest area, respectively (Roy et al., 2025). The Indian Sundarbans is divided into three parts, namely, the core zone, buffer zone, and transition zone. It is also characterized by innumerable rivers, rivulets, estuaries, coastal dunes, mud flats, creeks, inlets, mangrove swamps, and sandy beaches. The Sundarbans region consists of about 106 deltaic islands, of which 54 are inhabited by human settlements. Based upon the bio-geophysical attributes, the Sundarbans are distinctively classified into three divisions, namely, the swamp forest, mature delta, and sea face/beach. Mangrove trees are used for timber and construction

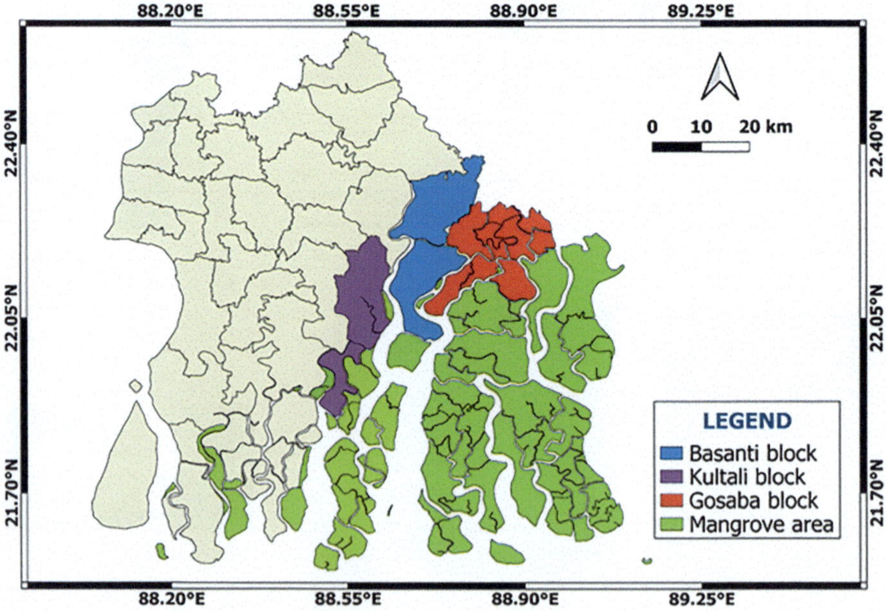

Fig. 2 The stretch and position of the Indian Sundarbans

materials (e.g. for houses, boats, and traps) as well as for fuel and charcoal production. Apiculture is widespread within the Sundarbans mangrove forests and provides honey and wax. Around 2000 people are engaged in beekeeping in the Indian Sundarbans, producing approximately 90% of India's natural honey. In addition, mangrove plants provide tannins for leather production and a wide array of medicinal uses. While crabs, molluscs, shrimps, and fish are caught in the ocean and the brackish waters surrounding the mangrove forests, the mangrove proper is the most important source for shrimp larvae, supplying the aquaculture. Mangroves are the nursery ground for many commercially important fish species.

A considerable decline of open mangrove forest stands and a reduction in biodiversity have been observed, mostly in the transitional zone of forest habitation in the Sundarbans. This transformation is attributed to the shifting of local people's interest from traditional forestry and subsistence farming towards alternative occupations like shrimp culture, coastal tourism, and commercial fishing, although environmentally hazardous livelihood activities like the collection of prawn seeds along the riverbanks are still persistent.

In Indonesia, mangrove seedling production programs have been supported by the local government (The Regency Office, The Marine and Fishery Department, and The Environmental Service of Indramayu) and the private sector (The *Pertamina* Oil and Gas National Company). These organizations provide the knowledge and skills for mangrove rehabilitation and replanting needed for establishing mangrove nurseries. Growing and purchasing mangrove seedlings has become a novel livelihood in the mangrove area of Indramayu. The support for mangrove nurseries from either private organizations or the government ceased in 2019. However, the local communities took the initiative to continue and maintain mangrove seedlings based on their experience, and they keep selling the seedlings without government and private sector support. Nonetheless, the State organizations are still focused on mangrove rehabilitation, and many mangrove seedlings are needed. In Sundarbans, the mangrove nursery development and seedling production program are regulated by Government Agencies and $$NGOs. Both the permanent nursery (for a long--term supply of seedlings) and the temporary nursery (for a short-term supply of seedlings) are observed here.

Based on the field data, the mangrove nurseries in the study regions of India and Indonesia are likely to produce good-quality seedlings. However, the farmers have not documented their nursery development methods, and therefore, such technology for mangrove seed production is not widely shared. Assessing the mangrove nursery design and propagation methods used in the study area are important steps to improving seedling production and subsequent successful rehabilitation of mangrove forests in these regions.

4 Mangrove Nursery Design

All types of nurseries were developed while considering the availability of seawater and shading. Seawater is prominent for mangrove growth since, naturally, they inhabit the intertidal zone where seawater in the ocean meets the land. The seedling growers adapted to relatively high salinity seawater, but seedlings grown at moderate salinity had better vegetative performance than seedlings grown at lower or higher salt concentrations (Moslehi et al., 2021). More than decades ago, Ball (1998) noticed that mangrove species richness was reduced in areas experiencing prolonged exposure to freshwater or hypersaline conditions. Shading of mangroves is very important for maintaining seedling quality (da Silva & Maia, 2019). Reducing 50% shade intensifies by shading net leads to the best growth of *R. mucronata*, while 75% shade intensifies supports the growth of *R. apiculata* seedlings (Basyuni et al., 2020). Shade intensity correlates with light intensity and, hence, the carbon fixation in photosynthesis (Ulqodry et al., 2014).

In Indonesia, the communities preferred to construct the nursery in the coastal area and pond. During the pre-COVID-19 period, the seedlings' demand for mangrove forest rehabilitation in West Java and East Java was high. The inhabitants extended nurseries not only in the wetland areas but also in the dryland areas. In the present study, four types of mangrove nursery development were noticed in Indramayu (Figs. 3 and 4). The nursery constructions are:

(a) In the tidal area on the coast area directly adjacent to the mainland, where natural shade from mangrove trees is utilized
(b) In the shrimp or fish pond (used to be mangrove forests) where artificial shading using a shading net is used
(c) In the garden around houses where existing trees are used for optimal shading
(d) In the dryland areas near the coast without seawater inundation with artificial shading

The farmers implemented a sound drainage system to avoid excessive waterlogging in the coastal nursery; they also controlled the water level by increasing the bed height from the ground (around 30 cm) in the ponds. They maintained the puddle by 2–5 cm so that the water would not flood the polybag where the mangrove propagule was grown. According to the farmers, tidal water never damages the seedlings. Most growers do not have a warehouse to store propagules before planting or fertilizer, organic matter, and nursery equipment. For all such farmers, mangrove nurseries are not the main source of livelihood because there is no regular demand that guarantees continuity of income.

In India, it is observed that the success in direct seeding is poor. Sometimes, direct seeding shows promising results for species with large propagules, such as *Rhizophora* spp., *Ceriops* spp., and *Bruguiera* spp. Forest nursery is the only means of seedling production for large-scale plantations. Two main types of nurseries are observed based on tidal location, namely, floating mangrove nurseries and flooded mangrove nurseries. A floating type of nursery is established in upland areas (above

Fig. 3 Mangrove nursery to support mangrove forest rehabilitation and replanting conducted in a the tidal area under mangrove trees (**a**), in fish ponds (**b**), in a home garden (**c**), and in dryland near natural mangrove forests (**d**)

Fig. 4 A remote sensing image of the mangrove nursery area in Indamayu, Indonesia, is shown by red lines

the highest tidal range), and it can be used for a long-term supply of seedlings. A flooded mangrove nursery is established in low intertidal areas regularly flooded by tidal waters. It provides a short-term supply of seedlings. The plastic pot and bags are mainly used to carry the seedlings to desirable areas (Fig. 5). The mangrove nursery area is shown in Fig. 6.

Fig. 5 Mangrove nursery in Sundarbans

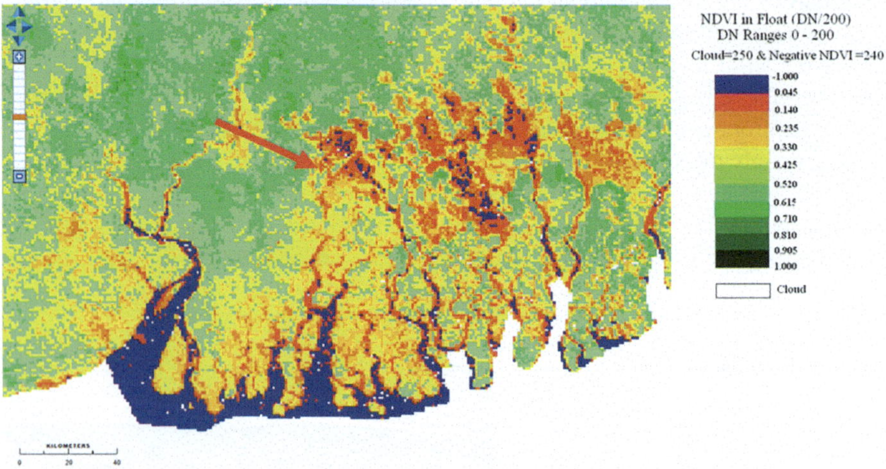

Fig. 6 NDVI map of June month of June in the Sundarbans region generated from OCM-2 (Ocean Colour Monitor 2) of the OCEANSAT-2 satellite, showing the Mangrove Nursery area through red arrow

5 Mangrove Species in the Nursery

The key mangrove species in Indramayu included *Avicenna marina* (Forssk) Vierh, *A. alba* Blume, *Bruguiera gymnorhiza*, *Rhizophora apiculata* Blume, *Rhizophora mucronata* Lam., and *Rhizophora stylosa* Griff. The *R. mucronata* population of mangrove forests in 2018 was 64% of the total plant number. Therefore, the demand for *R. mucronata* seedlings for rehabilitating or replanting mangrove areas is high. This species became dominant in the nursery. In small quantities, a few farmers grow *R. apiculata* and *R. stylosa*. Moreover, the propagules of *R. mucronata* are available mainly in the fruiting season of July to December. The inhabitants, especially housewives, take the propagule from mother plants that are well cared for. *R. mucronata* is easier to grow compared to other mangrove species. They also produce *R. stylosa* seedlings with shorter propagules that are more drought-tolerant.

Due to the low demand for *R. stylosa* seedlings, the growers have not preferred this species even though the survival rate of *R. stylosa* in the nursery was 80%, while that of *R. mucronata* was 70%. The *Avicennia marina* was rarely grown in the nursery because it is more challenging to grow this species in artificial conditions, even though the selling price is higher than other species. The survival rate of *A. marina* in the nursery was less than 50%. In Sundarbans, mangrove seedlings' survival rate is generally 77% after 4–7 months. However, *A. marina* had an 82% survival rate when transplanted as nursery-grown seedlings, while *R. mucronata* had a 55% survival rate using the drain and trench method (Chowdhury et al., 2018).

The Indonesian Ministry of Forestry-Japan International Corporation Agency--Sriwijaya University has published guidelines for rehabilitating mangroves in the ex-pond area (JICA, 2014). For nurseries, the manual has explained in detail about (1) nursery development; (2) nursery methods including fruit (cryptoviviparous)/propagule collection and selection, and fruit/seed cleaning and treatment, media preparation, planting, and maintenance of seedlings; and (3) Land preparation for rehabilitation/replanting. This guide also explains the importance of analysing soil characteristics, including water temperature, pH, and salinity. Indonesian university has released another technical guide for mangrove nurseries. The local government possibly has not yet disseminated such documents to mangrove seedling growers.

6 Propagation Method

The seedlings were grown in a 12-cm-height black polyethylene bag (polybag) with drainage holes containing local soil (Fig. 3a). In the wetland nursery, the soil was taken from the sediment, while in the dryland nursery, the soil is garden soil. Based on soil proximate analysis, the chemical characteristics of nursery soils are different (Table 1). The organic carbon (organic-C) of soil (sediment) in the wetland is higher than in the dryland due to slow enzymatic decomposition in a low-oxygen environment. The higher total nitrogen (total-N) in wetland areas is possibly caused by the water around the mangrove area being rich in organic and inorganic nutrients. Organic waste from shrimp/fish might have increased organic-C and total N. The soil reaction is neutral; most plant nutrients are available when soil pH is near neutral at pH 6.5–7.5.

The soil organic carbon is high in the western Indian Sundarbans because the area is adjacent to the protected mangrove forest. However, spatial variations in SOC are varied and decrease with the soil depth (Mitra et al., 2012). According to

Table 1 The content of organic carbon, plant macronutrients, and acidity of nursery soil

		Organic-C	Total-N	P	K		
Location	Depth (cm)	%				pH	EC
Wetland nursery	0–20	3.14	0.37	0.09	0.48	7.06	3.86
Dry land nursery	0–20	0.94	0.18	0.11	0.41	7.16	1.40

Mitra et al. (2012), the SOC in western and eastern Sundarbans was less than 0.9%, which is as low as SOC in Indonesian local soil, which is less than 1% (Table 1). The relatively low soil salinity is reported in the western sector of Sundarbans due to the freshwater release of the Farakka discharge through the main Hooghly channel, even though the increase of soil salinity due to the tidal inundation is reported (Biswas et al., 2017). In the Indian Sundarbans Mangrove Forest, the salinity in the land was 3.12–9.3 ds/m (Mandal et al., 2019), higher than that of the local soil of Indramayu.

The mature propagules/seeds are collected from mother plants for further growth. The picking time of propagules and seeds depends on the mangrove species (Table 2).

The growers need to treat the propagules treatment before planting. As many as 50–100 propagules were tied and stored horizontally in the shade for 7–14 days. Artificial shade is sometimes needed to reduce transpiration and balance the moisture throughout the propagule. Then, the propagule is plugged into the potted soil. Propagules planted directly in polybags without storage can be dried out and then bent so the seedlings are not straight. The height of the seedlings in the nursery depends on the species; generally, seedlings with two to four leaves can be transplanted to a rehabilitation/replanting location.

In Indramayu, the rhizophora nursery takes 6–7 months to produce 70-m-high seedlings ready to be transplanted. One pair of mature leaves will appear in the first month, and the 6- to 7-month-old seedlings have two to three pairs of mature leaves. Seedlings with two pairs of leaves (2 months old) can be used in rehabilitation for planting on the beachside. Plants taller than 1 m are not planted because their stems break easily (Fig. 7). For *Avicenna*, farmers have their own method for obtaining higher *Avicenna* seedlings. In 12-cm-height polybags, the propagule of Avicenna produces 30- to 40-cm-height seedlings, which are too low to avoid being submerged by rising tides. Therefore, the polybag must be larger, about 15-cm-height, and the height of 1-year-old seedlings can reach 60–70 cm, reducing the waves'

Table 2 Flowering and propagules/fruits harvesting of mangrove grown in Indamayu and Sundarbans

Species	Flowering season	Harvesting season
Indramayu area		
R. mucronata	May to August	July–December
R. apiculata	October–February	January–July
R. stylosa	October–February	January–July
Avicenna sp.	September–February	December–July (next year)
Sundarbans area		
Heritiera fomes	March–June	July–September
Xylocarpus mekongensis	April	August
X. granatum	June–July	August–October
Sonneratia apetala	March-–August	June–September
Bruguiera gymnorhiza	April–July	May–August

Fig. 7 (**a**) Mangrove seedlings were grown in the dryland with potted soil in Indramayu; (**b**) a grower with ready-to-transplant mangrove seedlings with two and three pairs of leaves

submergence when transplanting in the rehabilitation site. The delay in planting and limited buyers caused the mangrove seedling height of *R. mucronata* to become more than 1.5 m. Generally, farmers leave these plants in nurseries. Therefore, the nursery can turn into a mangrove forest. In the Sundarbans forest, seedling-raising nurseries last for 4–7 months; typically, seedlings reach a height of under 1 m.

7 Seedling Maintenance

Growers do not maintain plants intensively other than watering (dryland nursery) and managing drainage (wetland nursery). Fallen plants are also ignored. Mangrove researchers in India and Indonesia agree that fertilizer applications improve mangrove seedling growth, survival, and quality. Application of 6–12 g of nitrogen, phosphorous, and potassium (NPK) per plant grown in 10 kg growing media increased the number of leaves of mangrove *Rhizophora* seedlings up to two times more than that of mangrove without NPK (Trisnawati et al., 2019). The use of a mix of topsoil (70%) with either cow manure or chicken manure (30%) increased the height of *Rhizophora* seedlings to about five times higher than that of propagule seeds grown in muddy sand sediments in mangrove forests (Rizki & Novi, 2018). Growers believed that naturally, nurseries can produce good seedlings without good maintenance. The NPK strongly impacts the height and growth of mangrove seedlings in the Sundarbans. Nonetheless, the effectiveness of NPK on specific species, such *as Xylocarpus mekongensis* is low. However, recently, growers in Indramayu reported that they applied NPK fertilizer as much as about 3–5 g per plant to accelerate the growth due to the increased demand for seedlings. Moreover, growers applied insecticide to control the caterpillar that was eating up mangrove leaves.

In both countries, mangrove nurseries may generate income for growers. In 2022, the price of ready-to-grow seedlings in Indramayu ranged from Rp 1000 to Rp 2000 (approximately 0.07–0.14 USD), while cultivating a single seedling costs half the selling price. The revenue generated from these nurseries depends on the mangrove rehabilitation programs initiated by local or central government authorities and NGOs. A similar situation occurs in the Sundarbans; mangrove seedlings are primarily purchased by the government, NGOs, and local communities involved in mangrove restoration projects.

High-quality seedlings are not the only determining factor in the success of greening the mangrove area. In both study areas, climate change poses a challenge to ensuring the survival rate of seedlings. Climate resilience strategies for successful rehabilitation programs include selecting salt-tolerant mangroves through ion compartmentation, osmoregulation, and adaptation to salt influx (Parida & Jha, 2010). Growing mangroves in the intertidal zone is suggested to enhance plant survival rates. Early warning systems for extreme weather are essential for mangrove rehabilitation. The Indonesian Meteorology, Climatology, and Geophysics Office manage the early warning system for weather-related events (i.e. heavy rain, lightning, and strong winds) through various communication channels, including websites, mobile applications, SMS blasts, and social media, which are integrated with digital TV broadcasts. However, in general, farmers are not sufficiently aware of changes in the weather.

Both countries share similar mangrove rehabilitation policies aimed at restoring and enhancing mangrove ecosystems to reduce coastal erosion, boost the economies of coastal communities, and sometimes support ecotourism. The distinction lies in their use of remote sensing and GIS. India's policy emphasizes the use of remote sensing and GIS technologies for mangrove management and climate change resilience (Ghosh et al., 2015; Sunkur et al., 2024). The uses of this method in Indonesia are minor compared to India. Remote sensing and GIS are used in the research, i.e. mapping the degradation of mangrove areas.

8 Conclusion

In general, mangrove nursery development in Indramayu is not following the guidelines for mangrove nurseries in Indonesia, although they have been producing mangrove seedlings since 2017. The Directorate of Conservation Areas and the Development of Protected Forests and the Directorate General of Forest Protection and Nature Conservation at the Indonesian Ministry of Forestry have disseminated guidelines for mangrove nurseries together with JICA and Sriwijaya University in

South Sulawesi. The scope for rehabilitating mangroves in the ex-pond area is also explored.

The selection of the nursery location was appropriate; growers chose a flat topographic area, free from strong winds, close to the planting (rehabilitation) location, an accessible location, and close to the workers' residences. However, the nursery building design needs to be improved, especially those on dry land. For example, shading and bed design in dryland nurseries and organic matter amendment to enrich organic C in dryland soil. The Ministry of Forestry suggests the use of polybags in mangrove nurseries. Muddy sediment is appropriate for nurseries in the wetland because it agrees with the conditions for growing mangroves. However, for dryland nurseries, farmers need to mix the mineral soil with organic matter because, naturally, the topsoil of dryland does not meet the requirements for growing mangrove seedlings. At the work site, the mangrove area sediment contains relatively high organic matter (3.75%), while the dryland topsoil is only 0.9%. Therefore, organic matter amendment is required to increase organic-C content.

Mangrove nurseries are an additional livelihood for the coastal residents of Indramayu, especially before the COVID-19 pandemic. The mangrove nursery around the houses of local people shows the residents' enthusiasm for getting additional income from mangroves. Such a nursery method is not practiced in the Sundarbans area. The mangrove forest rehabilitation program and mangrove replanting in pond areas of Indramayu are likely to increase in the post-COVID era in West Java, Indonesia. Nursery and mangrove rehabilitation programs in India are encouraged due to consecutive cyclones, like Amphan, Bulbul, and Yaas. Mangrove plays an important role against cyclonic storms and saltwater intrusion during cyclonic activity. Earlier, especially before 2010, only one cyclone was observed in 3 years. But in the present decade, two cyclones per year are observed. Thus, the demand for qualified mangrove seedling growers is increasing, and it will positively influence people's income. The continuous involvement of mangrove experts, universities, the government, and mangrove-related institutions is needed to improve the quality of mangrove seedlings, ensuring the success of mangrove forest rehabilitation in Indramayu, Indonesia, and the Sundarbans, India.

Looking forward, several innovations are needed that can transform nursery science. Climate projection modelling will guide adaptive planting under rising sea levels and storm risks (Alongi, 2022). Biotechnological advances, including CRISPR-based approaches, may enhance tolerance to salinity and submergence (Parida & Jha, 2010). AI-driven drone monitoring can automate seedling counts and detect stress or pest outbreaks, reducing labour costs and response times (Bhadra et al., 2023). Finally, transboundary cooperation, as needed in shared ecosystems like the Sundarbans, will enable coordinated monitoring, financing, and ecosystem connectivity (Camacho et al., 2020). Together, these recommendations and innovations point to an integrated strategy where nurseries are not merely seedling suppliers but hubs of ecological resilience, technological innovation, and socio-economic empowerment.

Acknowledgement The support from the Asia Pacific Network Funded Project, entitled "Protecting Ecosystems and Livelihoods of the Sundarbans, a World Heritage Site: Assessing the Impact of Natural Hazards on Forest-Based Ecosystem Services" (Awarded as CRRP2019-851-Srivastava Year 2021–2022) is duly acknowledged.

References

Alongi, D. M. (2002). Present state and future of the world's mangrove forests. *Environmental Conservation, 29*(3), 331–349. https://doi.org/10.1017/S0376892902000231

Alongi, D. M. (2022). *Blue carbon: Coastal sequestration for climate change mitigation*. Springer.

Ball, M. C. (1998). Mangrove species richness in relation to salinity and waterlogging: A case study along the Adelaide River floodplain, northern Australia. *Global Ecology and Biogeography Letters, 7*(1), 73–82.

Basyuni, M., Yuriswan, W., & Bimantara, Y. (2020). Root development in mangrove plant *Rhizophora mucronata* under different shade intensities. *IOP Conference Series: Earth and Environmental Science, 452*, 012090. https://doi.org/10.1088/1755-1315/452/1/012090

Bayraktarov, E., Saunders, M. I., Abdullah, S., Mills, M., Beher, J., Possingham, H. P., Mumby, P. J., & Lovelock, C. E. (2016). The cost and feasibility of marine coastal restoration. *Ecological Applications, 26*(4), 1055–1074.

Bhadra, T., Nath, T., Ray, R., Sahu, N., & Raha, A. (2023). Monitoring the mangroves of Indian Sundarbans using geospatial techniques. *International Archives of the Photogrammetry, Remote Sensing and Spatial Information Sciences, 48*(1/W2), 405–412.

Biswas, S., Zaman, S., & Mitra, A. (2017). Soil characteristics of Indian Sundarbans: The designated world heritage site. *Scientific Journal of Biomedical Engineering & Biomedical Science, 1*(2), 53–59.

Camacho, L. D., Gevaña, D. T., Carandang, A. P., & Camacho, S. C. (2020). Resilience and sustainability of mangrove forest ecosystems. *Forests, 11*(9), 979. https://doi.org/10.3390/f11090979

Chowdhury, A., Naz, A., Bhattacharyya, S., & Sanyal, P. (2018). Cost–benefit analysis of "blue carbon" sequestration by plantation of few key mangrove species at Sundarban Biosphere Reserve, India. *Carbon Management, 9*(6), 575–586. https://doi.org/10.1080/17583004.2018.1518105

Das, B. (2025, April 30). Decision science could lead action in mangrove restoration. *Mongabay India*. Retrieved September 8, 2025, from https://india.mongabay.com

Ellison, A. M., Felson, A. J., & Friess, D. A. (2020). Mangrove rehabilitation and restoration as experimental adaptive management. *Frontiers in Marine Science, 7*, 327. https://doi.org/10.3389/fmars.2020.00327

Ericsson. (2021). *Connected Mangroves Project: Leveraging IoT for sustainability*. Ericsson.

FAO. (2015). *The world's mangroves 2000–2015*. FAO Forestry Paper No. 179. Food and Agriculture Organization of the United Nations.

FAO. (2020). *Asia-Pacific Forestry Sector Outlook Study III: Mangrove restoration practices*. FAO Regional Office.

Ghosh, S., Bakshi, M., Bhattacharyya, S., Nath, B., & Chaudhuri, P. (2015). A review of threats and vulnerabilities to mangrove habitats: With special emphasis on East Coast of India. *Journal of Earth Science & Climatic Change, 6*, 270. https://doi.org/10.4172/2157-7617.1000270

Gunawan, A., Hidayat, A., & Anggraini, E. (2018). Analisis kelembagaan rehabilitasi mangrove di Karangsong, Indramayu, Jawa Barat (Institutional analysis of mangrove rehabilitation in Karangsong, Indramayu, West Java). *Sodality: Jurnal Sosiologi Pedesaan, 6*(1), 1–7.

JICA (Japan International Cooperation Agency). (2014). *Technical guide for restoration in conservation areas—Mangrove ecosystems on former fish pond*. JICA-RECA Project. Retrieved March 11, 2021, from https://www.jica.go.jp/project/indonesian/indonesia/008/materials/c8h0vm00001etrc-att/pamphlet_06.pdf

Kanniah, K. D., Sheikhi, A., Cracknell, A. P., Goh, H. C., Tan, K. P., Ho, C. S., & Rasli, F. N. (2015). Satellite remote sensing for mangrove mapping and monitoring: Methods and applications. *Remote Sensing, 7*(5), 5447–5490.

Lewis, R. R. (2005). Ecological engineering for successful management and restoration of mangroves. *Ecological Engineering, 24*(4), 403–418.

Mandal, U. K., Burman, D., Bhardwaj, A. K., Nayak, D. B., Samui, A., Mullick, S., Mahanta, K. K., Lama, T. D., Maji, B., Mandal, S., Raut, S., & Sarangi, S. K. (2019). Waterlogging and coastal salinity management through land shaping and cropping intensification in climatically vulnerable Indian Sundarbans. *Agricultural Water Management, 216*, 12–26.

Mitra, A., Banerjee, K., & Sett, S. (2012). Spatial variation in organic carbon density of mangrove soil in Indian Sundarbans. *National Academy Science Letters, 35*(3), 147–154.

Moslehi, M., Pypker, T., Bijani, A., Ahmadi, A., & Hallaj, M. H. S. (2021). Effect of salinity on the vegetative characteristics, biomass and chemical content of red mangrove seedlings in the south of Iran. *Scientia Forestalis, 49*(132), e3748.

Parida, A. K., & Jha, B. (2010). Salt tolerance mechanisms in mangroves: A review. *Trees, 24*(2), 199–217.

Primavera, J. H., & Esteban, J. M. A. (2008). A review of mangrove rehabilitation in the Philippines: Successes, failures and future prospects. *Wetlands Ecology and Management, 16*, 345–358.

Rahmattunisa, R., Anna, Z., Yustiati, A., & Rizal, A. (2019). Economic valuation of mangrove forest in Pabean Ilir, Indramayu, Indonesia. *Global Scientific Journal, 7*(10), 1641–1649.

Rizki, R., & Novi, N. (2018). Respon pertumbuhan bibit mangrove *Rhizophora apiculata* Bl. pada media tanah topsoil. *Jurnal Bioconcetta, 3*(2), 41–54.

Roy, S. K., Mojumder, P., Chowdhury, A. A. & Md. Mahmudul Hasan (2025). Evaluating mangrove forest dynamics and fragmentation in Sundarbans, Bangladesh using high-resolution Sentinel-2 satellite images. *Global Ecology and Conservation, 58*, e03493). https://doi.org/10.1016/j.gecco.2025.e03493

Sasmito, S. D., Sillanpää, M., Hayes, M. A., Dargusch, P., & Friess, D. A. (2023). Challenges and opportunities for achieving sustainable development goals through restoration of Indonesia's mangroves. *Nature Ecology & Evolution, 7*, 62–70.

Silva, N. R. D., & Maia, R. C. (2019). Evaluation of the growth and survival of mangrove seedlings under different light intensities: Simulating the effect of mangrove deforestation. *Revista Árvore, 43*(3), e430308.

Sodikin, S. (2012). Persepsi masyarakat petani tambak terhadap kelestarian hutan mangrove di Desa Pabean Ilir Kecamatan Pasekan Kabupaten Indramayu (Pond farmers' perception on mangrove forest conservation in Pabean Ilir Village, Pasekan District, Indramayu Regency). *Jurnal Pendidikan Geografi, 12*(1), 41–48.

Sodikin, S., Hidayat, A., & Anggraini, E. (2023). Mangrove rehabilitation and eco-polybag innovation in Indonesia. *Polish Journal of Environmental Studies, 32*(4), 1234–1242.

Sunkur, R., Kantamaneni, K., Bokhoree, C., Rathnayake, U., & Fernando, M. (2024). Mangrove mapping and monitoring using remote sensing techniques towards climate change resilience. *Scientific Reports, 14*, 6949. https://doi.org/10.1038/s41598-024-57563-4

Syahid, L., Wahyunto, W., Sulaiman, A., & Jaya, I. N. S. (2019). GIS-based land suitability evaluation for mangrove ecosystem development in Southeast Asia. *IOP Conference Series: Earth and Environmental Science, 338*, 012044.

Toledo, G., Rojas, A., & Bashan, Y. (2001). Monitoring of black mangrove restoration with nursery-raised seedlings on an arid coastal lagoon. *Ecological Engineering, 18*(2), 149–158.

Trisnawati, I., Muryono, M. & Desmawati, I. (2019) The impact of herbivorous insect on leaves of mangrove species Rhizophora stylosa and its relation to leaf nutrient level. *Biodiversitas Journal of Biological Diversity, 20*(5):1409–1415. https://doi.org/10.13057/biodiv/d200533

Ulqodry, T. Z., Matsumoto, F., Okimoto, Y., Nose, A., & Zheng, S. H. (2014). Study on photosynthetic responses and chlorophyll fluorescence in *Rhizophora mucronata* seedlings under shade regimes. *Acta Physiologiae Plantarum, 36*(7), 1903–1917.

Waters, C. (2023, May 18). Why many mangrove planting efforts fail. *Smithsonian Magazine*.

Remote Sensing-Based Assessment of Mangrove Ecosystem with Special Reference to the Sundarbans of India

Dhananjay Barman, Sanjeev Kumar Srivastava, Gouranga Kar, Saon Banerjee, Debangana Banik, Bulu Basak, Mohammed A. S. Arfin-Khan, Sharif A. Mukul, Mochamad Candra Wirawan Arief, and Reginawanti Hindersah

Abstract Mangrove forests are unique ecosystems that exist on tropical and subtropical coastal wetlands throughout the world. They provide multiple ecosystem services which are ecologically and economically beneficial to the planet by providing food and fodder, fish and shrimp breeding, protection from cyclones, coastal erosion, and saltwater intrusion, providing medicinal ingredients, and attracting tourists. However, these mangrove ecosystems face multiple facets of climatic, ecological, and anthropogenic threats of extinction. The area under mangrove forest has drastically reduced over the time. For the protection and conservation of the mangrove forests, there is an urgent need for accurate mapping of mangrove area and assessment of their vulnerability to climate change, and other abiotic, biotic, and

D. Barman (✉) · G. Kar · D. Banik · B. Basak
ICAR—Central Research Institute for Jute and Allied Fibres, Kolkata, West Bengal, India

S. K. Srivastava
Geospatial Analytics for Conservation and Management, School of Science Technology and Engineering (SSTE), Sippy Downs, QLD, Australia

S. Banerjee
Bidhan Chandra Krishi Viswavidyalaya, Mohanpur, India

M. A. S. Arfin-Khan
Department of Forestry and Environmental Science School of Agriculture and Mineral Sciences, Shahjalal University of Science and Technology, Sylhet, Bangladesh

S. A. Mukul
United International University, Dhaka, Bangladesh

M. C. W. Arief
Center for Environment and Sustainability Science, Faculty of Fishery and Marine Science, Universitas Padjadjaran, Sumedang, West Java, Indonesia

R. Hindersah
Center for Environment and Sustainability Science, Faculty of Fishery and Marine Science, Universitas Padjadjaran, Sumedang, West Java, Indonesia

Faculty of Agriculture, Universitas Padjadjaran, Sumedang, West Java, Indonesia

S. K. Srivastava et al. (eds.), *Earth Observation for Monitoring Mangrove Ecosystems*, https://doi.org/10.1007/978-3-032-16668-5_5

anthropogenic threats. The remote sensing technique can play a vital role in assessing and providing spatiotemporal information on mangrove ecosystem distribution, species differentiation, health status, and ongoing changes of mangrove cover. Due to difficulty in accessing the mangrove areas by humans, various studies have been conducted by using different remote sensing techniques ranging from aerial photography to high- and medium-resolution optical imagery and from hyperspectral data to active microwave (SAR) data. Numerous scientific papers have reported studies on mangrove mapping, distribution, and health status over the last few decades. This chapter will primarily present how remote sensing techniques have been utilized by various researchers to address the multiple factors related to mangrove studies, including the Sundarbans mangrove of India, which is a UNESCO World Heritage site.

Keywords Mangrove forest · Remote sensing of mangrove · Coastal ecosystem mapping · Reflectance and backscattering properties of mangroves · Disaster mapping for mangrove

1 Introduction

Mangrove forests exist on tropical and subtropical coastal wetlands throughout the world (Chapman, 1976; Tomlinson, 2016). They are among the most biologically important ecosystems on the planet. Mangrove forests consist of trees, shrubs, and fern species (Duke et al., 1998) that have adapted to the harsh conditions of high salinity, anoxic soils, warm air and water temperatures, high tides, and muddy waters. These ecosystems act as a first line of defence against tropical and subtropical cyclones, hurricanes, and tsunamis by dissipating wave and wind energy. Moreover, the mangrove ecosystems provide tremendous ecological and economic services to mankind by providing food, fodder, fuel, medicine, recreation, etc. This also helps to regulate the environment by absorbing atmospheric carbon di-oxide, stabilizing coastlines, and protecting against storms. According to Barbier et al. (2011), the benefits from carbon sequestration depend on factors like the density of their biomass and the price of carbon in the market which can vary extensively ranging from around \$60 to more than \$1000 per hectare each year. Additionally, the storm protection it offers has been valued from \$300 to as high as \$9000 per hectare annually. In the Sundarbans, research showed that local communities gain substantial economic benefits from mangroves and are open to investing in their conservation. For instance, Iqbal (2020) used a choice experiment approach to estimate the marginal willingness to pay (MWTP) for improvements in ecosystem services. Also, Mukherjee et al. (2014) emphasized light on the ecological and economic trade-offs in the mangrove forests of South Asia, underscoring the importance of integrated valuation in policymaking and land-use planning. Even though there are obstacles like hypothetical bias, limited data, and challenges in capturing cultural services, these valuation efforts play a vital role in shaping Payment for Ecosystem Services (PES) initiatives, blue carbon strategies, and adaptive climate policies.

However, these mangrove ecosystems face multi-facets of climatological, ecological, and anthropogenic threats of extinction. Many parts of the ecosystems have immersed into sea water because of the sea-level rise due to global warming. The regeneration capacity of mangroves has been overtaken by the more frequent cyclonic events. Many parts of the mangrove forests have been degraded polluted and have lost biodiversity. Biodiversity in mangrove forests is increasingly at risk. In estimation, 1533 species are associated with mangroves, 15% of which are threatened with extinction. UNEP's report 'Decades of Mangrove Forest Change: What does it mean for nature, people and the climate?' found that nearly 50% of mangrove-associated mammals, 22% of fishes, 16% of plants, 13% of amphibians, and 8% of bird and reptile species are threatened with extinction. Worryingly, for 44% of the species, their extinction risk is increasing and of those already at risk, the situation is getting worse for 89%.

For addressing these issues and challenges, a proper and systematic assessment of the mangrove ecosystems is essential. As the manual survey and assessment is very difficult, and in some places, it is almost impossible due to the rough and tough characteristics of the ecosystems, satellite and unmanned aerial vehicle (UAV) can play a vital role in assessing the mangrove ecosystem. Many researchers worldwide have applied these techniques to assess the health of the mangrove ecosystem. These techniques in combination with modern-day artificial intelligence (AI) and machine learning (ML) techniques can provide even more real-time data in framing better management strategies for the mangrove ecosystems.

2 Global Distribution and Current State of the Mangrove Ecosystem

The global area of mangrove forest has been estimated to 147,359 km^2 (Global Mangrove Watch, 2020), 51% of which occurred in the Asia-Pacific, with 29% in the Americas, and 20% in Africa. Over the 24-year period from 1996 to 2020, it has been reported that the global mangrove forest cover has lost in the extent of 5245 km^2 (3.4%). However, in a latest assessment of UNEP using current satellite imagery, the global loss of mangroves has stabilized, and the mangrove cover increased in and around world's large rivers, estuaries, and deltas (UNEP, 2023).

According to the Global Forest Resource Assessment 2020 (FRA, 2020), 113 nations throughout the world have mangrove forests covering an estimated 14.72 million hectares. Asia has the highest documented mangrove area (5.33 million hectares), followed by South America (2.12 million hectares), Western and Central Africa (2.30 million hectares), North and Central America (2.55 million hectares), and Oceania that is claimed to have the least area of mangroves (1.26 million hectares). As per FRA 2020 report (Table 1), the global mangrove-cover between 1990 and 2020 declined from 15,759 hectares (157,590 km^2) to 14,717 hectares (147,170 km^2). Asia experienced the most significant loss, particularly in South and Southeast Asia (12.9%), where factors such as urbanization, aquaculture, and

Table 1 Global status of the mangrove cover

Region/subregion	Mangrove area (1000 ha)			
	1990	2000	2010	2020
Eastern and Southern Africa	929	902	883	905
Northern Africa	34	31	32	31
Western and Central Africa	2436	2400	2349	2304
Total Africa	3398	3332	3264	3240
East Asia	24	22	25	32
South and Southeast Asia	6117	6108	5713	5330
Western and Central Asia	190	190	190	184
Total Asia	6331	6320	5928	5545
Total Europe	0	0	0	0
Caribbean	787	789	774	891
Central America	492	482	483	466
North America	1152	1167	1190	1195
Total North and Central America	2431	2439	2447	2552
Total Oceania	1447	1150	1314	1255
Total South America	2152	2050	1976	2124
World	**15,759**	**15,292**	**14,928**	**14,717**

Source: FRA (2020), FAO

coastal development contributed to a loss of nearly 800 thousand hectares. Africa experienced a mild decrease, particularly in its western and central areas (4.6%), while Eastern and Southern Africa started to show a bit of recovery by 2020. Oceania, on the other hand, faced a net loss of 13.7%, even though there was some improvement noted in 2010. However, according to FSI Report 2023, India ranks as the world's third richest nation in terms of mangrove plant diversity, trailing behind Indonesia and Australia. More than 40% of the total area of mangroves was found to be in only four countries: Indonesia (19%), Brazil (9%), Nigeria (7%), and Mexico (6%). The status of world mangrove cover in 2024 shows that Southeast Asia has the largest share of mangrove cover, making up 33.6% of the total cover, followed by North and Central America and the Caribbean (14.4%), South America (13.2%), and West and Central Africa (15.5%). East and Southern Africa (5.3%), South Asia (6.6%), and Australia and New Zealand (7.0%) are other regions with moderate coverage. On the other hand, mangrove cover is comparatively low in the Pacific Islands (3.9%), the Middle East (0.2%), and East Asia (0.2%) (Fig. 1).

3 Climatic Variability of Mangrove Ecosystems

Climate change and its variability can significantly impact on mangrove ecosystems, which depend on species sensitivity and location (Randhir et al., 2024; Ward et al., 2016). Changes in temperature and rainfall due to climate change affect growth and development stages of mangrove species, leading to variations in leaf

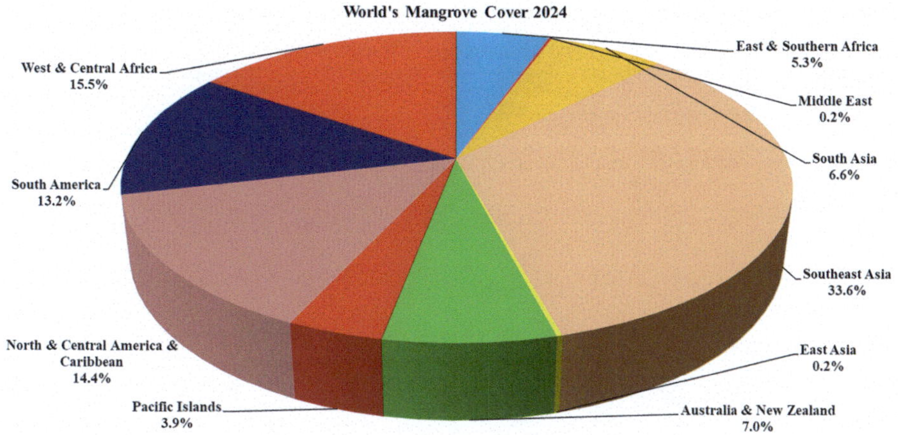

Fig. 1 The current status of the world's mangrove cover. [*Source*: The State of the World's Mangroves 2024]

area index (LAI) resulting changes in remotely sensed indices (e.g. Normalized Difference Vegetation Index, NDVI; Enhanced Vegetation Index, EVI) if data is not captured during peak biomass (Ganguly et al., 2012). Climate-driven shifts in species composition and structure alter spectral reflectance, reducing the accuracy of biomass models based on historical data (Clark et al., 2003). In tropical regions like the Sundarbans, increased cloud cover and atmospheric disturbances degrade optical sensor data quality, thus making it difficult for monitoring and assessment of the mangrove ecosystem (Baccini et al., 2017). The effects of climate change on three dominant mangrove species globally: red, black, and white mangroves, were studied using simulation, GIS, and statistical modelling and revealed that low precipitation can impact growth of all three mangrove species, where even 1 mm reduction in precipitation decreased the growth suitability of red, black, and white mangroves by 0.008%, 0.004%, and 0.008%, respectively (Randhir et al., 2024).The Intergovernmental Panel on Climate Change (IPCC) in 2013 reported that the global sea levels are likely to rise by between 0.28 and 0.98 m by 2100 which will cause coastal flooding in future. In recent decades, global sea-level rise has been reported at the rate of 3.2 ± 0.4 mm per year (Church & White, 2011). However, regional factors influence its variability from 1.9 mm per year in the Caribbean to 7.5 mm per year in parts of Indonesia (Nerem et al., 2010) and up to 9 mm per year in the lower Mississippi River Delta (NOAA, 2015). Sea-level rise can cause harm to mangrove by increasing inundation duration which leads to plant death (He et al., 2007) and shifting in species composition (Gilman et al., 2008). Sea-level rise and associated danger of sinking of mangrove area should be monitored periodically by using remote sensing (RS) data for taking appropriate climate action. Keeping in view the importance of the climate, an account of climatic situation of the major mangrove regions is presented in the Table 2.

Table 2 Climate of the mangrove regions of different countries of the world

Country	Area (in ha)	% global total	Rainfall (mm)	Temperature range (°C)	RH (%)	Tidal influence (m)
Indonesia	3,112,989	22.6	2000–4000	25–30	≥80	Semi-diurnal (1–3)
Australia	977,975	7.1	1000–3500	24–30	≥70	Intertidal (2–4)
Brazil	962,683	7.0	1500–2500	25–30	≥80	Semi-diurnal (2–4 and ≥4)
Mexico	741,917	5.4	1000–2500	25–30	≥80	Semi-diurnal (1–2 and ≥3)
Nigeria	653,669	4.7	1500–3000	26–30	≥80	Semi-diurnal (1–2)
Malaysia	505,386	3.7	2000–4000	25–32	≥80	Semi-diurnal (1–3)
Burma (Myanmar)	494,584	3.6	1500–5000	25–32	≥80	Semi-diurnal (2–4)
Papua new Guinea	48,121	3.5	2000–3500	26–30	≥80	Semi-diurnal (1.5–4)
Bangladesh	436,570	3.2	1500–3000	24–32	≥80	Semi-diurnal (3–6)
Cuba	421,538	3.1	1000–1500	24–30	≥80	Semi-diurnal (0.5–1.5)
India	368,276	2.7	1500–3000	25–32	75–85	Diurnal and semi-diurnal (2–6)
Guinea Bissau	338,652	2.5	1200–2000	25–32	70–80	Semi-diurnal (2–3)
Mozambique	318,652	2.3	1000–1500	25–30	70–80	Intertidal (1.5–3)
Madagascar	278,078	2.0	800–1500	25–30	≥80	Semi-diurnal (1.5–4)
Philippines	263,137	1.9	1500–3000	25–32	80–90	Intertidal (1–2)

4 Mangrove Area Mapping and Classification

The integration of satellite imagery with geospatial analysis has significantly improved the ability of researchers to detect changes in mangrove cover, evaluate species distribution, identify areas under ecological stresses, and distinguish mangroves from adjacent land cover types such as agricultural fields, mudflats, and inland vegetation (Kuenzer et al., 2011). Challenges in mangrove classification arose due to similar reflectance from other vegetation, rice fields, and water bodies. To overcome this, masking techniques can be used to separate mangrove areas from non-mangrove vegetation and water bodies, thus improving the accuracy of the classification (Salam et al., 2007). The unsupervised classification using 30 clusters in a study yielded an overall accuracy of 78.32% while supervised classification, supported by 272 ground-truth points, performed better with 81.62% accuracy. In the former method, the water bodies were classified accurately, but misclassifications occurred in identifying species like Gewa and grassland/bare land areas. Some of the classification errors in both methods could be attributed to environmental differences and a time gap between the acquisition of the satellite images and the creation of the mangrove inventory map. Despite this, both methods were considered reliable, with supervised classification slightly outperforming the unsupervised approach. However, some researchers reported that the use of machine learning algorithms such as Support Vector Machines (SVMs) and Random Forest (RF) has even been better to classify mangrove species and mapping and monitoring of mangrove forests (Campomanes et al., 2016; Jiang et al., 2021). The overall accuracy of RF can be 99.17%, whereas SVM showed 95.83% accuracy. Thus, these algorithms can be used for mapping and classification of mangrove ecosystems all over the world.

Numerous researchers used remote sensing techniques for mapping of Indian Sundarbans region (Biswas et al., 2024; Mondal & Bandyopadhyay, 2014; Mitra & Karmaker, 2010; Giri et al., 2007). Net erosion of approximately 3.7 km^2 per year across the Indian Sundarbans Delta between 1980 and 2021, with 243 km^2 of erosion and 91 km^2 of accretion over a 41-year period, was estimated using the NDVI (Biswas et al., 2024). Ghosh et al. (2015) employed an unsupervised (ISODATA) classification on a multi-temporal and multiscale analysis mapping colonial-era charts alongside 1960s Corona imagery and Landsat satellite data and found that the overall mangrove area in the Indian Sundarbans system remained relatively constant over time and there were significant internal changes in forest condition and stand structure, reflecting shifts in density and health of mangrove species. Giri et al. (2007) conducted a multi-temporal Landsat analysis from the 1970s to 2000s and reported that mangrove forest area in the Sundarbans decreased by approximately 1.2% over that period, with a more pronounced decline of approximately 2.5% between 1990s and 2000s. In another study, Datta and Deb (2012) employed multi-temporal Landsat data to investigate on land use/land cover (LULC) changes between 1975 and 2006 revealing a substantial decline of around 0.42% of the area's initial mangrove cover, primarily in the Sundarbans' forest-habitation interference zones. Kundu et al. (2021) classified Landsat imagery using NDVI from 1975 to 2018 into dense forest, sparse forest, water bodies, and wetland and found a net loss of approximately 2.1% ($\approx$ 47 km^2) of forest area, with varying trends in fragmentation and density. Mondal et al. (2018) used Landsat MSS and OLI multispectral satellite imagery to classify land use/land cover (LULC) on Sagar Island into categories such as forest vegetation, agriculture, built-up area, water bodies, and wetlands. Their classification achieved an overall accuracy of 79.53%, with a Kappa coefficient (K) of 0.7465.

4.1 Change Detection of Mangrove

The decrease in mangrove cover and its density is a threat to its flora and fauna. The monitoring of change in mangrove cover can be performed by adapting various change detection methods of the remote sensing (RS) technique. The traditional change detection methods like post-classification comparison (PCC), unsupervised conventional change detection, change vector analysis (CVA), and single threshold algorithm (STA) had limitations. To overcome this, Shimu et al. (2019) proposed a new method of using the Normalized Difference Vegetation Index (NDVI) to track changes in vegetation over a temporal scale, resulting in both progressive and regressive changes in mangrove vegetation.

The land use/land cover (LULC) change study in the South-Western Indian Sundarbans Biosphere Reserve from 1975 to 2006 using Landsat imagery showed a significant loss of open mangrove areas and biodiversity, particularly in forest--habitation zones, while dense mangrove cover increased in the reserved forests, indicating effective management (Datta & Deb, 2012). The study also showed that overall, the region lost about 0.42% of its original mangrove cover, and the rise of nonagricultural lands is linked to the growth of settlements, tourism, and alternative livelihoods like shrimp farming and commercial fishing, although harmful practices like prawn seed collection persisted.

Recent advances in remote sensing technologies have greatly improved LULC change mapping, particularly for mangrove ecosystems. High-resolution satellite images from sensors such as Sentinel and Landsat have enabled precise and accurate mapping of changes in mangrove cover over time (Kuenzer et al., 2011; Parida & Kumar, 2020). Samanta and Hazra (2016) employed a hybrid classification approach using Landsat TM/ETM+ imagery from 1986, 1996, 2002, and 2012 to assess mangrove cover changes in the Indian Sundarbans. By using the hybrid approach in combination of unsupervised (ISODATA) and supervised (Maximum Likelihood Classifier) classification techniques, they were able to classify the mangrove area into six land cover classes like dense forest, degraded forest, saline blanks, water body, sand (beaches/dunes), and mud flats. They obtained the classification accuracies between 85 and 90% and Kappa values of 0.79–0.88. Changes were analysed using post-classification comparison, highlighting spatial and temporal patterns of mangrove degradation and conversion, particularly in forest–habitation interference zones. Kumar et al. (2019) used five vegetation indices, viz. Mangrove Probability Vegetation Index (MPVI), Normalized Difference Wetland Vegetation Index (NDWVI), Shortwave Infrared Absorption Depth (SIAD), Normalized Difference Infrared Index (NDII) and Atmospherically Resistant Vegetation Index (ARVI) in the decision tree algorithm to improve the separability of mangrove forests from adjacent land cover classes, especially terrestrial vegetation using Support Vector Machine (SVM), Minimum Distance (MD), and Spectral Angle Mapper (SAM) and found that SVM performed better than MD and SAM with an overall precision of 99.08%over the Indian Sundarbans.

The impact analysis of LULC changes is essential for prioritizing the management practices. Thakur et al. (2021) analysed the impact of LULC changes in the Indian Sundarbans from 2000, 2010, and 2017, using Landsat TM, ETM+, and OLI imagery, and found a decline in mangrove swamps, plantations, and agricultural land, while sand beaches, mudflats, rivers, and agriculture increased. NDVI values dropped for mangroves from 0.441 to 0.229 and plantations from 0.266 to 0.195 but rose for rivers and aquaculture. Land Surface Temperature (LST) increased in settlements and sand beaches, showing a negative correlation with NDVI, likely due to the high evapotranspiration of mangroves. The results indicated the growth of non--vegetated areas and ecosystem stress. Similar results were reported by Kanjin and Alam (2024) where they showed changes in mangrove coverage from 1973 to 2023 using Landsat (Landsat 1–8) data and examined the relationship between vegetation health with NDVI and surface temperature using Moderate Resolution Imaging Spectroradiometer (MODIS) data in Bangladesh and Indian Sundarbans. Despite mangrove loss, vegetation health improved, indicating resilience. The study found a weak negative correlation between NDVI and LST, suggesting other factors, such as water bodies and land cover heterogeneity, also influenced temperature. The findings highlighted the need for conservation to prevent further degradation.

4.2 Disaster Mapping in Mangrove

Disasters are increasingly threatening populations and affecting sustainable development in the mangrove region. Satellite and unmanned aerial vehicle (UAV) technologies play a key role in disaster management by aiding in preparedness, warning, and response. However, challenges like limited access to high-resolution data and technological gaps remain, particularly between developed and developing countries. Recent advancements, including new spatiotemporal data sources and improved resolutions, offer different kind of solutions (Bello & Aina, 2014). SERVIR-Mekong initiative, in collaboration with NASA and USAID, is one such effort for disaster management carried out with the help of the Asian Disaster Preparedness Center and various regional partners in Cambodia, Laos, Thailand, Myanmar, and Vietnam (SERVIR Global, 2023). It plays a crucial role by offering free access to satellite data, dashboards, and customized decision-support tools designed for flood mapping, early warning systems, and monitoring of the mangrove ecosystem. By linking global data resources with local engagement and training, this showcases how countries can work together to fill gaps in data and capacity. This type of initiative empowers vulnerable coastal communities to manage disaster risks more effectively and boost their environmental resilience (SEI, 2023).

The entire global mangrove ecosystems, including the Indian Sundarbans, are exposed to various disasters like cyclones, hurricanes, typhoons, tsunamis and storm surges, flood inundation, embankment breaching, and coastal erosion (IUCN, 2025; Ghosh & Mistri, 2022). These disasters result severe ecological damage, which can be estimated by using Moderate Resolution Imaging Spectroradiometer

(MODIS) Global Disturbance Index (MGDI) given by Mildrexler et al. in 2009. For monitoring and management of cyclonic disasters, the MGDI was used by Dutta et al. (2015) in the Indian Sundarbans and found that the mangrove landmass decreased by approximately 329.45 km^2 at a rate of 7.48 km^2 per year from 1973 to 2017. Mandal et al. (2022) implemented the Analytic Hierarchy Process (AHP) along with Random Forest (RF) and Artificial Neural Network (ANN) to map multi--hazard risk of the Sundarbans. Their results showed that the southern and eastern parts of the study area faced high to extremely high risk, primarily driven by cyclonic hazards and embankment breaching.

5 Degradation and Deforestation of Mangroves

The degradation of mangroves is caused by multiple factors such as sea-level rise, development of fisheries, hyper-salinity, sediment movement, storm effects, shoreline erosion, and expansion of eutrophication in the surface depressions of swampy areas (Paul et al., 2017). About 40% of the tropical and 10.5% of the Sundarbans mangroves have been lost due to induced sea-level rise, extreme weather events, and human interferences (Padhy et al., 2022; Ghosh et al., 2015). Anthropogenic factors influencing the ecosystem have indicated that human-wildlife conflict (24.4%) was considered the biggest threat, followed by climate change (19.9%), salinity change (18.8%), livelihood pressures on natural resources (16.6%), shrimp seed collection (11.5%), and environment and water pollution (8.8%) (Vyas & Sengupta, 2012).

Integrating spectral data with other spatial datasets can improve insights, and documenting the accuracy of the results helps users assess the reliability of the findings (Coops et al., 2006). Landsat satellite data, particularly the Tasseled Cap (TC) transformation, is widely used for detecting forest disturbances. This study tested four TC-based data structures, such as brightness, greenness, wetness, and a novel Disturbance Index (DI), to identify stand-replacing disturbances in Russia and Washington State, USA. The TC transformations generally outperformed the original Landsat reflectance data in detecting changes. DI performed best where disturbances are more persistent than the areas where disturbances are more transient. The study highlights that both local forest recovery rates and the time interval between Landsat acquisitions should be considered when selecting a Tasseled Cap (TC) transformation for effective disturbance detection (Healey et al., 2005).

The MODIS Global Disturbance Index (MGDI) algorithm, which uses Aqua/MODIS data (LST and EVI), maps large-scale ecosystem disturbances (LSEDs) and their impact on the global carbon cycle. The algorithm detects changes in land-surface energy by analysing annual LST composites, minimizing natural variability. Applied to North America's woody ecosystems in 2005 and 2006, MGDI accurately identified wildfires, hurricane windfalls, and logging activities. The analysis found that 1.5% of North America's woody ecosystems were disturbed in 2005 and 0.5% in 2006, with results consistent with wildfire statistics (Mildrexler et al., 2009). Deforestation and forest degradation, driven by anthropogenic stress, climate change, and land conversion, were assessed in the Shendurney Wildlife Sanctuary, Kerala, through GIS, remote sensing, fragstat methods, and NDVI to evaluate forest degradation with spatial and nonspatial data, including anthropogenic and climate data. The disturbance index, ranging from 2.5 to 7.5, was categorized into four disturbance zones, viz. low, medium, high, and very high. This analysis supported the development of forest conservation and management strategies (Jose et al., 2011). These indices and methods can be used for assessing the mangrove degradation and deforestation globally.

6 Cyclonic Hazards in Mangrove

Due to climate change, the frequency and severity of cyclonic hazards have increased in the recent past, thereby significantly damaging the mangrove forest. Rising temperatures and rising sea levels exacerbated the damage, limiting the forest's ability to regenerate and threatening environmental sustainability (Gupta, 2021). The mangrove ecology of the Sundarbans has suffered significant damage due to super cyclones such as SIDR, Rashmi, and Aila, which has been effectively estimated by Dutta et al. (2015) using the MODIS Global Disturbance Index (MGDI). They analysed the pre- and post-cyclonic changes in EVI and LST to derive cyclonic impacts on vegetation greenness and surface temperature. The derived MGDI data were also transformed into percentage changes, and they compared them to pre- and post-cyclonic events to identify the cyclonic disturbance, with a threshold of 11% change from the temporal mean to define 'disturbed pixels'. This method can be useful to identify the cyclonic hazards in the mangrove ecosystem.

Cyclonic hazards generally occur due to the wind gusts, flooding, saltwater intrusion, storm surge, etc. (NOAA, 2023). These hazards in the mangrove have been reported by numerous researchers in the recent past (Mandal & Hosaka, 2020; Prakash et al., 2023; Mondal et al., 2024). Mandal and Hosaka (2020), using the classification and regression tree (CART), supervised classification technique on Landsat images to assess NDVI changes before and after 21 cyclones during 1988–2016 in the Google Earth Engine with 86% accuracy and a kappa coefficient of 0.80. They reported that the cyclone-affected forest area ranged from 0.5% to 24.1%, with the 18 cyclones impacting less than 10%, while the cyclone Sidr (24.1%) and one in 1988 (20.4%) caused the most damage. The cyclonic wind speed showed a significant positive correlation with the forest damage. Computing NDVI and EVI indices from the Landsat imagery, Prakash et al. (2023) found that the Amphan super cyclone caused a significant shift in mangrove cover from dense to sparse (approx. 50.18 km^2). They also reported that the barren land increased by 50.37% based on Normalized Difference Bareness Index (NDBaI), the water-bodies increased by 5.20% based on Normalized Difference Water Index (NDWI), and salinity increased based on Normalized Difference Salinity Index (NDSI).

Recently, Mondal et al. (2024) generated cyclonic flood-effective zones of the Sundarbans with the flood susceptibility spectral indices such as Normalized

Difference Vegetation Index (NDVI), Modified Normalized Difference Water Index (MNDWI), Normalized Difference Moisture Index (NDMI), Normalized Difference Built-up Index (NDBI), Bare Soil Index (BSI), and Normalized Difference Turbidity Index (NDTI) using Landsat 8 Operational Land Imager, Thermal Infrared Sensor, Resourcesat, LISS-III, and AWiFS sensors data. They found that the riverine and coastal areas in the northeast and southwest of the Sundarbans were highly susceptible to floods. Among the flood susceptibility indices, the MNDWI was found to be highly and significantly correlated with Normalized Difference Flood Index (NDFI).

7 Fire Hazards in Mangroves

Mangrove forest fires, though historically rare, have been increasingly recorded globally, particularly during natural droughts such as El Niño events, which dry out litter and peat layers, and when coupled with human activities like slash-and-burn agriculture, land conversion, and waste burning (Dookie et al., 2025). Fire hazards in mangrove forests are unlikely because of regular water inundation due to two high tides in a day. The typical swampy nature of the ecosystems is also protecting it from fire hazards. Extreme heat and high winds, especially during the dry season, further exacerbate the fire risk in the forests. The man-made fires are often attributed to discarded cigarettes from fishermen and honey collectors, clearing of paths through the mangrove forest or the flames from torches used to repel bees.

Bangladesh's Sundarbans face frequent fire hazards. Between 1970 and 2024, the Bangladesh forest department and residents reported a total of 40 fire incidents in the same area. Fire incidents in Bangladesh's East Sundarbans now occur nearly every year during the dry season. In many parts of the region, the canals that connect villages to the Sundarbans have dried up, and during summer, ponds within the mangroves also dry out. This increasing dryness leaves the forests highly susceptible to fires, which can spread rapidly due to a lack of manpower in the forest department.

Similar to the Bangladesh Sundarbans, many other mangrove ecosystems of the world are vulnerable to fire hazards. Remote sensing techniques like the Normalized Burn Ratio (NBR), Burn Area Index (BAI), and Mid-Infrared Burn Index (MIRBI) use reflectance differences in MIR and NIR bands to delineate recently burned regions versus unburned vegetation (Kurbanov et al., 2022). Using multi-temporal Landsat imagery and Normalized Burn Ratio (NBR) and differenced NBR (dNBR), a study detected that over 800 hectares of mangroves in South Sumatra experienced moderate to high burn severity (dNBR values exceeding 0.3) during the 2015 El Niño-driven fire season, particularly in inland zones, which are less influenced by tidal flooding (Crowe, 2020). Also, a fuel-type map and forest fire hazard rating for swampy mangrove forests can be developed using remote sensing data like Landsat TM imagery. By integrating this data with GIS layers such as fuel type, roads, and canals, a fire hazard rating map can be created (Razali et al., 2010). Fuzzy set theory integrated with decision-making algorithms within a GIS framework can be used to map potential forest fire risk and guide proactive fire management efforts (Vadrevu et al., 2010). However, in contrast to the Bangladesh Sundarbans, the Indian Sundarbans is clearly defined by natural boundaries, which help keep it completely free from encroachment and, thereby, protected from fire hazards. Because of the tidal fluctuations, the entire region is flooded twice daily, preventing any fire outbreaks. Also, due to the terrain, location, and diversity of species, the area is free from grazing activities.

8 Biomass Estimation

Accurate estimation of mangrove biomass is important for determining carbon stock, exploring climate mitigation potential, and directing restoration efforts in mangrove ecosystems. Simard et al. (2019) mapped the global mangrove canopy height and biomass estimate using ICESat Geoscience Laser Altimeter System (GLAS) LiDAR data and found a significant correlation between above-ground biomass (AGB) and canopy height. Using this method, the total global mangrove carbon stock (above and below-ground biomass and soil) is estimated at 5.03 Pg. Anand et al. (2020) integrated EO-1 Hyperion hyperspectral data with Bhitarkanika field data to estimate carbon stock and biomass. Ten mangrove species were classified using the SAM classifier with a Kappa value of 0.81. Between different regression models, a second-order polynomial based on EVI provided the greatest accuracy ($R^2 \approx 0.87$), followed by NDVI ($R^2 \approx 0.84$), predicting carbon stocks between 459,820 and 514,470 tonnes.

The multi-temporal Sentinel-1 and 2 data-derived variables were used by Ghosh et al. (2021) to estimate the above-ground biomass (AGB) using variables such as, Normalized Difference Vegetation Index (NDVI), Soil Adjusted Vegetation Index (SAVI), Green NDVI (GNDVI), Normalized Difference Index produced using band 4 and 5 of Sentinel-2 data (NDI45), Inverted Red-Edge Chlorophyll Index (IRCEI), Transformed Normalized Difference Vegetation Index (TNDVI), Enhanced Vegetation Index (EVI), and Two Bands Enhanced Vegetation Index (EVI2) in Bhitarkanika Wildlife Sanctuary by Random Forest (RF), Gradient Boosted Model (GBM), and Extreme Gradient Boosting (XGB) models. The best performance was achieved using an ensemble model with a reduced RMSE of 72.86 t/ha and a normalized RMSE of 11.38% which implies that ensemble machine learning techniques enhance the precision of biomass estimation in mangrove ecosystems and support improved blue carbon assessments. In 2017, Castillo et al. estimated and mapped above-ground biomass (AGB) of mangrove forests using a combination of satellite-derived predictors that included SAR backscatter coefficients (VH and VV), GLCM texture features, and vegetation indices such as NDVI and EVI from Sentinel-2. The resulting biomass maps achieved overall accuracy between 85.3 and 86.6% with RMSE values between 27.8 and 28.5 Mg/ha and correlation coefficients (r) ranging from 0.82 to 0.84, demonstrating potential for scalable blue carbon stock assessments and land-use change monitoring.

By using image processing and digital photogrammetry, a canopy height model (CHM) generated using drone-based RGB images could reliably estimate tree height, offering an efficient method for forest inventory. This approach has the potential to enhance forest management across various land tenures and forest types (Srivastava et al., 2022). UAV-based photogrammetry combined with allometric equations improved the accuracy for estimation of above-ground biomass in both natural and restored mangrove ecosystems using digital surface models (DSM) and digital terrain models (DTM) to generate canopy height models (CHMs). The mean total AGB in natural mangroves and restored mangroves was found to be 239 Mg/ha and 232 Mg/ha, respectively (Basyuni et al., 2025). Overall, these methods provide non-destructive, scalable, and economical biomass assessment techniques that are crucial for ecosystem monitoring, blue carbon accounting, and climate mitigation plans.

9 Species Discrimination in Mangrove

Accurate monitoring of mangrove species distribution is vital for assessing their health and supporting restoration efforts due to their enormous ecological and economic importance. Hyperspectral sensors could aid species identification, though field challenges may arise (Vaiphasa et al., 2005). Spectral data (2151 bands between 350 and 2500 nm) of crown canopy leaves from 16 Thai mangrove species were captured under laboratory conditions and tested statistically to assess species differentiation. Significant differences were found in 1941 bands ($p < 0.05$), with 477 bands at $p < 0.01$. Spectral separability, measured by the Jeffries-Matusita distance, showed most species could be separated, except for those in the Rhizophoraceae family. Using high-resolution multispectral remote sensing images, the location of species, *A. marina*, *A. alba*, and *A. officinalis* along the gradients of soil salinity were identified by Mitra and Karmaker (2010) having ecological optima at salinity values of 18.2 and 20.1 ppt, 17.5 and 18.6 ppt, and 13.02 and 25 ppt, respectively, indicating *A. marina* as the most salt-tolerant mangrove species in the Indian Sundarbans. Giri et al. (2014) employed Landsat TM and ETM+ images from 2010 and 1999, respectively, to detect changes in particular mangrove species zones, revealing that *Avicennia* sp. is the most dominant species, followed by *Excoecaria* sp. and *Phoenix* sp. Using Landsat satellite series images, Ghosh et al. (2016) were able to identify mangrove species composition and detect species-level over time, showing that *H. fomes* and *E. agallocha* both decreased by 9.9% in the Indian Sundarbans, while *Ceriops decandra*, *Sonneratia apelatala*, and *Xylocarpus mekongensis* increased by 12.9%, 38.4% and 57.3%, respectively.

Recently, a framework was developed using multisource spaceborne high-resolution imagery such as WorldView-2 (WV-2), Orbital HyperSpectral (OHS), and Advanced Land Observing Satellite-2 (ALOS-2) to classify mangrove species at a finer scale. Key variables for species classification were identified through the application of extreme gradient boosting with recursive feature elimination

(XGBoost-RFE) (Zhen et al., 2024). Anand et al. (2020) utilized Hyperion EO-1 hyperspectral satellite data (220 bands) to map dominant species in the Bhitarkanika mangroves through spectral angle mapping (SAM) and partial least squares regression (PLSR). According to the study, *Avicennia marina* had stronger reflectance in SWIR region because of leaf salt excretion, whereas *Rhizophora mucronata* had strong absorption in the NIR region because of its denser canopy structure and indicated overall classification accuracies of 80–85%.

10 Sundarbans Mangrove of India: A Case Study

The Indian Sundarbans, recognised as a World Heritage Site by UNESCO, is a biodiversity hotspot and a unique ecosystem by being the world's largest continuous mangrove ecosystem. This ecosystem plays a key role both ecologically and economically by protecting the area from cyclone-driven coastal erosion and for people's livelihoods by providing resources such as honey, fuelwood, and fish. The climate of the Sundarbans is subtropical. Temperature changes from 20 °C in December–January to 28 °C in June–July. The annual average rainfall is around 1963 mm, 75% of which occurs during June to September. This is in a tropical cyclone belt and often demonstrates a complex interaction between social and ecological systems.

The fluctuating water levels, temperature, and wind flow are affecting the topography of the area, making it prone to several natural hazards, leading to the destruction of forest resources, especially under climate change scenarios. The Sundarbans faces severe climate change challenges, including rising sea levels, increasing salinity, and frequent cyclones, which threaten both its ecosystem and local livelihoods. Some urgent actions, such as global emission reductions, promoting saline-resistant plants, and improving disaster response, are needed to protect the region from further damage and potential submergence (Mahadevia Ghimire & Vikas, 2012). Native mangrove species like *Avicennia marina*, *Rhizophora mucronata*, *Sonneratia apetala*, and *Ceriops decandra* are recognized for their high resistance to salt, ability to adapt to changing water conditions, and importance in stabilizing soil and supporting biodiversity. *Avicennia marina* can tolerate salinity levels over 60 parts per thousand (Parida & Jha, 2010). The forest area of the Sundarbans region is divided into two broad categories: (1) saltwater heritiera and (2) low mangrove. A large part of the 'reserve' forest areas known as 'tiger reserves' has been declared 'totally protected', and no forestry operation is allowed there. The most noted species of the wildlife of the region are the royal Bengal tiger, spotted deer, wild boar, and crocodiles.

The anthropogenic interference in the Sundarbans began in the late eighteenth century mainly through agricultural practices. The deforestation contributes 6–17% of anthropogenic CO_2 emission, for example, from 2000 to 2010, about 1.0 Pg C/year was emitted (Baccini et al., 2012). Further, the encroachment of seawater into the mangrove increases salinity, resulting in loss of high-value storm-stopping tree species such as Sundari (*Heritiera littoralis*), which gives this area the name Sundarbans. This is further affected by reduced freshwater and sediment supply to the system due to upstream dams on the rivers. There are many indicators of such deleterious effects, e.g. few mangrove islands have vanished over the last century, some have significantly shrunk over the last 50 years, houses are swept away due to repeated collapse of river embankments, abandoning rice farming for shrimping, and increased incidence of water-borne diseases. A comparison of the area with earlier survey records (e.g. early nineteenth-century Dampier and Hodges survey of forest extent) indicates that the mangrove has shrunk to almost half in size.

India's current mangrove cover is 4991.68 km^2, or 0.15% of its entire geographical area. According to the Forest Survey of India, the mangrove cover is categorized into Very Dense (canopy density of 70% and above), Moderately Dense (canopy density of 40% and more but less than 70%) and Open categories (canopy density of 10% and more but less than 40%). Very Dense Mangroves encompass 1463.97 km^2 (29.33% of the total mangrove area); Moderately Dense Mangroves cover 1500.81 km^2 (30.07%); and Open Mangroves cover 2026.87 km^2 (40.60%). India's mangrove cover has decreased from 2021 to 2023 by a net 7.43 km^2. The highest mangrove cover is found in the Sundarbans region of West Bengal (42.5% of the India's total mangrove cover) followed by Gujarat (23.3%), Andhra Pradesh (8.4%), Maharashtra (6.3%), and Odisha (5.2%). West Bengal has the largest expanse of extremely dense mangroves, covering about 981.63 km^2 out of the total mangrove area of 2119.16 km^2 followed by Gujarat, which contains 1164.06 km^2 of mangrove cover, predominantly featuring open mangroves (Table 3). The lowest mangrove cover is found in the Karaikal District of Puducherry. The states that show significant gain in mangrove cover are Andhra Pradesh (13 km^2) and Maharashtra (12 km^2). The rise in mangrove cover in Andhra Pradesh is mostly due to natural regeneration and plantation initiatives, conservation efforts and addition of leftover patches. Andhra Pradesh's districts of Krishna, Bapatla, and Kakinada have all seen a rise. In Maharashtra, the increase in mangrove cover is mostly attributable to natural regeneration and plantations done by the State government. The growth in mangrove cover has been found in the Raigarh and Palghar districts of Maharashtra. But there is a prominent decrease in mangrove cover observed in Gujarat by 36 km^2 (India State of Forest Report, Vol. 1, 2023).

The Land Use Land Cover (LULC) analysis and mapping of the Sundarbans region of the state of West Bengal in India were carried out to provide information to help users understand the problems and prospects of the unique ecosystem. The Landsat data for the years 1988 and 2022 were used for LULC classification using supervised classification in the QGIS environment (Fig. 2). The Sundarbans region was classified into seven LULC classes, such as forest, vegetation, river, water body, fallow land, agricultural land, and built-up areas for both years using a supervised classification method. The respective changes of these classes were −313.8, +139.0, −102.4, +259.0, +625.8, −1465.9, and +858.2 km^2. The positive sign (+) denoted the increase of the area under a particular class, and the negative sign (−) denoted its decrease in the year of 2022 from 1988. Over 34 years of time, the lands under the forest, river, and agriculture classes have decreased substantially. However, the

Table 3 State/Union Territories (UT)-wise Mangrove Cover Assessment 2023 (in km^2) in India

Sl. no.	State/UT	Very dense mangrove	Moderately dense mangrove	Open mangrove	Total mangrove cover
1	Andhra Pradesh	0.00	213.90	207.53	421.43
2	Goa	0.00	23.75	7.59	31.34
3	Gujarat	0.00	179.09	984.97	1164.06
4	Karnataka	0.11	3.15	10.94	14.20
5	Kerala	0.00	4.73	4.72	9.45
6	Maharashtra	0.00	89.82	225.27	315.09
7	Odisha	81.67	94.61	82.78	259.06
8	Tamil Nadu	1.19	25.07	15.65	41.91
9	West Bengal	981.63	703.79	433.74	2119.16
10	A & N Islands	399.37	162.64	46.28	608.29
11	D& NH and Daman & Diu	0.00	0.21	3.65	3.86
12	Puducherry	0.00	0.08	3.75	3.83
Total		**1463.97**	**1500.94**	**2046.87**	**5011.78**

Source: India's Biennial State of Forest Report, 2023

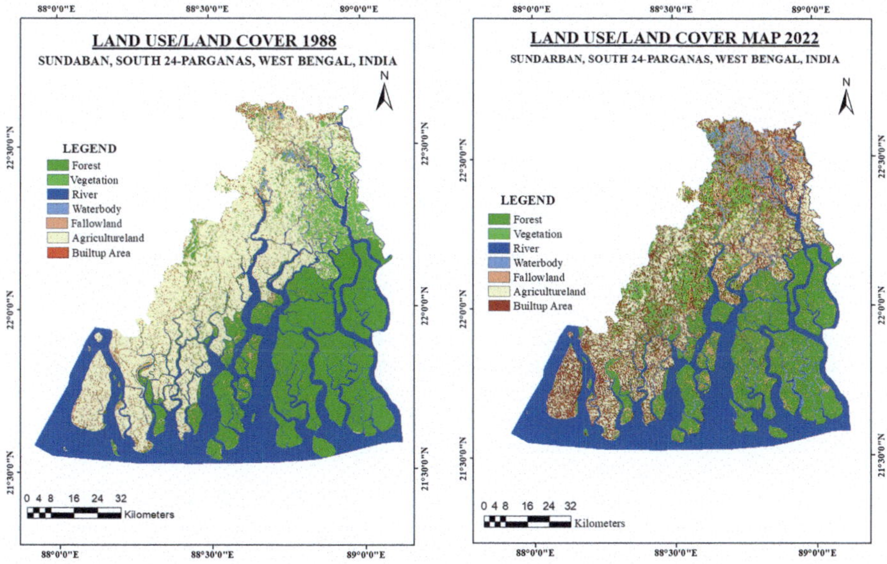

Fig. 2 Land use land cover map of the Indian Sundarbans region for the years 1988 and 2022

lands under other vegetation, water bodies (other than rivers), fallow, and built-up areas have increased in the Sundarbans region. This LULC mapping will enable the monitoring of temporal dynamics of agricultural ecosystems, forest conversions, surface water bodies, etc. on a timely basis for fulfilling the sustainable development goal in the Indian Sundarbans region.

11 Future Need for Remote Sensing Research on Mangrove

Mangrove forests are threatened by many human pressures, including conversion to aquaculture and agriculture, coastal development, overharvesting, agricultural and industrial pollutions, and climate change impacts. There is an urgent need for homogenized classification schemes and standardized data-processing methods for remote sensing-based assessment of mangrove ecosystems using common platforms like Google Earth Engine, Copernicus Global Land Services, etc. This will enable a coherent and seamless data accessibility to the next users for generating the new comparable data with the archive. Similarly, a homogenized transparent accuracy assessment will give confidence for prioritizing the issues. This can provide insights into researchable issues and the adoption of new techniques for the mangrove ecosystems. Developing a database management system for the world mangrove will promote further investigations on synergetic data use. Satellite and unmanned aerial vehicle (UAV) can play a major role in the evaluation of ecosystem services of mangrove by interdisciplinary and well-networked research teams at the regional, national, and global level.

Acknowledgements This chapter is an outcome of the Multinational Collaborative Project partially supported by Asia Pacific Network for Global Change Research (APN-GCR) and University of the Sunshine Coast (UniSC), Australia. The logistic support from the Indian Council of Agricultural Research for the study is duly acknowledged.

References

Anand, A., Pandey, P. C., Petropoulos, G. P., Pavlides, A., Srivastava, P. K., Sharma, J. K., & Malhi, R. K. M. (2020). Use of hyperion for mangrove forest carbon stock assessment in Bhitarkanika forest reserve: A contribution towards blue carbon initiative. *Remote Sensing, 12*(4), 597.

Baccini, A. G. S. J., Goetz, S. J., Walker, W. S., Laporte, N. T., Sun, M., Sulla-Menashe, D., Hackler, J., Beck, P. S. A., Dubayah, R., Friedl, M. A., & Samanta, S. (2012). Estimated carbon dioxide emissions from tropical deforestation improved by carbon-density maps. *Nature Climate Change, 2*(3), 182–185.

Baccini, A., Walker, W., Carvalho, L., Farina, M., Sulla-Menashe, D., & Houghton, R. A. (2017). Tropical forests are a net carbon source based on aboveground measurements of gain and loss. *Science, 358*(6360), 230–234.

Barbier, E. B., Hacker, S. D., Kennedy, C., Koch, E. W., Stier, A. C., & Silliman, B. R. (2011). The value of estuarine and coastal ecosystem services. *Ecological Monographs, 81*(2), 169–193.

Basyuni, M., Mubaraq, A., Amelia, R., Wirasatriya, A., Iryanthony, S. B., Slamet, B., Al Mustaniroh, S. S., Pradisty, N. A., Sidik, F., Hanintyo, R., & Sumarga, E. (2025). Mangrove aboveground biomass estimation using UAV imagery and a constructed height model in Budeng–Perancak, Bali, Indonesia. *Ecological Informatics, 86*, 103037.

Bello, O. M., & Aina, Y. A. (2014). Satellite remote sensing as a tool in disaster management and sustainable development: Towards a synergistic approach. *Procedia-Social and Behavioral Sciences, 120*, 365–373.

Biswas, S., Sarkar, K., & Das, T. K. (2024). Decadal changes (1980–2021) of shoreline and mangrove cover in Sundarban Delta, India, using remote sensing and GIS. *Journal of Physical Oceanography, 54*(8), 1655–1674.

Campomanes, F., Pada, A. V., & Silapan, J. (2016, September). Mangrove classification using support vector machines and random forest algorithm: A comparative study. In *6th International Conference on Geographic Object-Based Image Analysis, GEOBIA 2016: Solutions & Synergies*. University of Twente, Faculty of Geo-Information Science and Earth Observation (ITC).

Castillo, J. A. A., Apan, A. A., Maraseni, T. N., & Salmo, S. G., III. (2017). Estimation and mapping of above-ground biomass of mangrove forests and their replacement land uses in the Philippines using Sentinel imagery. *ISPRS Journal of Photogrammetry and Remote Sensing, 134*, 70–85.

Chapman, V. J. (1976). Mangrove vegetation. J. Cramer, Vaduz, Pp. 447.

Church, J. A., & White, N. J. (2011). Sea-level rise from the late 19th to the early 21st century. *Surveys in Geophysics, 32*(4), 585–602.

Clark, D. A., Piper, S. C., Keeling, C. D., & Clark, D. B. (2003). Tropical rain forest tree growth and atmospheric carbon dynamics linked to interannual temperature variation during 1984–2000. *Proceedings of the National Academy of Sciences, 100*(10), 5852–5857.

Coops, N. C., Wulder, M. A., & White, J. C. (2006). Identifying and describing forest disturbance and spatial pattern: Data selection issues and methodological implications. In *Forest disturbance and spatial pattern: Remote sensing and GIS approaches* (p. 264).

Crowe, M. (2020). Do mangroves burn? A remote sensed study of the impacts of forest fires on mangrove populations in South Sumatra, Indonesia. MSc degree in Climate Change Science and Policy, King's College London.

Datta, D., & Deb, S. (2012). Analysis of coastal land use/land cover changes in the Indian Sunderbans using remotely sensed data. *Geo-spatial Information Science, 15*(4), 241–250.

Dookie, S., Ansari, A. A., & Jaikishun, S. (2025). Forest-fire interactions, impacts, and implications: A focus on mangroves. *New Zealand Journal of Forestry Science, 55*.

Duke, N., Ball, M., & Ellison, J. (1998). Factors influencing biodiversity and distributional gradients in mangroves. *Global Ecology & Biogeography Letters, 7*(1), 27–47.

Dutta, D., Das, P. K., Paul, S., Sharma, J. R., & Dadhwal, V. K. (2015). Assessment of ecological disturbance in the mangrove forest of Sundarbans caused by cyclones using MODIS time-series data (2001–2011). *Natural Hazards, 79*(2), 775–790.

FAO, 2020. Global Forest Resources Assessment 2020: Main report. .

Ganguly, S., Nemani, R. R., Zhang, G., Hashimoto, H., Milesi, C., Michaelis, A., Wang, W., Votava, P., Samanta, A., Melton, F., & Dungan, J. L. (2012). Generating global leaf area index from Landsat: Algorithm formulation and demonstration. *Remote Sensing of Environment, 122*, 185–202.

Ghosh, S., & Mistri, B. (2022). Analyzing the multi-hazard coastal vulnerability of Matla–Bidya inter-estuarine area of Indian Sundarbans using analytical hierarchy process and geospatial techniques. *Estuarine, Coastal and Shelf Science, 279*, 108144.

Ghosh, A., Schmidt, S., Fickert, T., & Nüsser, M. (2015). The Indian Sundarban mangrove forests: History, utilization, conservation strategies and local perception. *Diversity, 7*(2), 149–169.

Ghosh, M. K., Kumar, L., & Roy, C. (2016). Mapping long-term changes in mangrove species composition and distribution in the Sundarbans. *Forests, 7*(12), 305.

Ghosh, S. M., Behera, M. D., Jagadish, B., Das, A. K., & Mishra, D. R. (2021). A novel approach for estimation of aboveground biomass of a carbon-rich mangrove site in India. *Journal of Environmental Management, 292*, 112816.

Gilman, E. L., Ellison, J., Duke, N. C., & Field, C. (2008). Threats to mangroves from climate change and adaptation options: A review. *Aquatic Botany, 89*(2), 237–250.

Giri, C., Pengra, B., Zhu, Z., Singh, A., & Tieszen, L. L. (2007). Monitoring mangrove forest dynamics of the Sundarbans in Bangladesh and India using multi-temporal satellite data from 1973 to 2000. *Estuarine, Coastal and Shelf Science, 73*(1–2), 91–100.

Giri, S., Mukhopadhyay, A., Hazra, S., Mukherjee, S., Roy, D., Ghosh, S., Ghosh, T., & Mitra, D. (2014). A study on abundance and distribution of mangrove species in Indian Sundarban using remote sensing technique. *Journal of Coastal Conservation, 18*(4), 359–367.

Global Mangrove Watch. (2020). *Mangrove extent dataset*. https://www.globalmangrovewatch.org

Gupta, S. (2021). Disturbance of mangrove forest due to climate change: The prospects of Sundarban. *International Journal of Multidisciplinary: Applied Business and Education Research, 2*(12), 1306–1313.

He, B., Lai, T., Fan, H., Wang, W., & Zheng, H. (2007). Comparison of flooding-tolerance in four mangrove species in a diurnal tidal zone in the Beibu Gulf. *Estuarine, Coastal and Shelf Science, 74*(1–2), 254–262.

Healey, S. P., Cohen, W. B., Zhiqiang, Y., & Krankina, O. N. (2005). Comparison of Tasseled Cap-based Landsat data structures for use in forest disturbance detection. *Remote Sensing of Environment, 97*(3), 301–310.

IPCC. (2013). Summary for policymakers. In T. F. Stocker, D. Qin, G.-K. Plattner, M. Tignor, S. K. Allen, J. Boschung, A. Nauels, Y. Xia, V. Bex, & P. M. Midgley (Eds.), *Climate change 2013: The physical science basis. Contribution of Working Group I to the Fifth Assessment Report of the Intergovernmental Panel on Climate Change*. Cambridge University Press.

Iqbal, M. H. (2020). Valuing ecosystem services of Sundarbans Mangrove Forest: Approach of choice experiment. *Global Ecology and Conservation, 24*, e01273.

IUCN. (2025). *The IUCN red list of threatened species*. Version 2025-1. Retrieved July 28, 2025.

Jiang, Y., Zhang, L., Yan, M., Qi, J., Fu, T., Fan, S., & Chen, B. (2021). High-resolution mangrove forests classification with machine learning using worldview and UAV hyperspectral data. *Remote Sensing, 13*(8), 1529.

Jose, S., Alex, C. J. C. J., Kumar, S., Varghese, A., & Madhu, G. (2011). Landscape metric modeling—A technique for forest disturbance assessment in Shendurney wildlife sanctuary. *Environmental Research, Engineering and Management, 58*(4), 34–41.

Kanjin, K., & Alam, B. M. (2024). Assessing changes in land cover, NDVI, and LST in the Sundarbans mangrove forest in Bangladesh and India: A GIS and remote sensing approach. *Remote Sensing Applications: Society and Environment, 36*, 101289.

Kuenzer, C., Bluemel, A., Gebhardt, S., Quoc, T. V., & Dech, S. (2011). Remote sensing of mangrove ecosystems: A review. *Remote Sensing, 3*(5), 878–928.

Kumar, T., Mandal, A., Dutta, D., Nagaraja, R., & Dadhwal, V. K. (2019). Discrimination and classification of mangrove forests using EO-1 Hyperion data: A case study of Indian Sundarbans. *Geocarto International, 34*(4), 415–442.

Kundu, K., Halder, P., & Mandal, J. K. (2021). Change detection and patch analysis of Sundarban forest during 1975–2018 using remote sensing and GIS data. *SN Computer Science, 2*(5), 364.

Kurbanov, E., Vorobev, O., Lezhnin, S., Sha, J., Wang, J., Li, X., Cole, J., Dergunov, D., & Wang, Y. (2022). Remote sensing of forest burnt area, burn severity, and post-fire recovery: A review. *Remote Sensing, 14*(19), 4714.

Mahadevia Ghimire, K., & Vikas, M. (2012). Climate change–impact on the Sundarbans, a case study. *International Scientific Journal: Environmental Science, 2*(1), 7–15.

Mandal, M. S. H., & Hosaka, T. (2020). Assessing cyclone disturbances (1988–2016) in the Sundarbans mangrove forests using Landsat and Google Earth Engine. *Natural Hazards, 102*(1), 133–150.

Mandal, P., Maiti, A., Paul, S., Bhattacharya, S., & Paul, S. (2022). Mapping the multi-hazards risk index for coastal block of Sundarban, India using AHP and machine learning algorithms. *Tropical Cyclone Research and Review, 11*(4), 225.

Mildrexler, D. J., Zhao, M., & Running, S. W. (2009). Testing a MODIS global disturbance index across North America. *Remote Sensing of Environment, 113*(10), 2103–2117.

Mitra, D., & Karmaker, S. (2010). Mangrove classification in Sundarban using high resolution multi-spectral remote sensing data and GIS. *Asian Journal of Environment and Disaster Management, 2*(2).

Mondal, I., & Bandyopadhyay, J. (2014). Coastal zone mapping through geospatial technology for resource management of Indian Sundarban, West Bengal, India. *International Journal of Remote Sensing Applications, 4*(2), 103–112.

Mondal, I., Thakur, S., Ghosh, P., De, T. K., & Bandyopadhyay, J. (2018). Land use/land cover modeling of Sagar Island, India using remote sensing and GIS techniques. In *Emerging technologies in data mining and information security: Proceedings of IEMIS 2018* (Vol. 1, pp. 771–785). Springer Singapore.

Mondal, B. K., Mahata, S., Basu, T., Das, R., Patra, R., Abdelrahman, K., Fnais, M. S., & Praharaj, S. (2024). Analysis of the post-cyclonic physical flood susceptibility and changes of mangrove forest area using multi-criteria decision-making process and geospatial analysis in Indian Sundarbans. *Atmosphere, 15*(4), 432.

Mukherjee, N., Sutherland, W. J., Khan, M. N. I., Berger, U., Schmitz, N., Dahdouh-Guebas, F., & Koedam, N. (2014). Using expert knowledge and modeling to define mangrove composition, functioning, and threats and estimate time frame for recovery. *Ecology and Evolution, 4*(11), 2247–2262.

Nerem, R. S., Chambers, D. P., Choe, C., & Mitchum, G. T. (2010). Estimating mean sea level change from the TOPEX and Jason altimeter missions. *Marine Geodesy, 33*(S1), 435–446.

NOAA. (2015). National Oceanic and Atmospheric Administration, National Ocean Service, Center for Operational Oceanographic Products and Services, Tides and Currents. http://tidesandcurrents.noaa.gov/

NOAA. (2023). *Cyclone hazards & safety*. https://www.noaa.gov/jetstream/tc-hazards

Padhy, S. R., Dash, P. K., & Bhattacharyya, P. (2022). Challenges, opportunities, and climate change adaptation strategies of mangrove-agriculture ecosystem in the Sundarbans, India: A review. *Wetlands Ecology and Management, 30*(1), 191–206.

Parida, A. K., & Jha, B. (2010). Salt tolerance mechanisms in mangroves: A review. *Trees, 24*(2), 199–217.

Parida, B. R., & Kumar, P. (2020). Mapping and dynamic analysis of mangrove forest during 2009–2019 using landsat–5 and sentinel–2 satellite data along Odisha Coast. *Tropical Ecology, 61*(4), 538–549.

Paul, A. K., Ray, R., Kamila, A., & Jana, S. (2017). Mangrove degradation in the Sundarbans. In *Coastal wetlands: Alteration and remediation* (pp. 357–392). Springer International Publishing.

Prakash, S., Varghese, A., Amrutha, A. S., & Baiju, K. R. (2023). Assessment of ecological disturbance on Indian sundarbans with special reference to Amphan Cyclone by using geospatial technology. *Journal of Geospatial Surveying, 3*(1), 26.

Randhir, T. O., Toffling, K., & Griffin, C. R. (2024). Impacts of climate change and variability on the growth potential of global mangrove distribution. *Sustainable and Resilient Infrastructure, 9*(1), 63–71.

Razali, S. M., Nuruddin, A. A., Malek, I. A., & Patah, N. A. (2010). Forest fire hazard rating assessment in peat swamp forest using Landsat thematic mapper image. *Journal of Applied Remote Sensing, 4*(1), 043531.

Salam, M. A., Ross, L. G., & Beveridge, C. M. C. (2007). The use of GIS and remote sensing techniques to classify the Sundarbans Mangrove vegetation. *Journal of Agroforestry and Environment, 1*(1), 7–15.

Samanta, K., & Hazra, S. (2016). Mangrove forest cover changes in Indian Sundarban (1986–2012) using remote sensing and GIS. In *Environment and earth observation: Case studies in India* (pp. 97–108). Springer International Publishing.

SEI (Stockholm Environment Institute). (2023). *NASA data and images help increase climate resilience in the Mekong region*. https://www.sei.org/features/nasa-data-images-help-increase-climate-resilience-mekong-region/

SERVIR Global. (2023). *SERVIR-Mekong*. https://www.servirglobal.net/Global/Regions/SERVIR-Mekong

Shimu, S. A., Aktar, M., Afjal, M. I., Nitu, A. M., Uddin, M. P., & Al Mamun, M. (2019, December). NDVI based change detection in Sundarban Mangrove Forest using remote sensing data. In

2019 4th International Conference on Electrical Information and Communication Technology (EICT) (pp. 1–5). IEEE.
Simard, M., Fatoyinbo, L., Smetanka, C., Rivera-Monroy, V. H., Castañeda-Moya, E., Thomas, N., & Van der Stocken, T. (2019). Mangrove canopy height globally related to precipitation, temperature and cyclone frequency. *Nature Geoscience, 12*(1), 40–45.
Srivastava, S. K., Seng, K. P., Ang, L. M., Pachas, A. N. A., & Lewis, T. (2022). Drone-based environmental monitoring and image processing approaches for resource estimates of private native forest. *Sensors, 22*(20), 7872.
Thakur, S., Maity, D., Mondal, I., Basumatary, G., Ghosh, P. B., Das, P., & De, T. K. (2021). Assessment of changes in land use, land cover, and land surface temperature in the mangrove forest of Sundarbans, northeast coast of India. *Environment, Development and Sustainability, 23*(2), 1917–1943.
Tomlinson, P. B. (2016). *The botany of mangroves*. Cambridge University Press.
UNEP. (2023). *State of the world's mangroves 2023*. United Nations Environment Programme. https://www.unep.org/resources/report/state-worlds-mangroves-2023
Vadrevu, K. P., Eaturu, A., & Badarinath, K. (2010). Fire risk evaluation using multicriteria analysis—A case study. *Environmental Monitoring and Assessment, 166*(1), 223–239.
Vaiphasa, C., Ongsomwang, S., Vaiphasa, T., & Skidmore, A. K. (2005). Tropical mangrove species discrimination using hyperspectral data: A laboratory study. *Estuarine, Coastal and Shelf Science, 65*(1–2), 371–379.
Vyas, P., & Sengupta, K. (2012). Mangrove conservation and restoration in the Indian Sundarbans. In *Sharing lessons on mangrove restoration* (p. 93).
Ward, R. D., Friess, D. A., Day, R. H., & Mackenzie, R. A. (2016). Impacts of climate change on mangrove ecosystems: A region by region overview. *Ecosystem Health and Sustainability, 2*(4), e01211.
Zhen, J., Mao, D., Shen, Z., Zhao, D., Xu, Y., Wang, J., Jia, M., Wang, Z., & Ren, C. (2024). Performance of xgboost ensemble learning algorithm for mangrove species classification with multisource spaceborne remote sensing data. *Journal of Remote Sensing, 4*, 0146.

Use of Archived Data and Geospatial Methods for Characterising Sundarbans

Sanjeev Kumar Srivastava, Prashant K. Srivastava, Harikesh Singh, Sharif A. Mukul, Mohammed A. S. Arfin-Khan, Mukunda Dev Behera, and Dhananjay Barman

Abstract The Sundarbans, designated as a UNESCO World Heritage Site and recognised as the world's largest contiguous mangrove forest, is an ecologically critical region. This unique ecosystem faces significant vulnerability to a range of natural hazards, including cyclones, coastal erosion, land loss, and pollution—all of which contribute to the degradation of its biodiversity and the disruption of essential ecosystem services. Addressing these multifaceted challenges demands an integrated, transboundary approach that harmonises ecological conservation with socioeconomic development. This involves not only safeguarding forest habitats but also enhancing livelihoods and reducing poverty among local communities. To enable a comprehensive analysis of both the Indian and Bangladesh parts of the Sundarbans, it is vital to leverage temporal insights from archived geospatial data sets—such as historical topographic maps, declassified satellite-based photographs, and satellite imagery from various Earth observation platforms. In this chapter, we examine a

S. K. Srivastava (✉) · H. Singh
Geospatial Analytics for Conservation and Management, School of Science Technology and Engineering (SSTE), Sippy Downs, QLD, Australia
e-mail: sanjeev.srivastava@usc.edu.au; ssrivast@usc.edu.au

P. K. Srivastava
Remote Sensing Laboratory, Institute of Environment and Sustainable Development, Banaras Hindu University, Varanasi, India

S. A. Mukul
Department of Environment and Development Studies, United International University, Dhaka, Bangladesh

M. A. S. Arfin-Khan
Department of Forestry and Environmental Science, Shahjalal University of Science and Technology, Sylhet, Bangladesh

M. D. Behera
Centre for Ocean, River, Atmosphere and Land Sciences (CORAL), Indian Institute of Technology Kharagpur, Kharagpur, India

D. Barman
ICAR—Central Research Institute for Jute and Allied Fibers, Kolkata, West Bengal, India

S. K. Srivastava et al. (eds.), *Earth Observation for Monitoring Mangrove Ecosystems*, https://doi.org/10.1007/978-3-032-16668-5_6

variety of data sources that can offer more in-depth insights into the dynamics of this ecosystem. We also compare the utility of contemporary remote sensing data with archived information dating back to the mid-eighteenth century.

Keywords Archived data · Geospatial data · Remote sensing · Mangrove forest · Sundarbans

1 Introduction

The Sundarbans, the world's largest contiguous mangrove forest and a designated UNESCO World Heritage Site, is situated in the tidally active lower deltaic plain of the Ganges-Brahmaputra-Meghna basin along the Bay of Bengal (UNESCO, 2025). Spread approximately 10,000 km^2 across India and Bangladesh, this unique ecosystem supports the livelihoods of around 7.2 million people (Nishat et al., 2019). It provides critical ecosystem services, including acting as a natural buffer against cyclones, coastal erosion, and other environmental hazards (Mukul et al., 2019). The Sundarbans play a key role both ecologically, by protecting the area from coastal erosion, and for people's livelihoods, by providing resources such as honey, fuelwood, and fish (Kibria et al., 2018).

In the Sundarbans region, agricultural land is being abandoned in favour of shrimp aquaculture, and due to increasing salinity level and freshwater scarcity, the region is experiencing a rise in water-borne diseases (Dasgupta et al., 2020). A World Bank study has projected large-scale climate-induced migration from the fringes of the Sundarbans (Nishat et al., 2019).

This world's largest mangrove forest is composed of a network of low-lying conglomeration islands created from sediment loads deposited from the Himalayan rivers (Ghosh et al., 2015). These islands are of mudflats with salt pans and natural levees that are separated by tidal creeks (Siddiqi, 2001) and provide the world's largest remaining habitat for the globally endangered Bengal tiger (*Panthera tigris* tigris) (Mukul et al., 2019). This ecosystem, located in a tropical cyclone belt, demonstrates a complex interaction between social and ecological systems (Kibria et al., 2018). The fluctuating water levels, temperature, and wind flow are affecting the topography of the area, making it prone to several natural hazards, leading to the destruction of forest resources (Jandl et al., 2007). Studies have recommended an integrated approach that envisages the ecological and adjacent socio-cultural systems to be reconciled for natural hazards protection, livelihood improvement, and poverty alleviation (Sayer et al., 2013). The Indian vision emphasises community wellbeing, ecosystem conservation and phased out-migration. The Bangladesh vision emphasises the management and conservation of the delta (IUCN, 2014). Studies have indicated that the majority of the forested grids in the region are likely to experience a shift in forest types due to climate change scenarios (Ravindranath et al., 2006). A study on the comparison of the area with earlier survey records (e.g. early nineteenth-century Dampier and Hodges survey of forest extent) indicates that the mangrove has shrunk to almost half in size (DasGupta & Shaw, 2017). Previous studies have indicated that the anthropogenic interference began in the region in the late eighteenth century, mainly through agricultural practices (Ghosh et al., 2015), and deforestation is contributing 6–17% of anthropogenic CO2 emission (2000–2010, about 1.0 Pg C/year was emitted) (Baccini et al., 2012).

As mentioned above, the Sundarbans is increasingly threatened by a combination of anthropogenic pressures and natural forces (Sievers et al., 2020). The degradation of forest cover and associated resources is driven by deforestation, land-use change, and the rising frequency of climate-induced events such as cyclones, floods, and droughts (Karsch et al., 2023). Many villages have already witnessed the disappearance of protective mangrove belts, which serve as the region's first line of defence against climate change impacts. The intrusion of seawater has led to increased salinity, resulting in the decline of high-value, storm-resistant tree species such as Sundari (*Heritiera fomes*), from which the Sundarbans derives its name. This ecological stress is further exacerbated by reduced freshwater and sediment inflows due to upstream damming and river regulation (Dasgupta et al., 2020). The consequences are visible and alarming: several mangrove islands have vanished over the past century, others have significantly shrunk in the last 50 years, and repeated embankment failures have displaced communities and destroyed homes.

In this chapter, we discuss key data sources that can be utilised for long-term assessment of the ecological and socioeconomic conditions of the Sundarbans.

1.1 Study Area

This study investigates the current extent of the Sundarbans mangrove ecosystem and its adjacent regions that used to be mangrove forests (Fig. 1). We demonstrate the utility of archived historic geospatial data sets to highlight regions where there is a loss of land or of mangroves over the last 175 years.

1.2 Analytical Framework

The study followed a four step workflow: (1) georeferencing of archived topographic maps (root mean square error (RMSE) < 50 m) using identifiable ground control points, (2) mosaicking and orthorectification of CORONA KH-4B film scans, (3) overlay and change detection with modern satellite imagery in ArcGIS Pro and QGIS, and (4) quantitative estimation of mangrove loss and land conversion using supervised classification and manual visual interpretation. This framework ensured comparability across disparate data sets collected over a span of nearly 250 years.

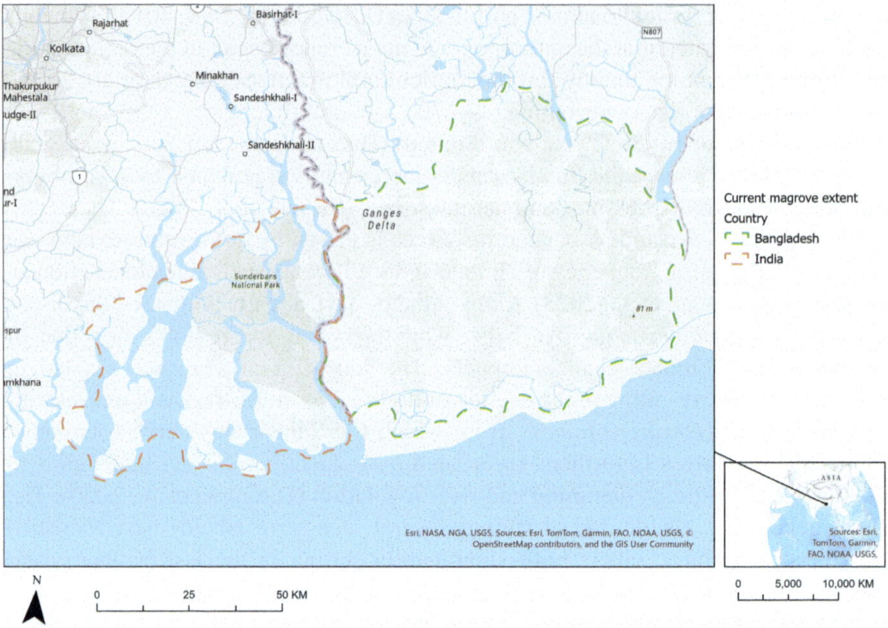

Fig. 1 The study area showing the current extent of mangroves in India and Bangladesh

1.3 Remote Sensing Data

A wide range of satellite imagery is now available for mapping the current extent of the Sundarbans mangrove ecosystem, with near-daily acquisitions from multiple platforms. In this study, we utilise imagery from Landsat and PlanetScope satellites. The Landsat programme has been providing global-scale remote sensing data since 1972, while PlanetScope has offered near-daily coverage in recent years (Planet Team, 2019). These data sets are widely used for various Earth resource analyses, including vegetation mapping and monitoring in the Sundarbans. Given the availability of well-established algorithms for vegetation classification and condition assessment using these missions, this chapter focuses on the use of historic, high-resolution, declassified satellite imagery from the CORONA mission to complement contemporary analyses.

The USA's Corona mission, which was the first satellite-based remote sensing mission operational during 1960–1972 (CIA, 2025). These satellites, code-named as CORONA, ARGON, and LANYARD, collected more than 860,000 images of the Earth's surface covering 1900 million square kilometres. The USA's National Archives and Records Administration (NARA) was given the responsibility for the original film and provided access to a duplicate copy for public viewing of the film. The USGS was also provided a copy to support science products. The data set can be accessed from the USGS's Earth Explorer website (USGS, 2025a, 2025b).

A number of camera systems were used for collecting photographic films. The earlier missions carried a single panoramic camera (KH-1, KH-2, KH-3, and KH-6) or a single frame camera (KH-5). The later systems carried two panoramic cameras with a separation angle of 30°, with one camera looking forward and the other looking aft (KH-4, KH-4A, and KH-4B) (USGS, 2025a, 2025b).

For this study, we used images acquired in 1972 by the CORONA satellite with the KH-4B camera that had a spatial resolution of 1.8 m. The images were greyscale single band, and the interpretation was based on the tone, texture, patterns, shadow, shape, and size of various features. The images were scanned using high-performance photogrammetric film scanner to create digital products at 7 micron or 14 micron resolutions, and the scanned images can be subjected to photogrammetry for 3d information extraction (USGS, 2025a, 2025b).

The images are available in different parts for each panoramic scene, and for merging all the image parts, a specialised stereogrammetry software was used. The open-source software from NASA, AMES Stereo Pipeline (ASP), was used to stitch the individual film scans to generate a whole image (Beyer et al., 2018) (Fig. 2).

The images collected from two different angles were downloaded (DA and DF series), and a mosaic was created for the study area using ArcGIS Pro software (Figs. 3 and 4). The KH4A had a dual panoramic camera system with fore (forward) and aft (backward) looking cameras having a 15° off-nadir view angle, resulting in a convergence angle of 30° and a baseline-to-height ratio of 0.54. The images were named as DF and DA series for and alt looking cameras (Figs. 3 and 4).

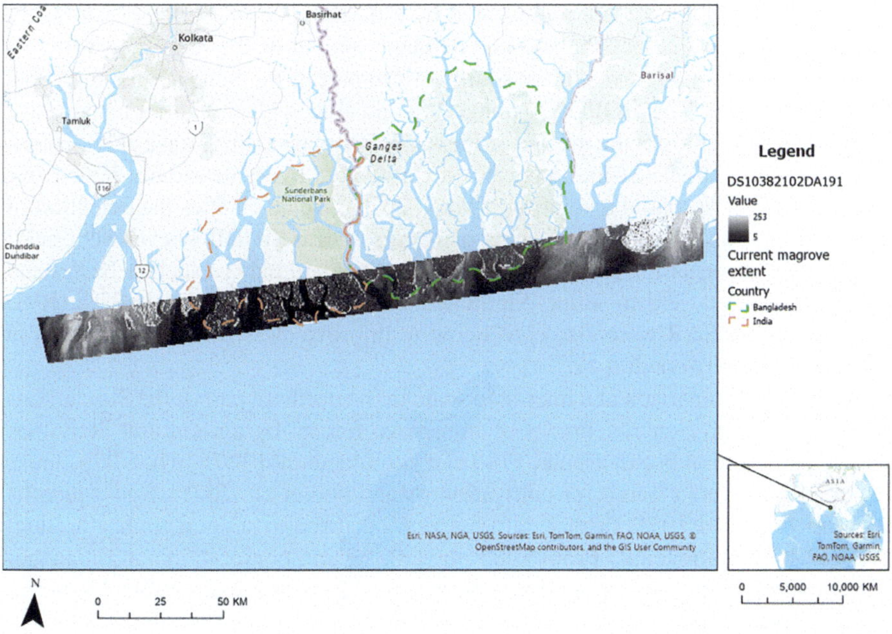

Fig. 2 An example stitched panoramic scene taken from the CORONA satellite

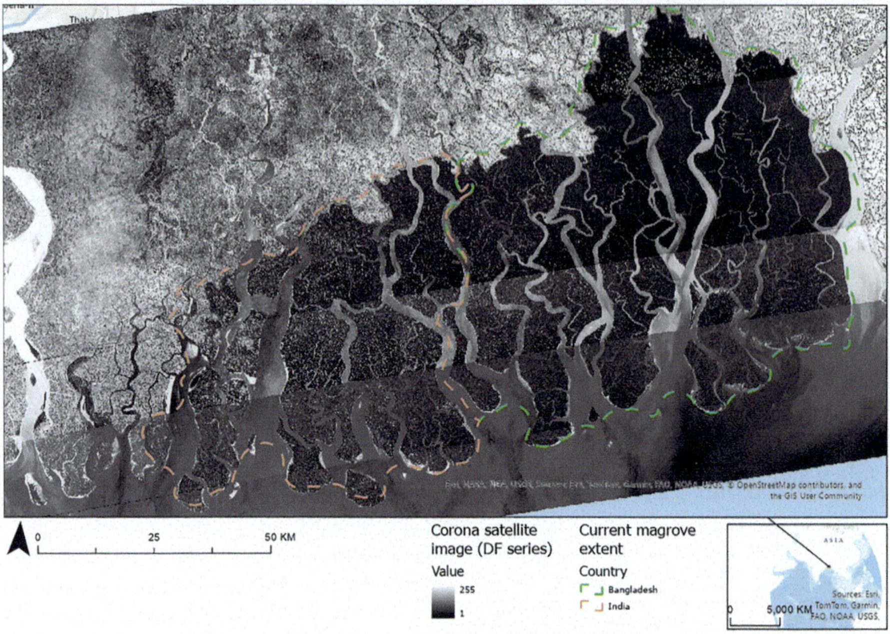

Fig. 3 The mosaic was created for the study area using images collected in the DF series. The DF series had the forward coverage with the camera set in panoramic forward mode

The mangrove areas were represented by a darker tone in the image and were identified by their distinct shapes. The detailed version of the map shows loss of land masses with mangroves in the southwestern part of the Sundarbans in the last 50 years (Fig. 5). The CORONA image is compared with the current mangrove boundaries derived from the mosaic of PlanetScope images (Srivastava et al., 2025). The selection of PlanetScope imagery was motivated by its near-daily revisit frequency and 3–5 m spatial resolution (Planet, 2025), which enables the detection of fine-scale land-use changes that coarser platforms such as MODIS or even Sentinel-2 may miss. For vegetation condition assessment, standard indices such as NDVI and EVI were derived, and in saline-dominated tracts, radar vegetation indices (RVI) from Sentinel-1 SAR were also considered to improve classification robustness in persistently cloudy conditions.

A detailed comparison at a finer map scale in the northern part of the Sundarbans reveals progressive encroachment of mangrove forests by agricultural activities. This transformation began around 1780 and continued until 1875, when large areas of mangroves were cleared for cultivation (Mahmood et al., 2021). Subsequently, between 1876 and 1951, further deforestation occurred to support timber production and generate revenue. From 1952 onward, several conservation and management plans were introduced to protect the region (Mahmood et al., 2021). However, satellite imagery from 1972 to 2020 reveals continued agricultural expansion into mangrove areas, underscoring the ongoing pressures on this fragile ecosystem (Fig. 6).

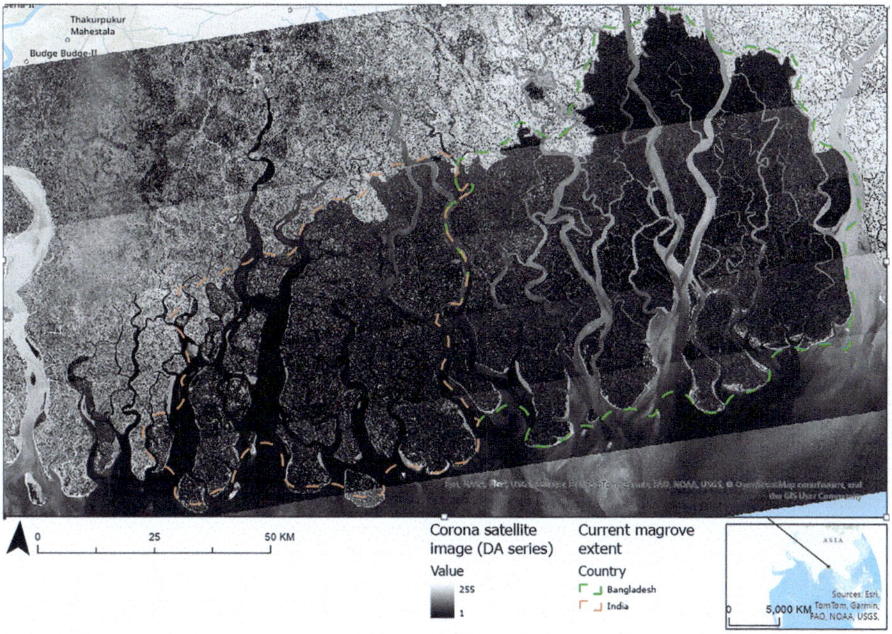

Fig. 4 The mosaic was created for the study area using images collected in the DA series. The DA series collected images in aft (backward) panoramic mode

1.4 Topographic Maps

Topographic maps offer a comprehensive overview of the landscape by depicting key natural and human-made features with precision. Their accuracy stems from data collected through systematic ground surveys conducted by trained professionals, followed by meticulous cartographic plotting. As part of national mapping programmes, these maps adhere to standardised specifications for representing surface features, which are consistently tied to the scale of the map series.

The Survey of India, established in 1767, laid the foundation for mapping the Indian subcontinent. However, systematic mapping of Earth's surface features began only in the latter half of the nineteenth century (Coldstream, 1919). Early maps were produced at coarse scales, limiting their detail and utility. In the early twentieth century, a structured topographic survey scheme was introduced, beginning with maps at a 1:1 million scale (covering 4° latitude × 4° longitude). These were subsequently subdivided into 16 sheets at a 1:250,000 scale, each covering 1° latitude × 1° longitude (Edney, 1991).

For this study, the earliest map utilised was from the 1:250,000 scale series, published in 1913, based on survey data collected between 1895 and 1900, with some hydrological information dating back to 1878 (Fig. 5). Additional maps used in this analysis were drawn from the 1:50,000 to 1:63,000 series (Brown, 2023), most of which were published in the 1920s and 1940s. The 1920s maps incorporated data

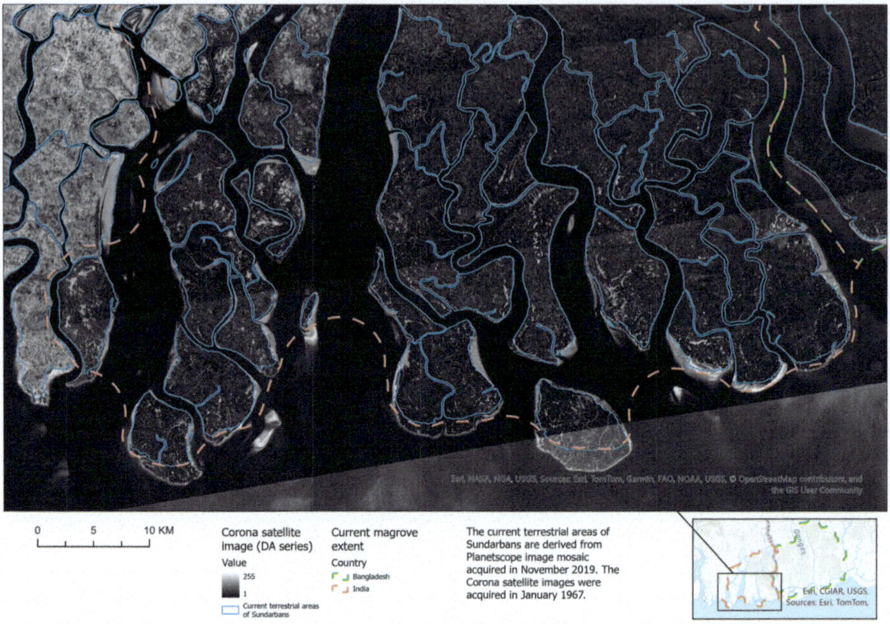

Fig. 5 The southern part of the Sundarbans shows a loss of land over the past 50 years. The blue polygons represent the current extent of Sundarbans mangroves, which are derived from PlanetScope data

collected primarily between 1904 and 1908, while the 1940s maps combined earlier survey data with new information gathered in the 1920s and updates derived from aerial photography conducted in the 1940s. Collectively, these maps offer a valuable historical perspective of the region's landscape (Fig. 6).

The largest-scale maps available for the region were produced at approximately 1:50,000 scale and were published primarily during the 1920s and 1940s (Fig. 7). For this study, the earliest map utilised was from the 1:250,000 scale series, published in 1913, based on survey data collected between 1895 and 1900, with some hydrological information dating back to 1878. Additional maps used in this analysis were drawn from the 1:50,000 to 1:63,000 series (Brown, 2023), most of which were published in the 1920s and 1940s. The 1920s maps incorporated data collected primarily between 1904 and 1908, while the 1940s maps combined earlier survey data with new information gathered in the 1920s and updates derived from aerial photography conducted in the 1940s. Collectively, these maps offer a valuable historical perspective of the region's landscape (Fig. 8).

Several historic topographic maps are available at coarse resolutions. Notably, the region was surveyed by British surveyors in the early eighteenth century, culminating in the publication of a complete map of the study area in 1761, based on survey data collected in 1760 (Fig. 9). Many of these early maps are poorly drawn and lack precise geographic coordinates, making georeferencing a significant

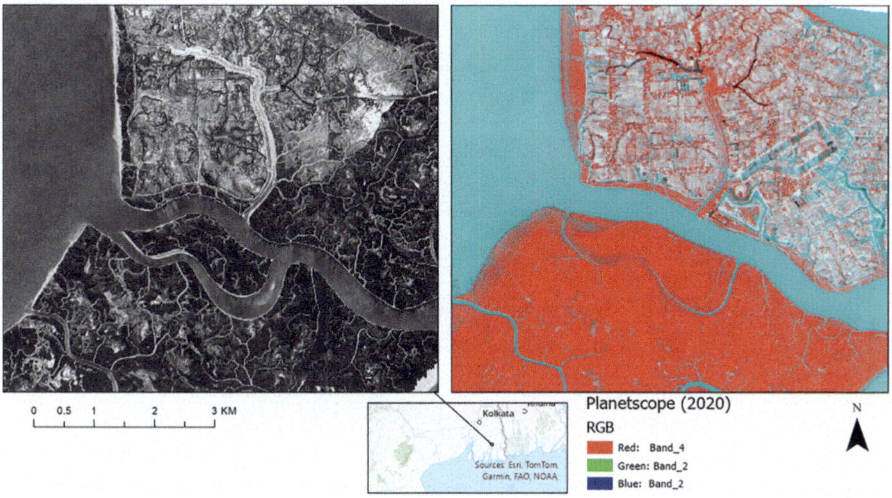

Fig. 6 A local scale comparison between CORONA (1972) and PlanetScope (2020) images showing the conversion of mangrove forests to agricultural lands

challenge. Their limited spatial accuracy and incomplete metadata further complicate integration with modern geospatial data sets.

Another significant topographic map, available from the mid-nineteenth century, was produced at an approximate scale of 1:250,000. This map includes latitude and longitude coordinates, enabling accurate georeferencing using GIS software. Published in 1853, it was based on survey data collected between 1847 and 1851 (Fig. 10) and primarily covers the western portion of the Sundarbans. A comparative analysis with recent Landsat 8/9 imagery reveals notable mangrove forest loss, particularly in the northern and western regions of the study area (Fig. 10).

The western part of the Sundarbans has undergone significant transformation, with large areas converted into human settlements, including villages, small towns, agricultural lands, and road networks (Fig. 11). These regions now show no visible signs of mangrove forests. Mangrove ecosystems are well-known for their role in mitigating the adverse impacts of cyclones and sea-level rise. However, the western Sundarbans is now highly susceptible to coastal erosion (Mohammed et al., 2024). The replacement of mangroves with agricultural and residential land use increases the vulnerability of these areas to extreme weather events and long-term environmental degradation.

The northern part of the region, once dominated by mangrove forests, has also experienced significant encroachment by human settlements and is now largely characterised by aquaculture activities. In the Landsat false-colour composite imagery, dark blue patches clearly indicate the presence of aquaculture ponds and facilities (Fig. 12). This transformation reflects a shift in land use that has implications for ecosystem health and resilience. The destructive aquaculture practices of the past are well-known to destroy mangrove forests in the region (Parvin et al., 2023).

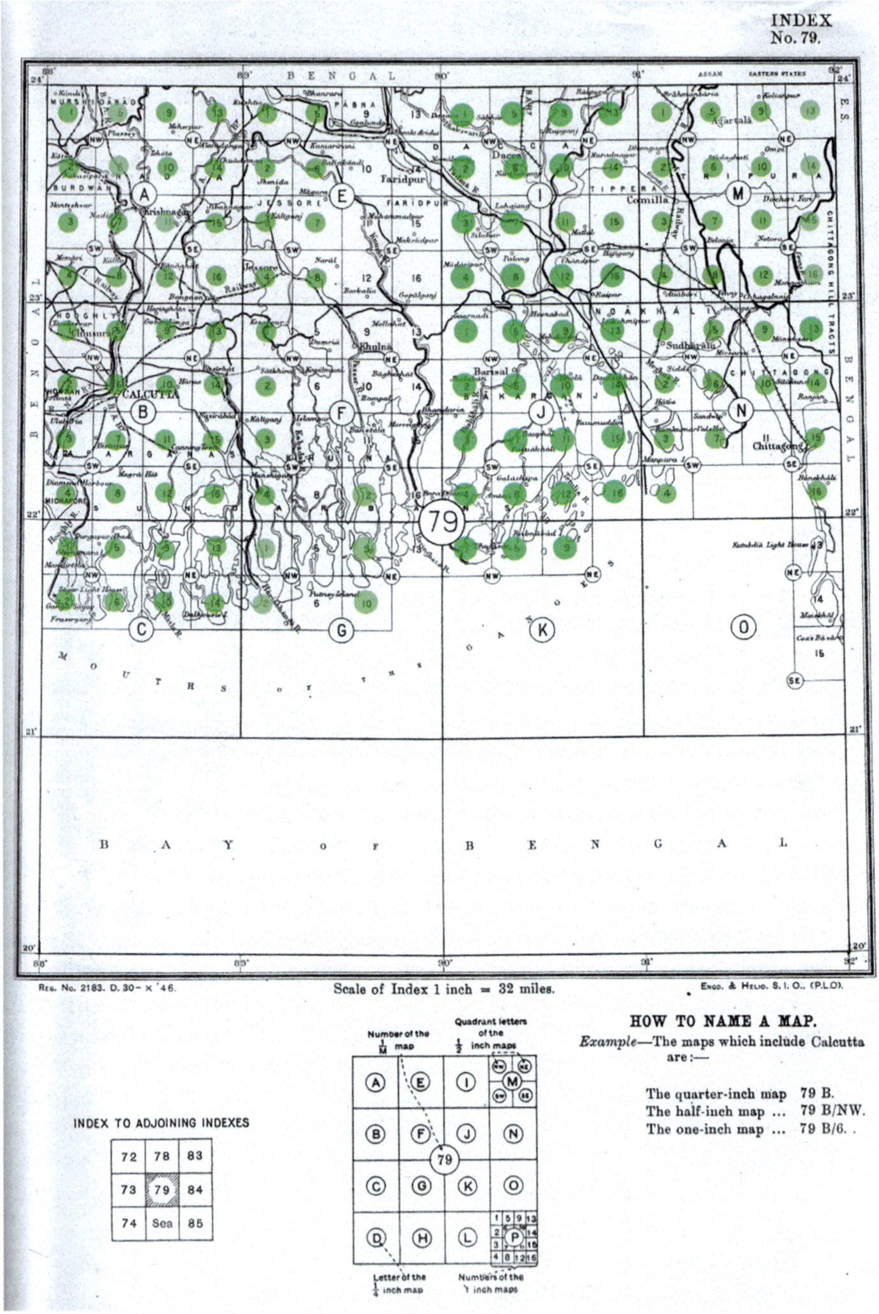

Fig. 7 The coverage of India and the surrounding regions by topographic maps (1:50–63,000) in the Sundarbans region (source: Brown, 2023)

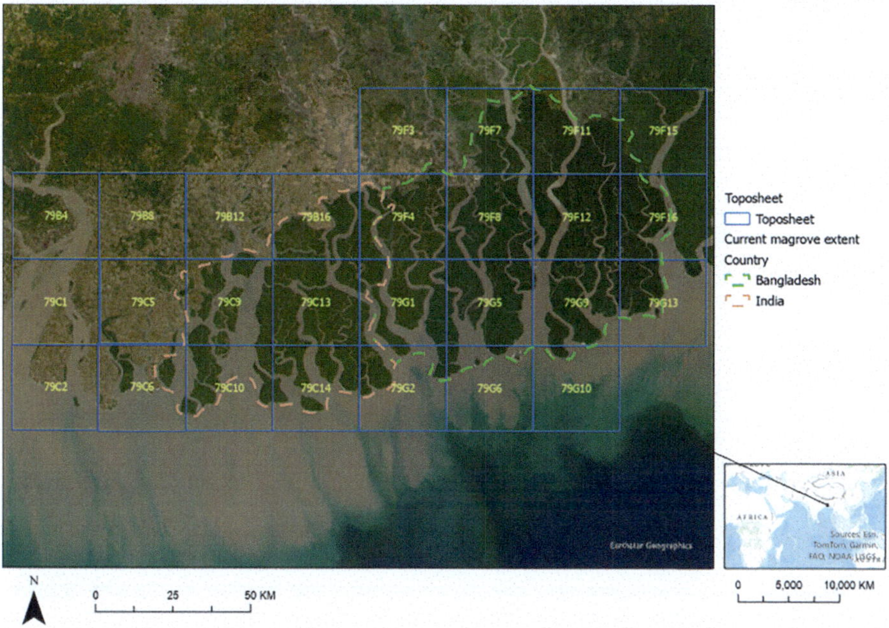

Fig. 8 The current extent of the mangrove vegetation is distributed across India and Bangladesh. The earlier extent was much larger, which can be seen in topographic maps published in the early twentieth century. These toposheets are available at roughly 1:63,000 map scale and were published in the early and mid-twentieth century. The recent topographic maps can be obtained from the survey departments of India and Bangladesh

A well-established human settlement in the region can be seen in the photographs taken by the authors during a field visit to the area, featuring aquaculture ponds, road networks, residential structures, agricultural fields, and railway tracks (Fig. 13). While exploring Google Street View, we noticed several other anthropogenic structures such as railway tracks and mobile towers.

1.5 *Combining Historic Topographic Maps with Declassified Images and Recently Acquired Remote Sensing Data*

Integrating multiple archival data sets offers valuable insights into the long-term ecological dynamics of ecosystems. Such an approach is particularly critical for future management, as climate projections indicate an increased frequency of extreme weather events, which could severely impact ecosystem health and, consequently, local livelihoods (Sarkar et al., 2024). This strategy becomes even more important in regions like the Sundarbans, where large areas are inaccessible and inhabited by wildlife, making remote sensing-based desktop analyses the most practical and effective method for monitoring and assessment (Srivastava et al., 2025).

Fig. 9 The oldest topographic map of Sundarbans available at the scale of approximately 1:300,000. The map was published in 1761, and the area was surveyed in 1760. This map has geographic area and map scale, but key coordinate information is missing, making georeferencing a challenge (source: Brown, 2023)

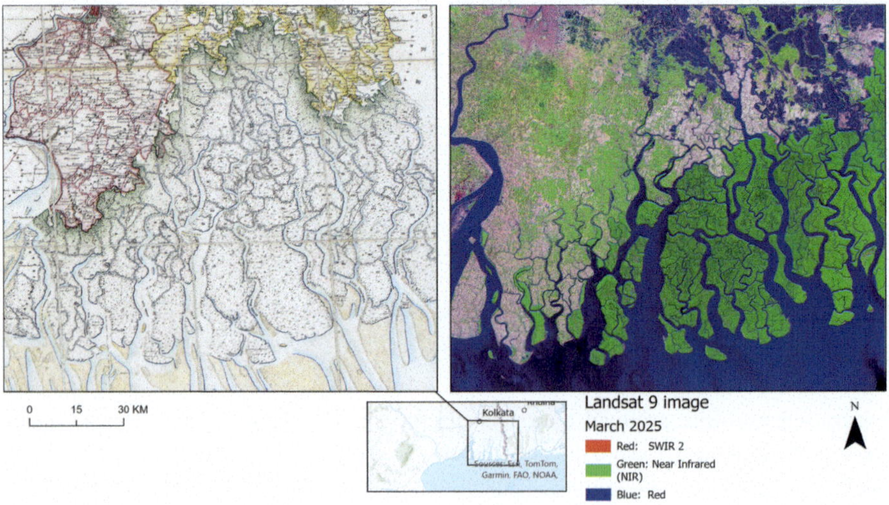

Fig. 10 The topographic map, prepared in 1853 from survey data collected between 1847 to 1851 clearly shows a much larger northern extent of the Sundarbans mangroves when compared with a recently acquired Landsat 9 image. The dark green colour in the Landsat image indicates mangrove forests. These mangrove forests are replaced by urban areas and agricultural fields on the western and northern sides

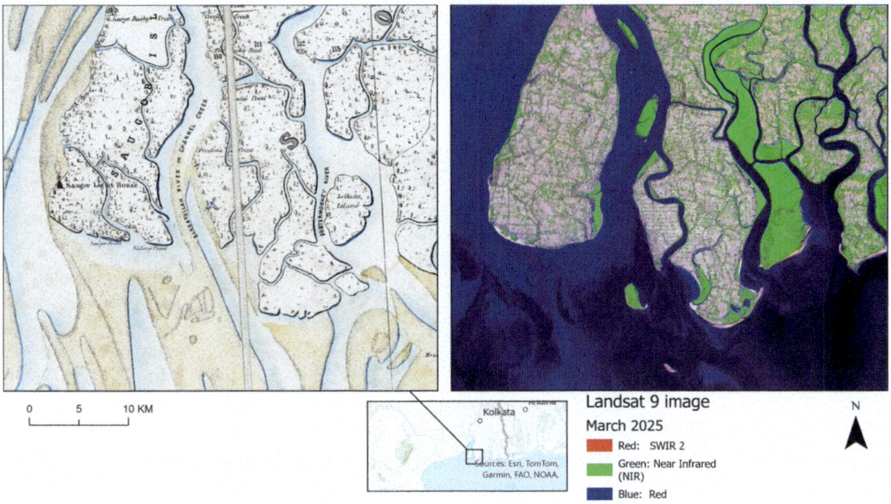

Fig. 11 The detailed version of Fig. 8 focuses on the western part of the Sundarbans, which is now fully converted to small towns, villages, and agricultural fields with road networks

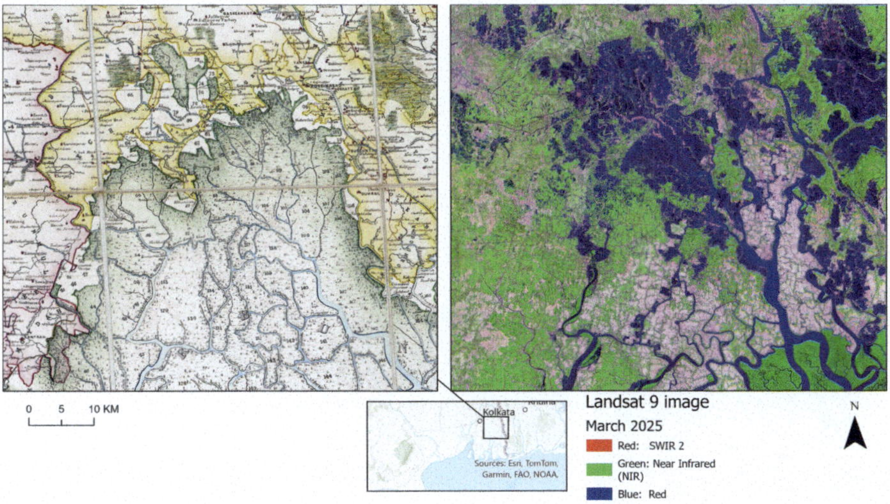

Fig. 12 The detailed version of Fig. 8 focuses on the northern part of the Sundarbans, which is now fully converted to small towns, villages and agricultural fields with road networks

The topographic maps published in 1922, based on survey data collected in 1905 primarily for revenue purposes, provide valuable insights into the ecological conditions of that period (Fig. 14). When compared with the 1942 topographic map, clear evidence of anthropogenic encroachment and land loss emerges. Further comparison with remote sensing imagery from the early 1970s reveals continued changes in both landmass and mangrove cover. A recent Landsat image highlights significant

Fig. 13 Photographs taken in the northern part of the regions mentioned in Fig. 10. This area was a mangrove forest in the past, but now it is an established human settlement with aquaculture ponds, houses, power lines, and roads (source: photographs taken by authors, February 2023)

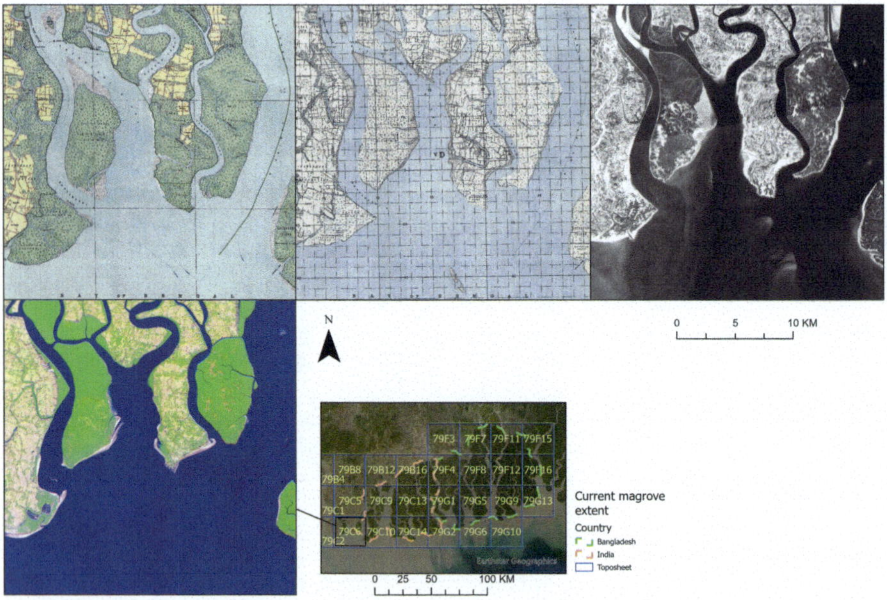

Fig. 14 A comparison of geospatial information collected over 120 years. The information is from early and mid-twentieth-century topographic maps, CORONA satellite data collected in the 1970s, and a recent Landsat 9 image. An anthropogenic conversion of a mangrove ecosystem into human settlement can be seen in one of the islands that were dominated by mangroves in the year 1905

transformations, particularly in the southern part of the region, underscoring the extent of ecological change over time (Fig. 15).

Recent Google Earth Street View imagery of the same area reveals that land once covered by mangrove forests has been transformed into a well-established human

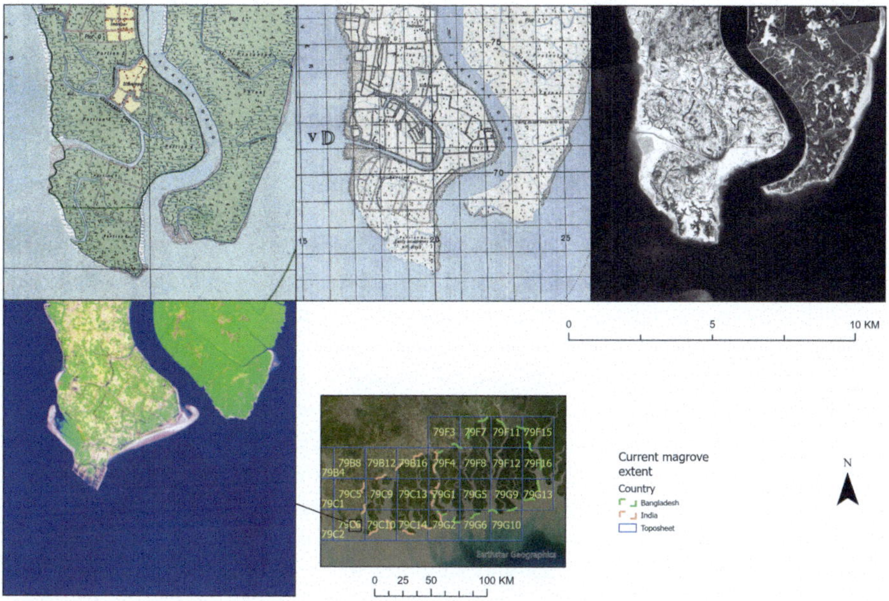

Fig. 15 A detailed view of the southern end of the island, where mangrove forests are totally lost. Land loss over time can also be noticed in the southern end, mostly in the last 50 years

Fig. 16 Examples of mangrove area converted into human settlements showing commuters, power lines, roads, and houses (source: photographs taken by authors, February 2023)

settlement. The area now features a connected network of roads, residential buildings, agricultural fields, communication towers, power lines, and people commuting for work (Fig. 16). The recent conversions of agricultural fields to aquaculture farms have been a concern for this ecosystem (Sumon et al., 2025).

2 Conclusions and Recommendations

Archived data sets play a vital role in long-term ecological monitoring, offering critical insights that inform future planning and conservation strategies for sensitive ecosystems. Many human settlements are increasingly vulnerable to extreme weather events, particularly in areas where natural ecosystems—once capable of buffering such impacts—have been degraded or lost. With advancements in data processing and image analysis technologies, it is now possible to extract meaningful information from historical landscape records. This enables the identification of high-risk regions and supports more targeted efforts to preserve and restore ecologically significant areas.

A variety of additional sources can be utilised to enrich long-term ecological assessments, including archived aerial photographs of the region. Beyond topographic maps and satellite imagery, nonspatial historical records containing geographic descriptions also offer valuable context for understanding ecosystem changes over time. Integrating these diverse data sources using modern geospatial technologies and recent high-resolution data sets enables a more comprehensive reconstruction of past landscapes and supports more informed conservation and management strategies.

Future research should increasingly integrate high-resolution LiDAR and UAV photogrammetry for canopy structure mapping, SAR time-series (Sentinel-1, ALOS PALSAR) for salinity and inundation monitoring, and machine learning–based change detection algorithms for automated analysis of multi-temporal data sets. Such innovations, when combined with archived historical baselines, can provide a holistic monitoring system that not only tracks ecological integrity but also anticipates areas most vulnerable to sea-level rise and human encroachment.

Acknowledgement The book chapter is written from the study under the Multinational Collaborative Project partially supported by Asia Pacific Network for Global Change Research (APN-GCR) and University of the Sunshine Coast (UniSC), Australia. The logistic support from Indian Council of Agricultural Research, Department of Agricultural Research and Education, Govt. of India for the study is duly acknowledged.

References

Baccini, A., Goetz, S., Walker, W., Laporte, N., Sun, M., Sulla-Menashe, D., Hackler, J., Beck, P., Dubayah, R., & Friedl, M. (2012). Estimated carbon dioxide emissions from tropical deforestation improved by carbon-density maps. *Nature Climate Change, 2*(3), 182–185.

Beyer, R. A., Alexandrov, O., & McMichael, S. (2018). The Ames Stereo Pipeline: NASA's open source software for deriving and processing terrain data. *Earth and Space Science, 5*(9), 537–548.

Brown, J. (2023). *79 Block—63k or 50k scale Maps of South Asia*. Zenodo 2023. https://doi.org/10.5281/zenodo.7869274

CIA. (2025). *CORONA: America's First Imaging Satellite Program* 2025 [cited August 2025]. https://www.cia.gov/legacy/museum/exhibit/corona-americas-first-imaging-satellite-program/
Coldstream, W. M. (1919). *Survey of India maps. Vol. 7, Records of the Survey of India*. Photo Litho Office.
DasGupta, R., & Shaw, R. (2017). Perceptive insight into incentive design and sustainability of participatory mangrove management: a case study from the Indian Sundarbans. *Journal of Forestry Research, 28*(4), 815–829.
Dasgupta, S., Wheeler, D., Sobhan, M. I., Bandyopadhyay, S., & Paul, T. (2020). Coping with climate change in the Sundarbans: Lessons from multidisciplinary studies. World Bank Publications.
Edney, M. H. (1991). The atlas of India 1823–1947/the natural history of A topographic map series. *Cartographica, 28*(4), 59–91.
Ghosh, A., Schmidt, S., Fickert, T., & Nüsser, M. (2015). The Indian Sundarban mangrove forests: History, utilization, conservation strategies and local perception. *Diversity, 7*(2), 149–169.
IUCN. (2014). *Bangladesh Sundarban Delta Vision 2050: A first step in its formulation*. International Union for Conservation of Nature.
Jandl, R., Lindner, M., Vesterdal, L., Bauwens, B., Baritz, R., Hagedorn, F., Johnson, D. W., Minkkinen, K., & Byrne, K. A. (2007). How strongly can forest management influence soil carbon sequestration? *Geoderma, 137*(3–4), 253–268.
Karsch, G., Mukul, S. A., & Srivastava, S. K. (2023). Annual mangrove vegetation cover changes (2014–2020) in Indian Sundarbans National Park using Landsat 8 and Google Earth Engine. *Sustainability, 15*(6), 5592.
Kibria, A., Costanza, R., Groves, C., & Behie, A. M. (2018). The interactions between livelihood capitals and access of local communities to the forest provisioning services of the Sundarbans Mangrove Forest, Bangladesh. *Ecosystem Services, 32*, 41–49.
Mahmood, H., Ahmed, M., Islam, T., Uddin, M. Z., Ahmed, Z. U., & Saha, C. (2021). Paradigm shift in the management of the Sundarbans mangrove forest of Bangladesh: Issues and challenges. *Trees, Forests and People, 5*, 100094.
Mohammed, Sultana, F., Khan, A., Ahammed, S., Saimun, M. S. R., Bhuiyan, M. S., Srivastava, S. K., Mukul, S. A., & Arfin-Khan, M. A. (2024). Assessing vulnerability to cyclone hazards in the world's largest mangrove forest, The Sundarbans: A geospatial analysis. *Forests, 15*(10), 1722.
Mukul, S. A., Alamgir, M., Sohel, M. S. I., Pert, P. L., Herbohn, J., Turton, S. M., Khan, M. S. I., Munim, S. A., Reza, A., & Laurance, W. F. (2019). Combined effects of climate change and sea-level rise project dramatic habitat loss of the globally endangered Bengal tiger in the Bangladesh Sundarbans. *Science of the Total Environment, 663*, 830–840.
Nishat, B., Rahman, A., & Mahmud, S. (2019). *Landscape narrative of the Sundarban: Towards collaborative management by Bangladesh and India*. World Bank Group.
Parvin, S., Sakib, M. H., Islam, M. L., Brown, C. L., Islam, M. S., & Mahmud, Y. (2023). Coastal aquaculture in Bangladesh: Sundarbans's role against climate change. *Marine Pollution Bulletin, 194*, 115431.
Planet. (2025). *Planet Explorer* 2025 [cited January 2025]. https://www.planet.com/explorer/
Planet Team. (2019). *Planet application program interface: In space for life on Earth* 2025 [cited December 2019]. https://api.planet.com/
Ravindranath, N. H., Joshi, N. V., Sukumar, R., & Saxena, A. (2006). Impact of climate change on forests in India. *Current Science, 90*(3), 354–361.
Sarkar, P., Banerjee, S., Biswas, S., Saha, S., Pal, D., Naskar, M. K., Srivastava, S. K., Barman, D., Kar, G., & Mukul, S. A. (2024). Contribution of mangrove ecosystem services to local livelihoods in the Indian Sundarbans. *Sustainability, 16*(16), 6804.
Sayer, J., Sunderland, T., Ghazoul, J., Pfund, J.-L., Sheil, D., Meijaard, E., Venter, M., Boedhihartono, A. K., Day, M., & Garcia, C. (2013). Ten principles for a landscape approach to reconciling agriculture, conservation, and other competing land uses. *Proceedings of the National Academy of Sciences, 110*(21), 8349–8356.

Siddiqi, N. (2001). Mangroves of Bangladesh Sundarbans and accretion areas. In *Mangrove ecosystems functions management*. Springer-Verlag.

Sievers, M., Chowdhury, M. R., Adame, M. F., Bhadury, P., Bhargava, R., Buelow, C., ... & Connolly, R. M. (2020). Indian Sundarbans mangrove forest considered endangered under Red List of Ecosystems, but there is cause for optimism. *Biological conservation, 251*, 108751.

Srivastava, S. K., Mukul, S., Banerjee, S., Barman, D., Kar, G., Hindersah, R., Arief, M. C. W., & Khan, M. A. S. A. (2025). A hierarchical and inclusive zonation of Sundarbans for monitoring and management of the world heritage area. *APN Science Bulletin, 15*(1), 1–18.

Sumon, K. A., Kanok, N. J. R., Sadat, M. A., Mainuddin, M., Wahid, S. M., & Karim, F. (2025). A comprehensive review on the negative impacts on Sundarbans fisheries: Insights from the hydrological changes modulated by climate change and anthropogenic activities. *Marine Pollution Bulletin, 220*, 118409.

UNESCO. (2025). *The Sundarbans*. UNESCO World Heritage Convention 2025 [cited August 2025]. https://whc.unesco.org/en/list/798/

USGS. (2025a). *USGS Earth Explorer* 2025 [cited January 2025]. https://earthexplorer.usgs.gov/

USGS. (2025b). *USGS EROS archive—Declassified data—Declassified satellite imagery—1* 2025 [cited August 2025]. https://www.usgs.gov/centers/eros/science/usgs-eros-archive-declassified-data-declassified-satellite-imagery-1

Cyclone Attacks and Their Impact in Eastern India with Special Reference to Sundarbans

Piyali Sarkar, Saon Banerjee, Sarathi Saha, Pintoo Bandopadhyay, Sanjeev Kumar Srivastava, Dhananjay Barman, and Gouranga Kar

Abstract The Bay of Bengal (BoB) region is one of the most cyclone-prone areas in the world, accounting for a large portion of tropical cyclone-related fatalities. The Indian Sundarbans and the adjoining deltaic coasts are mainly vulnerable due to its topography, fragile ecosystem, and an extremely high population density. This region is also a good habitat for a huge range of flora and fauna. This chapter examines the occurrence and intensity of tropical cyclones in eastern India. It also looks at their social and ecological impacts, with a focus on the Sundarbans. The analysis is based on historical records and some case studies of significant cyclones. These sources highlight both immediate damage and long-term ecological stress. To reduce vulnerability in areas hit by cyclones, it's crucial to combine ecological resilience with better forecasting and community preparedness. With the advancements in technology, the accuracy for cyclone forecasts has enhanced manifold. People of the coastal areas have become more aware of the risks and cyclone preparedness has also improved. At present special emphasis is given to disseminate the cyclone advisory to the coastal people around Indian Sundarbans.

Keywords Tropical cyclone · Sundarbans · Climate change · Coastal ecosystem

P. Sarkar · S. Banerjee (✉) · S. Saha
Department of Agricultural Meteorology and Physics, Bidhan Chandra Krishi Viswavidyalaya, Mohanpur, West Bengal, India

P. Bandopadhyay
Department of Agronomy, Bidhan Chandra Krishi Viswavidyalaya, Mohanpur, West Bengal, India

S. K. Srivastava
Geospatial Analytics for Conservation and Management, School of Science Technology and Engineering (SSTE), University of the Sunshine Coast, Sippy Downs, QLD, Australia

D. Barman · G. Kar
ICAR—Central Research Institute for Jute and Allied Fibers, Kolkata, West Bengal, India

S. K. Srivastava et al. (eds.), *Earth Observation for Monitoring Mangrove Ecosystems*, https://doi.org/10.1007/978-3-032-16668-5_7

1 Introduction

Eastern coastal region of India, particularly the Bay of Bengal (BoB), is recognized as one of the most hazardous areas in the world. It experiences several devastating tropical cyclones (TC) every year. The warm shallow water, high humidity, and funnel-shaped bay contribute to the rapid strengthening and landfall of cyclonic systems (Emanuel, 2005; Sahoo & Bhaskaran, 2016). From 1891 to 2020, over 200 cyclonic storms formed in the North Indian Ocean, with about 58% impacting the coasts of Odisha and West Bengal (Balaguru et al., 2014). World's largest contiguous mangrove forest, the Sundarbans extending across the southern section of West Bengal and adjoining Bangladesh lies directly in the cyclone trajectory. It consists of many low-lying, saline islands mixed with tidal channels and is home to wide range of flora and fauna (Zaman & Chowdhury, 2024). This fragile ecosystem protects coastal regions from storm surges and erosion. It also supports the wellbeing and livelihoods of local communities by providing different ecosystem services (Halder et al., 2021; Mohammed et al., 2024; Sarkar et al., 2024). However, ongoing embankment breaches, soil salinization, and tidal flooding have increased ecological weakness and put around five million residents at risk (Das, 2023).

Recent analyses show a continuous threat and a rise in the prevalence of intense tropical cyclones (Category 4 and above) in the BoB since the beginning of twenty-first century (Balaguru et al., 2014). Tropical cyclones in this region mainly form during two distinct periods in the year: the pre-monsoon season (April to June) and the post-monsoon season (October to December). However, post-monsoon cyclones tend to be more intense owing to more favourable environmental conditions (Girishkumar & Ravichandran, 2012). Recent decades have witnessed a marked rise in ferocity of post-monsoon cyclones in BoB, particularly during 1981–2010, with projections indicating a further escalation in the upcoming days (Balaguru et al., 2014). Long-term analyses also reveal a similar pattern. Based on 122 years of historical cyclone records from the India Meteorological Department (IMD), a study by Singh et al. in 2000 indicated a rise in the frequency of very intense cyclones. It also noted a twofold increase in less severe systems over the Bay of Bengal. Over the past century, nearly a 20% rise in the transformation of cyclonic disturbances into severe cyclones has been recorded. More recent work by Bhardwaj and Singh (2020), which assessed cyclone activity across 1972–2017, reported pronounced interannual variability in cyclone frequency, intensity, duration, and indices like accumulated cyclone energy (ACE) and power dissipation index (PDI).

Future projection studies suggest a potential increase of 2–11% in the globally averaged mean intensity of tropical cyclones, indicating a shift towards more intense storms. This raises concerns about heightened risks, including loss of life and property, due to flash floods and storm surges (Thomas et al., 2014). Given these projections of increasing cyclone intensity and associated risks, it becomes crucial to examine regional preparedness and adaptive capacities. In India, efforts to monitor cyclones and reduce risks started with the formation of the India Meteorological Department (IMD) in 1875. Today, significant investments in remote sensing technologies, better early warning systems, and community-based adaptations have greatly improved India's cyclone preparedness and lowered disaster risks (Sahana et al., 2023). The IMD has progressively expanded its observational infrastructure, modelling capacity, and forecasting systems, achieving near-zero fatalities during severe cyclones in recent years (Mohapatra et al., 2025). Along India's east coast, high-resolution inundation forecasting using RADARSAT-2 SAR data and ADCIRC modelling has improved flood prediction accuracy (Mandal et al., 2020). Nonetheless, household-level analyses of the Sundarbans reveal that insufficient early warning awareness and preparedness delays continue to undermine community resilience (Sahana et al., 2023).

However, the Sundarbans remain highly vulnerable due to their geographical exposure, weak socio-economic conditions, and ecological sensitivity, as highlighted by recent studies on cyclone hazards in the region (Mohammed et al., 2024). This indicates that, even with recent improvements in forecasting and preparedness, the eastern coastal zone continues to face significant risks from recurring cyclonic disasters. Tackling these challenges needs more focused research and stronger policy frameworks for lasting solutions (Ghosh & Mistri, 2023). This chapter combines historical and scientific views to look at cyclonic activity in the Bay of Bengal. It pays special attention to physical processes, trends in intensity and frequency, and the ecological and social impacts in the Sundarbans and the adjoining region.

2 Tropical Cyclone Formation and Prediction

2.1 Structure and Characteristics of Cyclone

A tropical cyclone (TC) is a large low-pressure system with a warm core. It forms over tropical or subtropical oceans and has organized thunderstorm activity along with a noticeable wind circulation at the surface. In the Northern Hemisphere, winds circulate counterclockwise, while in the Southern Hemisphere, they rotate clockwise due to Earth's rotation. A mature cyclonic system can reach up to 10 km in height and 500–700 km in width, moving at an average speed of 16–24 km/h but occasionally accelerating to more than 60 km/h. According to wind strength, weather systems develop from tropical depressions to tropical storms and finally to tropical cyclones or hurricanes or typhoons, depending on the region, once sustained winds exceed 120 km/h (World Meteorological Organization, 2023).

2.2 Precondition for Cyclone Formation

There are certain pre-requisites required to transform the mild low-pressure system with clouds into a tropical cyclone (Mohapatra & Sharma, 2019). The most important conditions are as follows:

1. Sea surface temperatures must exceed 27 °C to create latent heat through evaporation.
2. High humidity and an unstable atmosphere are essential for the development of cumulonimbus convection.
3. Low vertical wind shear is critical to keep storms aligned vertically.
4. A sufficient Coriolis force is needed for rotation, which is usually found at latitudes above 5°.
5. Surface winds must converge to push air upwards and organize it into a rotating system.

2.3 Cyclone Prediction

Predicting tropical cyclones (TCs) is important but difficult because of the many complex factors that affect storm development and movement. Although improvements in traditional numerical weather prediction (NWP) models have significantly helped with track forecasts, they still struggle with changes in intensity and rapid intensification. Nonlinear air-sea interactions and inner-core dynamics play major roles in these challenges (Leroux et al., 2018; Rogers, 2021).

Recent advancements in observation methods have boosted forecasting accuracy. Better monitoring of storm structures and landfall events has come from high-resolution geostationary satellites, low-earth orbiting sensors, and aircraft reconnaissance. Leroux et al. (2018) pointed out that these datasets have improved forecasts for track shifts near complex topography and have led to probabilistic forecasts for both track and intensity. Rogers (2021) highlighted the importance of boundary layer processes, interactions with vertical wind shear, and eyewall replacement cycles in affecting intensity changes and noted ongoing gaps in observations. At the same time, artificial intelligence (AI) and machine learning (ML) have become useful additions to NWP. Bi et al. (2023) showed that Pangu-Weather, a 3D deep learning model trained on 40 years of global data, runs faster than traditional systems. It provides more accurate medium-range cyclone tracking compared to ECMWF's high-resolution model. Reviews by Chen et al. (2020) and Wang et al. (2022) have shown that ML techniques are increasingly used for forecasting cyclone formation, tracking, intensity, rainfall, and storm surge. These methods are particularly good at using multisource datasets, such as satellite, radar, and surface observations, to identify patterns and extract features. However, challenges remain related to how well models can be understood, how stable they are, and their accuracy in capturing the complex nonlinear dynamics of TCs (Wang et al., 2022).

3 Climate Change, Tropical Cyclones, and Coastal Resilience

Climate change is expected to significantly affect tropical cyclones, with evidence showing an increase in the frequency and intensity of severe storms (Hochard et al., 2021). Global warming is likely to cause more frequent and severe tropical cyclones including a rise in Category 4 and 5 storms, particularly in the Western North Pacific region (Mohd Anuar et al., 2020). This warming also leads to more extreme rainfall and storm activity especially during monsoon periods in areas like Southeast Asia (Pérez-Alarcón et al., 2023; Knutson et al., 2020). A major result of climate change is rising global sea levels. This will likely worsen storm surges and make coastal communities and property more vulnerable to flooding (Mohammed et al., 2024; Roy et al., 2023; Chakrabortty et al., 2025). Mangrove ecosystems are known for their protective role by reducing the physical impacts of storms by decreasing water flow pressure, lowering storm surge height, dampening wind speeds, and even limiting saltwater intrusion (Sunkur et al., 2023; Das & Vincent, 2009; Hochard et al., 2019). Empirical studies from Odisha's Bhitarkanika region show that villages protected by mangroves experienced significantly less damage to houses, crops, and livestock than those without this protection or those that depended only on embankments. The damage costs were US $33.31 per household compared to US $153.74 (Badola & Hussain, 2005). Other analyses in hundreds of villages showed significantly fewer deaths where there was more mangrove cover, even when considering other factors (Das & Vincent, 2009). Modeling also indicates that a 10% increase in mangrove width could lower human casualties by about 12.5% and reduce structural collapse by around 2% (Das, 2007). Finally, the estimated storm protection value is US $4335 per hectare or US $68,586 per kilometre width. This amount far exceeds both land prices and returns from other land uses. It strongly supports mangrove conservation as a practical resilience strategy (Das, 2022).

While regional evidence highlights the important protective role of mangroves against cyclones in India, a series of recent disastrous events worldwide has sparked scientific debate on tropical cyclone trends related to climate change. Events like Typhoon Haiyan and Hurricanes Harvey, Irma, and Maria have increased discussions among major scientific organizations about the patterns and future direction of tropical cyclone activity (WMO, 2021). Some studies suggest that tropical cyclone activity has increased in certain regions, while others argue that natural variability on regional, annual, and inter-decadal scales makes it hard to see a clear signal driven by climate (Chand et al., 2022; Knutson et al., 2010; Shan et al., 2025). Although there is no clear consensus on whether global warming is currently having any measurable impact on tropical cyclones, climate models indicate that there may be an increase in tropical cyclone intensity in the future (Wang et al., 2017; Raavi & Walsh, 2021; Vecchi et al., 2019). At the same time, model projections suggest that the frequency of cyclones may either be stable or decline (Chand et al., 2022; Sobel et al., 2021).

In the Bay of Bengal and Sundarbans region, empirical evidence shows worsening effects from recent events, particularly the severe impact of Cyclone Amphan in

2020 on the Sundarbans. A remote sensing-based approach combining satellite data from optical (Sentinel-2), radar (Sentinel-1), and LiDAR as shown by Mondal et al. (2022) revealed that Amphan flooded at least 6821 km^2 of land in 39 districts of coastal India and Bangladesh for over 10 days after making landfall. Within the Sundarbans, mangrove health declined sharply. NDVI values dropped by more than 0.2 units over 3.45 km^2 of forest. This shows both immediate ecological stress and difficulties for long-term recovery.

4 Cyclonic Activities in Eastern India in Recent Past

The coastline of Eastern India has faced a series of powerful cyclones, each leaving unique hazard patterns. These include strong winds, storm surges, excessive rainfall, and additional effects such as landslides and saltwater intrusion. In the following section, some important insights about the major events that have occurred since 1999 have been discussed.

4.1 Odisha Super-Cyclone (1999)

The Odisha cyclone of 1999 was the strongest and most destructive tropical cyclone ever recorded in the northern Indian Ocean. It hit Odisha on October 29, 1999, causing extensive damage to lives and property in the coastal districts of Odisha, including Balasore, Bhadrak, Kendrapara, Jagatsinghpur, Puri, Khurda, and Ganjam (Figs. 1 and 2). A study by Das and Vincent (2009) demonstrated the key protective role of mangroves where villages with larger mangrove areas saw much lower mortality rates, with each additional 100 m of mangrove width decreasing the chance of death by nearly 30%. Fanchiotti et al. (2020) reported a loss of 18,420 km^2 of agricultural land, destruction of nearly 2 million tonnes of rice, and the deaths of 400,000 livestock, along with lasting socioeconomic challenges.

4.2 Cyclone Sidr (2007)

Cyclone Sidr hit Bangladesh in November 2007, being classified as a Category 4 storm, and resulted in extensive destruction. A field study conducted after the event indicated that at least 3447 people died, mainly due to the storm surge, and pointed out deficiencies in both cyclone shelters and the distribution of warnings as significant contributing factors to the high human toll (Islam et al., 2011). The ecological and economic impacts were also severe. A case study in Patuakhali area revealed that nearly 16.8 million homestead trees were destroyed, including various fruit and

Fig. 1 Damages caused by 1999 Odisha Super cyclone. (Source: Orissapost, 30 October 2021)

timber species, which severely affected the livelihoods and subsistence of rural families (Haq et al., 2012).

4.3 Cyclone Aila (2009)

In May 2009, Cyclone Aila made landfall, causing significant damage to coastal Bangladesh and India as well as the Sundarbans. In coastal Bangladesh, Moniruzzaman et al. (2014) investigated the effects of Cyclone Aila at the household level. The sectors most impacted were shelter (41.4%), homestead forest (19.3%), lighting (15.2%), and livestock (10.6%). Recovery was still not uniform 4 years later; while sanitation and drinking water had completely recovered, the homestead forest, food preparation, and livestock sectors had fallen far behind.

Rahman et al. (2015) evaluated the effects of Cyclone Aila on household-level disasters and recovery in Bangladesh using the Modified Domestic Assets Index (MDAI). They discovered that the MDAI accurately reflected differences in homestead assets, livestock, food preparation, shelter, water, sanitation, and other factors using data from 67 households. Even years after the cyclone, recovery was uneven across sectors, and the results showed significant losses in household assets and

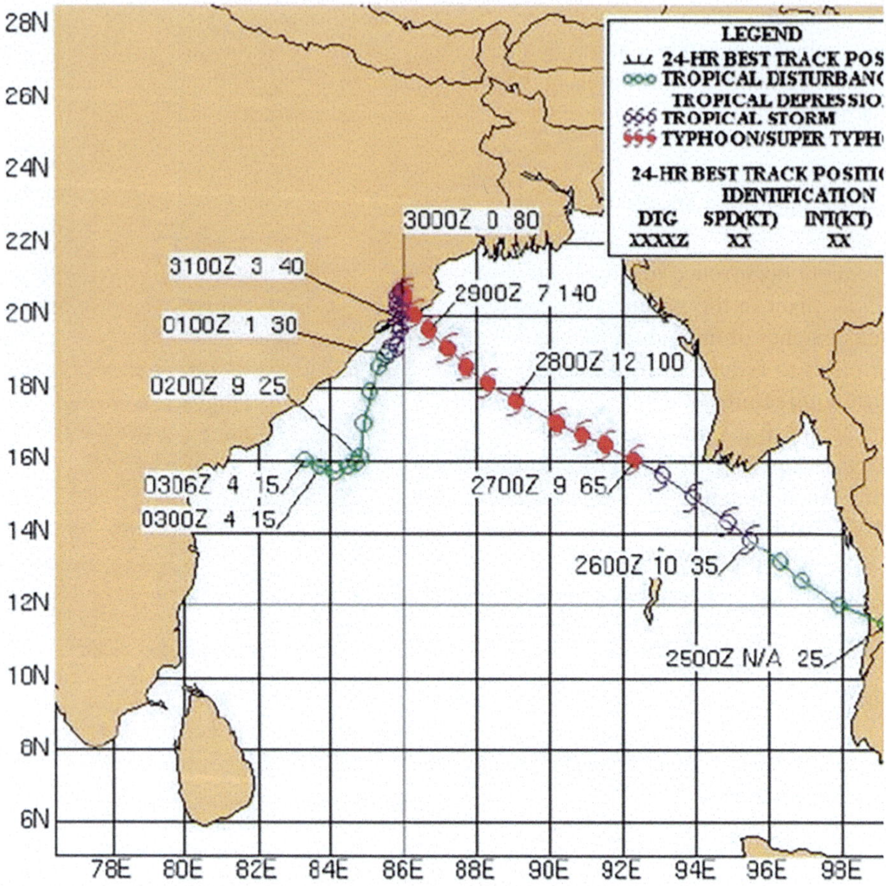

Fig. 2 Track of Orissa super cyclone 1999. (Source: Das et al., 2017)

shelter. The study proved that the MDAI is a useful methodological tool for assessing the effects of disasters and long-term recovery.

4.4 Cyclone Hudhud (2014)

Tropical Cyclone Hudhud made landfall on India's coast of Andhra Pradesh, near the city of Visakhapatnam, as a Category '3 storm' with a speed of 185 km per hour (115 mph) on 12 October 2014. The storm caused extensive damage to the city's infrastructure, including the airport, power grid, and residential areas (Figs. 3 and 4). Based on reports from the media and government sources, more than 248,000 people in 320 villages have been affected (Red Cross, 2014). A remote sensing analysis by Konda et al. (2018) showed loss of vegetation of 30.67 km^2 and

Fig. 3 Damages caused by cyclone Hudhud. (Source: India.com, 12 October 2014)

43.37 km^2 In Visakhapatnam and Bheemunipatnam, respectively. Consequently, the city's green spaces saw nearly 40% decline in canopy cover.

4.5 Cyclone Titli (2018)

On 10 October 2018, Titli made landfall on the coast of Gopalpur in Andhra Pradesh (Fig. 5). Titli being the most intense tropical cyclone in the Bay of Bengal of 2018 caused huge damages to live, property, and the standing crops. It affected almost 80.1 lakhs of population and disrupted road and telecommunications networks and power and water supply (Sphere India, 2019).

4.6 Cyclone Fani (2019)

Cyclone Fani, one of the most powerful pre-monsoon cyclones to hit Odisha in recent decades, made landfall in May 2019 with sustained wind gusts exceeding 200 km/h, which is equivalent to a Category 4 hurricane (Figs. 5 and 6). The Indian Red Cross Society conducted a damage assessment in Odisha state on May 8 and found that 18,168 villages in 14 districts, Angul, Balasore, Bhadrak, Cuttack, Dhenkanal, Ganjam, Jagatsinghpur, Jajpur, Kendrapara, Keonjhar, Khordha, Mayurbhanj, Nayagarh, and Puri, saw significant damage. More than 16 million

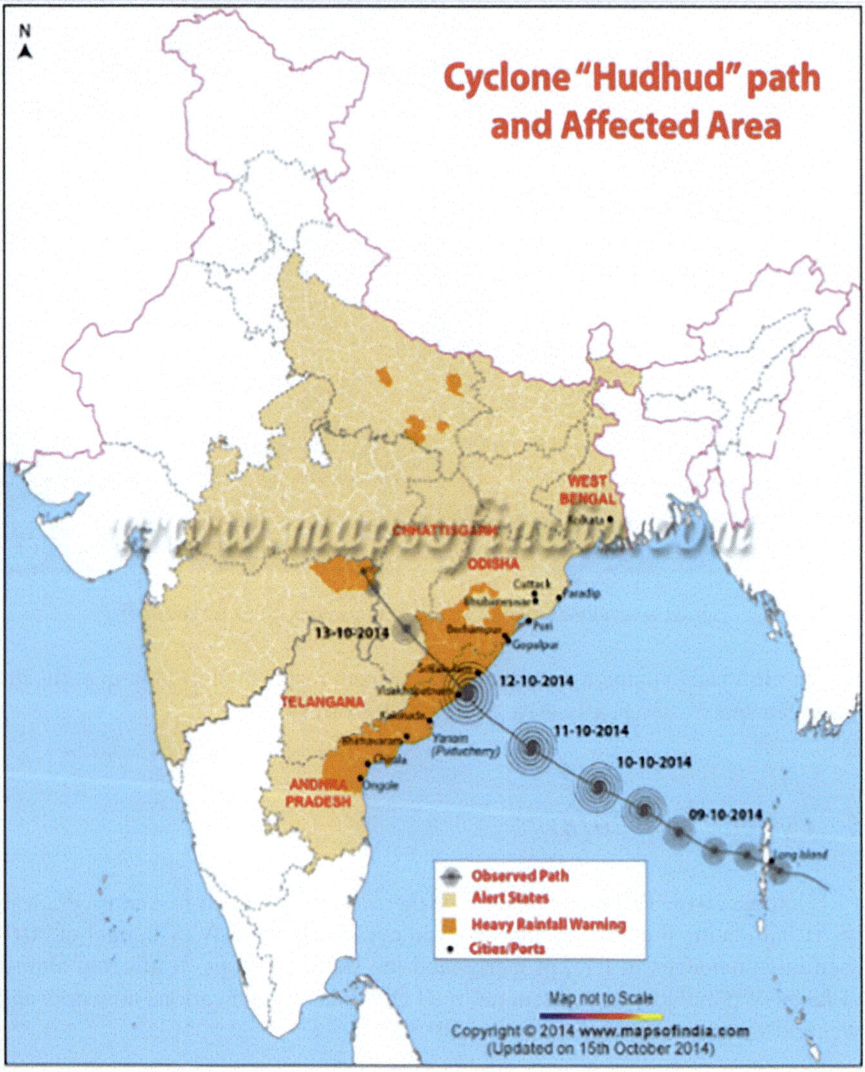

Fig. 4 Map showing Cyclone Hudhud path. (Source: Maps of India.com)

people were impacted, and over 189,000 homes and about 153,000 hectares of crops were damaged in these 14 districts. A geospatial study of Balukhand Wildlife Sanctuary using Sentinel-2A data revealed canopy loss and vegetation stress, with species like *Eucalyptus globulus* and *Casuarina equisetifolia* being particularly impacted (Nandi et al., 2020). Apart from its environmental consequences, Cyclone Fani significantly altered the hydrodynamics of the Bay of Bengal. Wahiduzzaman and Paul (2023) used a semi-implicit finite difference model to simulate storm surge dynamics along the Bangladesh coast which showed a strong correlation with actual

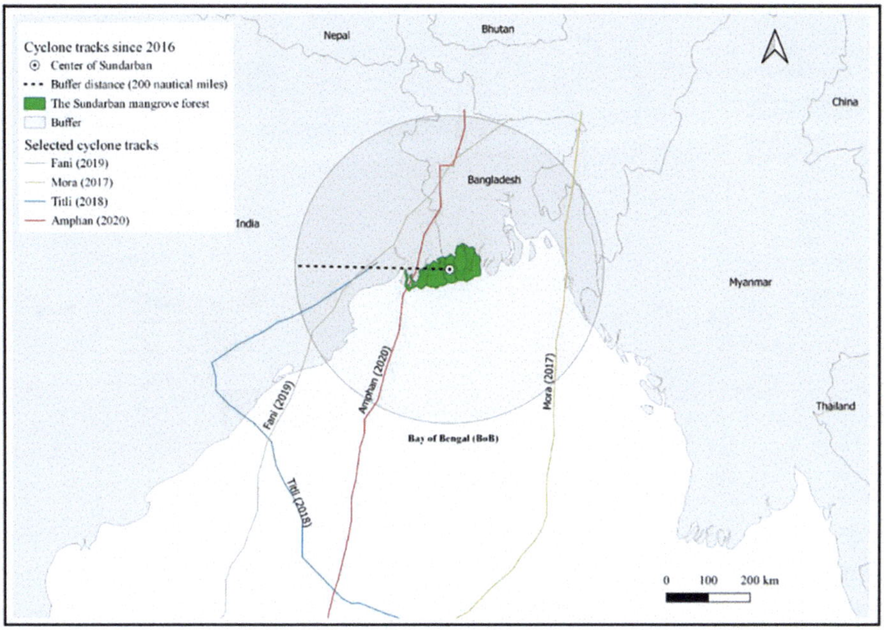

Fig. 5 Tracks of selected cyclones since 2016 over the BoB region. (Source: Sharma et al., 2022)

water levels and the expected maximum surge heights in many places exceeded 5 m. These surges led to extensive coastal flooding and showed how vulnerable offshore islands are.

4.7 *Cyclone Bulbul (2019)*

The incredibly strong Bulbul, the second cyclone of 2019, hit coastal west Bengal on November 9 particularly affecting Sundarban delta (IMD, 2021). According to a Sentinel-1 SAR data assessment, South 24 Parganas saw the biggest rise in flooding, going from 99.92 km^2 before the storm to 149.22 km^2 after it, whereas East Medinipur and North 24 Parganas saw less change (Biswakarma et al., 2021). In coastal Bangladesh, complementary geospatial research by Sonet et al. (2024) revealed that Bulbul changed 45.08 km^2 of land cover, with significant decreases in thick vegetation (4.56 km^2), agriculture (4.75 km^2), and settlements (7.72 km^2), while water bodies grew by 22.49 km^2 as a result of flooding.

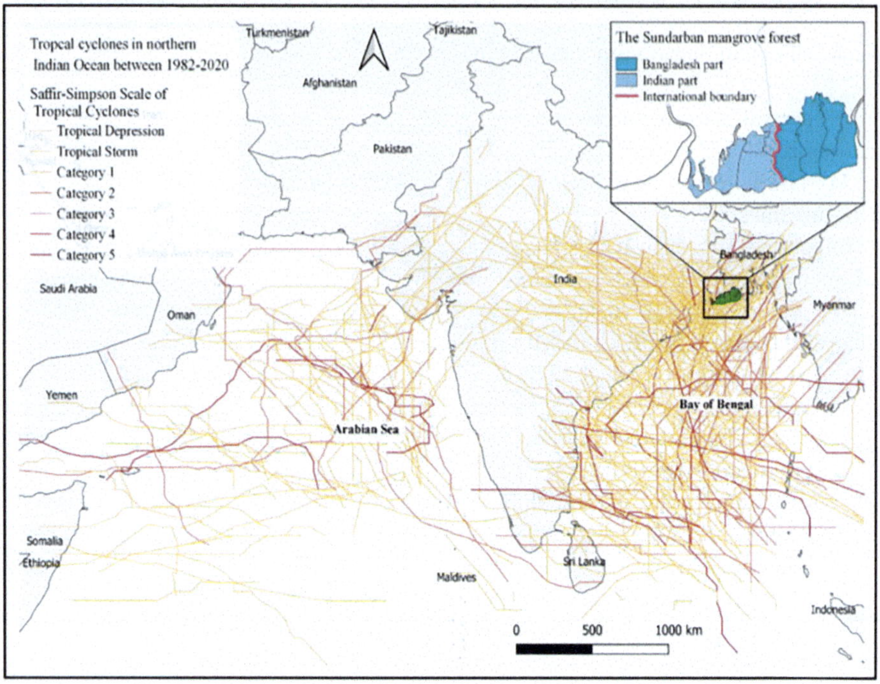

Fig. 6 Number of TC observed since 1982 over the northern Indian Ocean. (Source: Sharma et al., 2022)

4.8 *Cyclone Amphan (2020)*

Cyclone Amphan was the largest storm to hit the Ganges Delta since the 1999 Odisha Super Cyclone, with winds of 185 km/h when it hit the Sundarbans and coastal West Bengal on 20 May 2020 (Fig. 7). 70% of the population of West Bengal, the epicentre of the landfall, was immediately impacted, with at least 86 fatalities and ₹1.02 trillion in damages reported (Red Cross, 2021). Odisha additionally reported 4.4 million affected individuals and ₹3.2 billion in losses. Amphan has been extensively researched for its effects and is known to be the most expensive cyclone in the history of the North Indian Ocean (US $14 billion; WMO, 2021). Using Sentinel-1 SAR data along with a Random Forest classifier, Behera et al. (2022) achieved an inundation mapping accuracy of 82.22%, identifying more than 2198 km^2 of flooded regions across four districts in Odisha and 17 in West Bengal. The districts most severely impacted were Tamluk (302.96 km^2), Basirhat (277.86 km^2), Kanthi (254.26 km^2), and Kendrapara (168.94 km^2). The cropland was the hardest hit, resulting in substantial losses to livelihoods. Additionally, Mishra et al. (2021) recorded Cyclone Amphan's effects on Sundarban mangroves, noting that dense cover decreased from 77 to 34%, sparse cover increased from 20 to 63%, and more

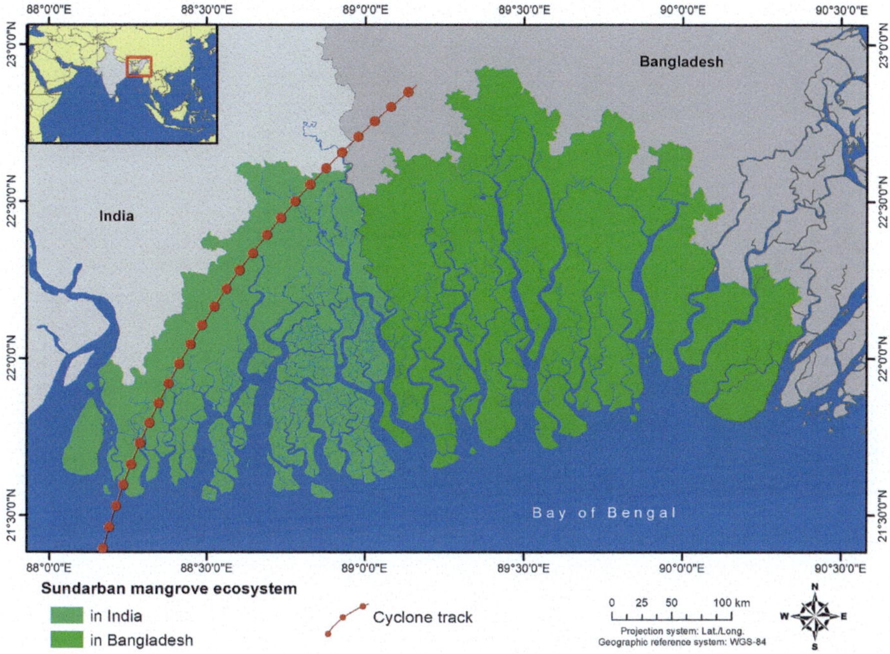

Fig. 7 Map of Sundarbans region showing the route of the Cyclone Amphan, 2020. (Source: Mishra et al., 2021)

than 68% of shoreline transects experienced erosion, with an average retreat of 31 m. The cyclone caused the uprooting of salt-sensitive species like Sundari while inundating both agricultural fields and aquaculture ponds.

4.9 Cyclone Yaas (2021)

On 26 May 2021, very severe cyclonic storm Yaas hit the Odisha coastline, bringing wind speeds of 130–145 km/h. It caused heavy rainfall and storm surges that affected Odisha, West Bengal, and Jharkhand (Figs. 8 and 9). Just before the cyclone, a rare outbreak of tornadoes occurred in the North 24 Parganas and Hooghly districts, increasing the level of destruction.

Several geospatial studies assessed the flooding and damage caused by Yaas. A study by Khatun et al. (2022) mapped inundation in Purba Medinipur, West Bengal, using Sentinel-1 SAR data on Google Earth Engine (GEE). They reported that about 239.69 km^2, or 5% of the district, was flooded, with cropland making up 95% of the affected area. A broader Sentinel-based assessment estimated that 2528.7 km^2, or 6.8% of the coastal land, across Odisha and West Bengal was impacted. This included over 24% of croplands in West Bengal (Geological Journal, 2025).

Fig. 8 Damages caused by Cyclone Yaas. (Source: BBC, 26 May 2021)

Complementing this, Paul and Chowdhury (2021) used several space-borne data sets to show that Cyclone Yaas flooded several coastal blocks of West Bengal. These include Sagar (37.10 km^2), Namkhana (78.12 km^2), Pathar Pratima (58.74 km^2), Ramnagar I and II (34.86 km^2), and Khejuri (22.27 km^2). During the event, the area received nearly 1195 mm of rainfall. Beyond agriculture, Yaas also devastated the aquaculture sector. A study on shrimp farming (Bhanja et al., 2021) reported that nearly 3000 hectares of shrimp farms in coastal districts of West Bengal and Odisha were inundated. The overall damage was estimated between 2000 and 5000 hectares. The cyclone resulted in the loss of approximately 12,000 tons of shrimp, which stalled shrimp production in the region for the rest of the year. This led to significant financial losses for farmers. These damages not only disrupted local livelihoods but also hindered the economic and social development of shrimp-farming communities.

4.10 Cyclone Gulab (2021)

Cyclone Gulab hit coastal Andhra Pradesh and Odisha in September 2021. RMSI (2021) estimated the total losses at INR 2000 crore. Seventy percent of the damages were in agriculture, especially in Srikakulam, Vizianagaram, Ganjam, and Gajapati districts. The storm revealed the uneven vulnerability of farming communities, who experience repeated crop and income losses.

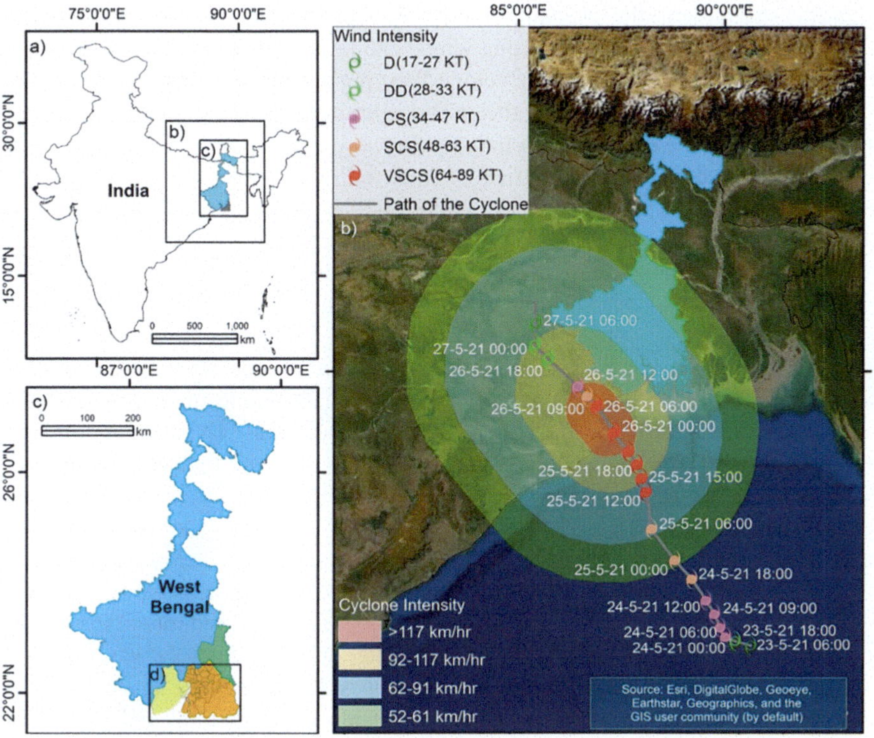

Fig. 9 Map showing cyclone trajectory during the period 23 May–27 May 2021, variability of wind intensity and its influence zone. (Source: Paul & Chowdhury, 2021)

4.11 Cyclone Jawad (2021)

Though relatively weak, Cyclone Jawad caused significant water-related issues in December 2021. Paradeep received 201 mm of rain in 1 day. This heavy rainfall disrupted port operations and flooded farmland. Paddy and potato crops suffered greatly across Odisha and West Bengal. This shows that even low-intensity cyclones can lead to serious agricultural problems when paired with heavy rainfall (IMD, 2021).

5 Advisories and Preparedness Measures for Cyclones

During the 1999 Odisha Super Cyclone, the India Meteorological Department (IMD) played an important role in issuing timely warnings. Six bulletins were released between 27 and 31 October, starting with advice for fishermen to stay out of the sea. When a depression formed over the Andaman Sea on 25 October and began to strengthen, the Andaman and Nicobar Administration was quickly informed about the potential for strong winds and heavy rainfall. After it developed into a cyclonic storm, IMD's Area Cyclone Warning Centre in Kolkata started sending bulletins from 26 October to the island administration and state authorities. Similar preparation was evident in later cyclones. For example, before Hudhud in 2014, alerts spread across nine districts of Andhra Pradesh. This allowed the evacuation of over 700,000 people, including about half a million from within the state, before landfall caused extensive damage to crops like paddy, groundnut, sugarcane, and pulses. During Amphan in 2020, advisories led to nearly 300,000 people relocating in West Bengal, backed by disaster relief units, Coast Guard search-and-rescue teams, and pre-positioned National Disaster Response Force (NDRF) contingents. For Yaas in 2021, IMD issued an orange alert on 25 May for districts such as Mayurbhanj, Cuttack, and Puri. At the same time, Kolkata Port Trust raised a red alert and halted all shipping operations. Evacuations already started on 24 May in East Midnapore and Jhargram, moving thousands of at-risk residents. Even for the weaker Jawad in 2021, proactive alerts led to the evacuation of more than 11,000 people in West Bengal, along with steps like trimming tree branches, checking power lines, and setting up 188 evacuation shelters. In Kolkata, precautionary measures were taken in urban areas, while in the coastal subdivision of Kakdwip in South 24 Parganas, over 20,000 people were moved inland to prepare for potential storm surge flooding. These examples show how timely warnings and coordinated evacuation plans have saved many lives. However, they also emphasize the ongoing need for strong embankments, durable infrastructure, and long-term policies to boost disaster risk reduction and improve community recovery efforts.

6 Conclusions

From the brief history of recent cyclonic activities, it can be concluded that the numbers of cyclonic events are increasing day by day. After the landfall of Hudhud in 2014, the next devastating cyclone was seen in 2018. But after that, almost each year cyclonic events devastated the eastern coast of India. In 2019, two cyclonic events (Bulbul and Fani) were observed; while in 2021, people of eastern coast in India encountered three cyclonic events (Yaas, Gulab, and Jawad). With the advancement of satellite technology, the cyclone warning system is strong enough in recent years. Hence, proper advisory can be issued at least 4 days before the landfall of a cyclone. Consequently, the death tolls can be minimized. However, the loss of crops, livestock, and biodiversity cannot be avoided many times. Moreover, the tropical cyclone has a huge impact on Sundarbans and other coastal ecosystems to a great extent.

Acknowledgement The help received from APN funded Project on 'Protecting ecosystems and livelihoods of the Sundarbans, a World Heritage site: Assessing the impact of natural hazards on forest-based ecosystem services' (Project Code: CRRP2020-08MY-Srivastava) is duly acknowledged.

References

Badola, R., & Hussain, S. A. (2005). Valuing ecosystem functions: An empirical study on the storm protection function of Bhitarkanika mangrove ecosystem, India. *Environmental Conservation, 32*(1), 85–92.

Balaguru, K., Taraphdar, S., Leung, L. R., & Foltz, G. R. (2014). Increase in the intensity of post-monsoon Bay of Bengal tropical cyclones. *Geophysical Research Letters, 41*(10), 3594–3601.

BBC. (2021). *Cyclone Yaas: Severe storm lashes India and Bangladesh.* https://www.bbc.com/news/world-asia-india-57237953

Behera, M. D., Prakash, J., Paramanik, S., Mudi, S., Dash, J., Varghese, R., Roy, P. S., Abhilash, P. C., Gupta, A. K., & Srivastava, P. K. (2022). Assessment of tropical cyclone amphan affected inundation areas using sentinel-1 satellite data. *Tropical Ecology, 63*(1), 9–19.

Bhanja, A., Payra, P., Das, M., Mandal, S., & Mandal, B. (2021). Impact of Yaas cyclone on shrimp farming along the coastal belt of West Bengal and Odisha. *Bulletin No., 4*, p. 9.

Bhardwaj, P., & Singh, O. (2020). Climatological characteristics of Bay of Bengal tropical cyclones: 1972–2017: P. Bhardwaj, O. Singh. *Theoretical and Applied Climatology, 139*(1), 615–629.

Bi, K., Xie, L., Zhang, H., Chen, X., Gu, X., & Tian, Q. (2023). Accurate medium-range global weather forecasting with 3D neural networks. *Nature, 619*(7970), 533–538.

Biswakarma, P., Singh, M., Sarma, A. K., & Joshi, V. (2021). Assessment of the water inundation area due to the tropical cyclone Bulbul (2019) in the selected districts of West Bengal, India with the application of RS and GIS tools. *Proceedings of the Indian National Science Academy, 87*(4), 628–639.

Chakrabortty, R., Ali, T., Atabay, S., Roy, P., & Pande, C. B. (2025). Impact of climate change scenario on sea level rise and future coastal flooding in major coastal cities of India. *Scientific Reports, 15*(1), 28689.

Chand, S. S., Walsh, K. J., Camargo, S. J., Kossin, J. P., Tory, K. J., Wehner, M. F., Chan, J. C., Klotzbach, P. J., Dowdy, A. J., Bell, S. S., & Ramsay, H. A. (2022). Declining tropical cyclone frequency under global warming. *Nature Climate Change, 12*(7), 655–661.

Chen, R., Zhang, W., & Wang, X. (2020). Machine learning in tropical cyclone forecast modeling: A review. *Atmosphere, 11*(7), 676.

Das, S. (2007). *Storm protection by mangroves in Orissa: An analysis of the 1999 super cyclone*. SANDEE.

Das, S. (2022). Valuing the role of mangroves in storm damage reduction in coastal areas of Odisha. In *Climate change and community resilience* (p. 257).

Das, G. K. (2023). Climate change and coastal hazards in Sunderbans. In *Coastal environments of India: A coastal West Bengal perspective* (pp. 175–198). Springer International Publishing.

Das, S., & Vincent, J. R. (2009). Mangroves protected villages and reduced death toll during Indian super cyclone. *Proceedings of the National Academy of Sciences, 106*(18), 7357–7360.

Das, Y., Mohanty, U. C., & Jain, I. (2017). Numerical simulation on Bay of Bengal's response to cyclones using the Princeton Ocean model. *Brazilian Journal of Oceanography, 65*(2), 128–145.

Emanuel, K. (2005). Increasing destructiveness of tropical cyclones over the past 30 years. *Nature, 436*(7051), 686–688.

Fanchiotti, M., Dash, J., Tompkins, E. L., & Hutton, C. W. (2020). The 1999 super cyclone in Odisha, India: A systematic review of documented losses. *International Journal of Disaster Risk Reduction, 51*, 101790.

Ghosh, S., & Mistri, B. (2023). Cyclone-induced coastal vulnerability, livelihood challenges and mitigation measures of Matla–Bidya inter-estuarine area, Indian Sundarban. *Natural Hazards, 116*(3), 3857–3878.

Girishkumar, M. S., & Ravichandran, M. (2012). The influences of ENSO on tropical cyclone activity in the Bay of Bengal during October–December. *Journal of Geophysical Research: Oceans, 117*(C2).

Halder, N. K., Merchant, A., Misbahuzzaman, K., Wagner, S., & Mukul, S. A. (2021). Why some trees are more vulnerable during catastrophic cyclone events in the Sundarbans mangrove forest of Bangladesh? *Forest Ecology and Management, 490*, 119117.

Haq, M. Z., Robbani, M., Ali, M., Hasan, M. M., Hasan, M. M., Uddin, M. J., Begum, M., Teixeira da Silva, J. A., Pan, X. Y., & Karim, M. R. (2012). Damage and management of cyclone Sidr-affected homestead tree plantations: A case study from Patuakhali, Bangladesh. *Natural Hazards, 64*(2), 1305–1322.

Hochard, J. P., Hamilton, S., & Barbier, E. B. (2019). Mangroves shelter coastal economic activity from cyclones. *Proceedings of the National Academy of Sciences, 116*(25), 12232–12237.

Hochard, J. P., Barbier, E. B., & Hamilton, S. E. (2021). Mangroves and coastal topography create economic "safe havens" from tropical storms. *Scientific Reports, 11*(1), 15359.

IMD (2021). Cyclonic Storm Jawad over southeast Bay of Bengal (02–05 December 2021). India Meteorological Department Ministry of Earth Sciences, Government of India. https://rsmcnewdelhi.imd.gov.in/uploads/report/26/26_f08cd4_detailed%20report%20JAWAD.pdf

India.com. (2014). *Cyclone Hudhud: Three killed as Cyclone hits Andhra Pradesh coast.* https://www.india.com/news/india/cyclone-hudhud-three-killed-as-cyclone-hits-andhra-pradesh-coast-170636/

Islam, A. S., Bala, S. K., Hussain, M. A., Hossain, M. A., & Rahman, M. M. (2011). Performance of coastal structures during Cyclone Sidr. *Natural Hazards Review, 12*(3), 111–116.

Khatun, M., Garai, S., Sharma, J., Singh, R., Tiwari, S., & Rahaman, S. M. (2022). Flood mapping and damage assessment due to the super cyclone Yaas using Google Earth Engine in Purba Medinipur, West Bengal, India. *Environmental Monitoring and Assessment, 194*(12), 869.

Knutson, T. R., McBride, J. L., Chan, J., Emanuel, K., Holland, G., Landsea, C., Held, I., Kossin, J. P., Srivastava, A. K., & Sugi, M. (2010). Tropical cyclones and climate change. *Nature Geoscience, 3*(3), 157–163.

Knutson, T., Camargo, S. J., Chan, J. C., Emanuel, K., Ho, C. H., Kossin, J., Mohapatra, M., Satoh, M., Sugi, M., Walsh, K., & Wu, L. (2020). Tropical cyclones and climate change assessment. *Bulletin of the American Meteorological Society, 101*(3), E303–E322.

Konda, V. G. R. K., Chejarla, V. R., Mandla, V. R., Voleti, V., & Chokkavarapu, N. (2018). Vegetation damage assessment due to Hudhud cyclone based on NDVI using Landsat-8 satellite imagery. *Arabian Journal of Geosciences, 11*(2), 35.

Leroux, M. D., Wood, K., Elsberry, R. L., Cayanan, E. O., Hendricks, E., Kucas, M., Otto, P., Rogers, R., Sampson, B., & Yu, Z. (2018). Recent advances in research and forecasting of tropical cyclone track, intensity, and structure at landfall. *Tropical Cyclone Research and Review, 7*(2), 85–105.

Mandal, A. K., Ramakrishnan, R., Pandey, S., Rao, A. D., & Kumar, P. (2020). An early warning system for inundation forecast due to a tropical cyclone along the east coast of India. *Natural Hazards, 103*(2), 2277–2293.

Mishra, M., Acharyya, T., Santos, C. A. G., da Silva, R. M., Kar, D., Kamal, A. H. M., & Raulo, S. (2021). Geo-ecological impact assessment of severe cyclonic storm Amphan on Sundarban mangrove forest using geospatial technology. *Estuarine, Coastal and Shelf Science, 260*, 107486.

Mohammed, Sultana, F., Khan, A., Ahammed, S., Saimun, M. S. R., Bhuiyan, M. S., Srivastava, S. K., Mukul, S. A., & Arfin-Khan, M. A. (2024). Assessing vulnerability to cyclone hazards in the world's largest mangrove forest, The Sundarbans: A geospatial analysis. *Forests, 15*(10), 1722.

Mohapatra, M., & Sharma, M. (2019). Cyclone warning services in India during recent years: A review. *Mausam, 70*(4), 635–666.

Mohapatra, M., Sharma, M., & Pattanaik, D. R. (2025). Evolution of cyclone monitoring and forecasting system in India: A review. *Mausam, 76*(1), 103–124.

Mohd Anuar, N., Shafiai, S. H., & Hashim, A. M. (2020). Climate change impact on tropical cyclone evolution and storm surge severity in the east coast of Peninsular Malaysia. In *IOP Conference Series: Materials Science and Engineering* (Vol. 736, No. 7). Institute of Physics Publishing.

Mondal, P., Dutta, T., Qadir, A., & Sharma, S. (2022). Radar and optical remote sensing for near real-time assessments of cyclone impacts on coastal ecosystems. *Remote Sensing in Ecology and Conservation, 8*(4), 506–520.

Moniruzzaman, M., Rahman, M. A., & Siddik, M. A. (2014). Assessing the damage and recovery at household level: A case study of cyclone Aila affected area. *ASA University Review, 8*(1), 263–274.

Nandi, G., Neogy, S., Roy, A. K., & Datta, D. (2020). Immediate disturbances induced by tropical cyclone Fani on the coastal forest landscape of eastern India: A geospatial analysis. *Remote Sensing Applications: Society and Environment, 20*, 100407.

Paul, S., & Chowdhury, S. (2021). Investigation of the character and impact of tropical cyclone Yaas: A study over coastal districts of West Bengal, India. *Safety in Extreme Environments, 3*, 219–235.

Pérez-Alarcón, A., Fernández-Alvarez, J. C., & Coll-Hidalgo, P. (2023). Global increase of the intensity of tropical cyclones under global warming based on their maximum potential intensity and CMIP6 models. *Environmental Processes, 10*(2), 36.

Raavi, P. H., & Walsh, K. J. (2021, December). Response of tropical cyclone frequency to sea surface temperatures using aqua-planet simulations. In *Oceans* (Vol. 2, No. 4, pp. 785–810). MDPI.

Rahman, M. A., Moniruzzaman, M., Siddik, M. A., New, D. O. H. S., & Mohakhali, D. (2015). Assessing the impact and recovery of cyclone AILA on households: The modified domestic assets index (MDAI) approach. *Jagannath University Journal of Life and Earth Sciences, 1*, 17–25.

Red Cross. (2014, October 14). *India: Cyclone HudHud—Information bulletin no. 2*. International Federation of Red Cross and Red Crescent Societies, TC-2014-000141-IND. https://reliefweb.int/report/india/india-cyclone-hudhud-information-bulletin-n-2

Red Cross. (2021, March 1). *India: Cyclone Amphan final report*. International Federation of Red cross and Red Crescent Societies, TC-2020-000135-IND.

RMSI. (2021, September 29). *Cyclone Gulab*. Cyclone Advisory, RMSI. https://www.rmsi.com/ri/advisory/PIER_Cyclone_Advisory_Gulab_Sept_29_2021.pdf

Rogers, R. F. (2021). Recent advances in our understanding of tropical cyclone intensity change processes from airborne observations. *Atmosphere, 12*(5), 650.

Roy, P., Pal, S. C., Chakrabortty, R., Chowdhuri, I., Saha, A., & Shit, M. (2023). Effects of climate change and sea-level rise on coastal habitat: Vulnerability assessment, adaptation strategies and policy recommendations. *Journal of Environmental Management, 330*, 117187.

Sahana, M., Patel, P. P., Rehman, S., Rahaman, M. H., Masroor, M., Imdad, K., & Sajjad, H. (2023). Assessing the effectiveness of existing early warning systems and emergency preparedness towards reducing cyclone-induced losses in the Sundarban Biosphere Region, India. *International Journal of Disaster Risk Reduction, 90*, 103645.

Sahoo, B., & Bhaskaran, P. K. (2016). Assessment on historical cyclone tracks in the Bay of Bengal, east coast of India. *International Journal of Climatology, 36*(1), 95.

Sarkar, P., Banerjee, S., Biswas, S., Saha, S., Pal, D., Naskar, M. K., Srivastava, S. K., Barman, D., Kar, G., & Mukul, S. A. (2024). Contribution of mangrove ecosystem services to local livelihoods in the Indian Sundarbans. *Sustainability, 16*(16), 6804.

Shan, K., Song, F., Lin, Y., Chu, W., He, C., Chu, P. S., Wu, L., & Yu, X. (2025). How does globally accumulated tropical cyclone energy vary in response to a changing climate? *Science Bulletin, 70*(6), 943–950.

Sharma, S., Suwa, R., Ray, R., & Mandal, M. S. H. (2022). Successive cyclones attacked the world's largest mangrove forest located in the Bay of Bengal under pandemic. *Sustainability, 14*(9), 5130.

Singh, O. P., Ali Khan, T. M., & Rahman, M. S. (2000). Changes in the frequency of tropical cyclones over the North Indian Ocean. *Meteorology and Atmospheric Physics, 75*(1), 11–20.

Sobel, A. H., Wing, A. A., Camargo, S. J., Patricola, C. M., Vecchi, G. A., Lee, C. Y., & Tippett, M. K. (2021). Tropical cyclone frequency. *Earth's Future, 9*(12), e2021EF002275.

Sonet, M. S., Hasan, M. Y., Chakma, S., & Al Kafy, A. (2024). Assessing tropical cyclone impacts in coastal Bangladesh: A change detection analysis on cyclone Bulbul using geospatial analysis and remote sensing techniques. *International Journal of Disaster Risk Reduction, 112*, 104726.

Sphere India. (2019). *Humanitarian Snapshot Report 12: Aftermath of Cyclone Titli in AP and Odisha*. October–8 November 2018, Sphere India, 9th November 2018.

Sunkur, R., Kantamaneni, K., Bokhoree, C., & Ravan, S. (2023). Mangroves' role in supporting ecosystem-based techniques to reduce disaster risk and adapt to climate change: A review. *Journal of Sea Research, 196*, 102449.

Thomas, A., Samala, B. K., & Kaginalkar, A. (2014). Simulation of north Indian Ocean tropical cyclones using RAMS numerical weather prediction model. *Tropical Cyclone Research and Review, 3*(1), 44–52.

Vecchi, G. A., Delworth, T. L., Murakami, H., Underwood, S. D., Wittenberg, A. T., Zeng, F., Zhang, W., Baldwin, J. W., Bhatia, K. T., Cooke, W., & He, J. (2019). Tropical cyclone sensitivities to CO2 doubling: Roles of atmospheric resolution, synoptic variability and background climate changes. *Climate Dynamics, 53*(9), 5999–6033.

Wahiduzzaman, M., & Paul, G. C. (2023). Assessing the impact of extensive severe cyclonic storm Fani on the coastal communities of Bangladesh: A case study. *Saudi Journal of Civil Engineering, 7*(03), 16–28.

Wang, C., Liang, J., & Hodges, K. I. (2017). Projections of tropical cyclones affecting Vietnam under climate change: Downscaled HadGEM2-ES using PRECIS 2.1. *Quarterly Journal of the Royal Meteorological Society, 143*(705), 1844–1859.

Wang, Z., Zhao, J., Huang, H., & Wang, X. (2022). A review on the application of machine learning methods in tropical cyclone forecasting. *Frontiers in Earth Science, 10*, 902596.

WMO. (2021, December 2). *WMO provisional report on the state of the global climate 2020*. World Meteorological Organization, United Nations.

WMO. (2023). *Classification of tropical cyclones*. https://wmo.int/content/classification-of-tropical-cyclones

Zaman, M. S., & Chowdhury, T. H. (2024). The Sundarbans, the world's largest tidal halophytic mangrove Forest: Its economic and ecological significance. *Bangla Journal of Interdisciplinary Sciences, 2*(1), E1–E15.

Advanced Remote Sensing and GIS Techniques for Mangrove

Hyperspectral Remote Sensing in Mangrove Mapping

Bhawana Sharma, Prashant K. Srivastava, Sanjeev Kumar Srivastava, and Mukunda Dev Behera

Abstract Mangrove forests possess one of the most affluent biodiversity in the coastal region around the globe. Due to their rich floral and faunal diversity, they hold higher ecological and economic importance. Identification of species composition in mangrove ecosystem is very crucial for better decision-making in conservation and planning restoration practices. Hyperspectral Remote Sensing has made remarkable contribution in mangrove species identification in contrast to other remote sensing techniques. Nowadays, integrated approach of traditional classification techniques with machine learning algorithms is prevalent in achieving highly accurate species identification in mangrove forests. This chapter gives a brief account on the role of HRS in mangrove species identification, methodologies involved in producing robust discrimination results, and a few case studies to assess the real-time application of HRS in mangrove species identification.

Keywords Hyperspectral Remote Sensing (HRS) · Supervised classification algorithm (SCA) · Unsupervised classification algorithm · Mangrove species · Mangrove forests (MFs)

B. Sharma · P. K. Srivastava (✉)
Remote Sensing Laboratory, Institute of Environment and Sustainable Development, Banaras Hindu University, Varanasi, India
e-mail: prashant.iesd@bhu.ac.in

S. K. Srivastava
Geospatial Analytics for Conservation and Management, School of Science Technology and Engineering (SSTE), University of the Sunshine Coast, Sippy Downs, QLD, Australia

M. D. Behera
Centre for Ocean, River, Atmosphere and Land Sciences (CORAL), Indian Institute of Technology Kharagpur, Kharagpur, India

S. K. Srivastava et al. (eds.), *Earth Observation for Monitoring Mangrove Ecosystems*, https://doi.org/10.1007/978-3-032-16668-5_8

1 Introduction

1.1 Hyperspectral Remote Sensing Technology

Advanced technologies of remote sensing are upgrading day-by-day to give high-end results in multiple aspects of earth observations. Hyperspectral Remote Sensing (HRS) technology is continuously progressing in various disciplines for environmental applications (Singh et al., 2022). The terms "image spectrometry" and "hyperspectral" were firstly introduced by Goetz et al. (1985) in his work on "Earth Remote Sensing." This evolutionary work increased the spectral bands from just seven to hundreds in an instant. The contiguous spectrum data from the visible to near-infrared (NIR) wavelength (350–2500 nm) region largely assisted in precise detection and identification of vast number of earth features (Goetz et al., 1985). HRS is utilized in various interdisciplinary subjects, from mineral exploration in geology (Mishra et al., 2021), vegetation discrimination (Singh et al., 2024), disease identification (Rayhana et al., 2023), and stress detection (Srivastava & Rawat, 2023) in agriculture and forestry to biodiversity mapping and monitoring in ecology (Shukla & Kot, 2016).

1.2 Why Are Mangrove Species Required to Be Identified?

Mangroves are woody vegetation of coastal regions which forms a highly efficient intertidal ecosystem. They are salt tolerant and well-suited to grow in low-oxygenated soil. Mangroves are highly sensitive to frost which explains their distribution in tropical and subtropical regions of the world (Fig. 1). They act as a barrier which prevents soil erosion and promotes sedimentation during the regular change in tidal waves. This dynamic ecosystem provides food and shelter to many aquatic and terrestrial faunal populations (NOAA, (Heumann, 2011; Friess et al., 2019)). Water purification, nutrient cycling, tourism, provision of timber and non-timber forest produces, and recreation are few other ecosystem services that mangroves forests continuously facilitate for everyone. Additionally, the mangrove forests (MFs) are rightfully considered as an important contributor of carbon sequestration as it can sequester vast amount of carbon in limited coastal habitats when compared with other terrestrial ecosystem types. Carbon sink in MFs referred to as "blue carbon." The wider term "ocean carbon sink" can also be used to describe the carbon sequestered in the coastal ecosystem of mangroves, seagrass beds, and salt marshes (Choudhary et al., 2024). Many studies were conducted in the last decade to estimate the carbon sink in mangroves at global level (Hamilton & Casey, 2016; Inoue, 2018; Goldberg et al., 2020). These studies also demonstrated the critical conditions and destructions caused to this specific ecosystem across the world (Adame et al., 2021). Despite having socio-economic and ecological importance, anthropogenic activities play key role in rapid exploitation and conversion of MFs for various

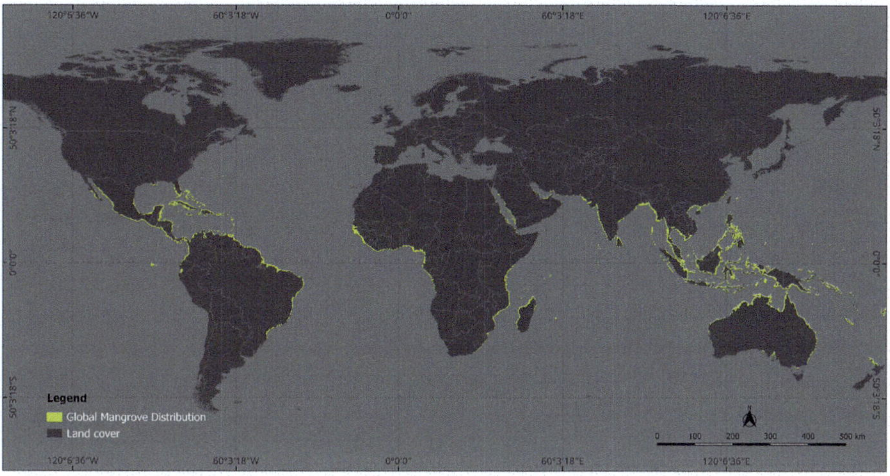

Fig. 1 MFs distribution around the globe (data source: Global Mangrove Watch, version 3.0, 2020)

agricultural and aquaculture practices. Deforestation, pollution, climate change, and encroachment cases heavily contributing to mangrove degradation. These instances demand the need for large-scale study to consistently monitor the dynamics of mangrove biodiversity (Saoum & Sarkar, 2024). Remote sensing technology not only provides large-scale data, but it is also inexpensive; it has all time available historical data archive of varying spatial, spectral, and temporal resolution; and above all, it is easily accessible to users around the world (Singh et al., 2024). Nowadays, remote approach plays significant role in mangrove restoration, decision-making, and execution of management plans in the real world.

1.3 Contributions of HRS in Mangrove Forest Analysis

Hyperspectral imaging widely utilized in discriminating mangroves from other vegetations (Yang et al., 2022), biodiversity mapping (Wang et al., 2022), biomass estimations (Pandey et al., 2019; Pham et al., 2019), health monitoring (Vidhya et al., 2014; Chaube et al., 2019; Kumar et al., 2020), detection of specific pest or disease infestations (Jiang et al., 2022), and species-level classifications (Held et al., 2003; Vaiphasa et al., 2005; Cao et al., 2021). Intensive studies like biochemical parameter (nitrogen and chlorophyll content) mapping (Axelsson et al., 2013; Flores-de-Santiago et al., 2013) and biophysical parameter (Chen et al., 2023) modeling using HRS technique also conducted frequently over MFs. Globally, many studies have proven that HRS technology significantly improve the classification accuracy and species-level discrimination (Pandey et al., 2019; Kumar et al., 2020; Li et al., 2021; Jiang et al., 2022) in MFs.

1.4 Role of HRS in Mangrove Species Identification

The hyperspectral sensors provide contiguous reflectance data of higher spectral resolution in VNIR and SWIR region of electromagnetic spectrum respectively. Thus, providing complete spectrum of earth features (Folkman et al., 2001) in more than 200 bands of narrow bandwidth. This attribute of hyperspectral technique facilitates highly detailed imageries (Salghuna & Pillutla, 2017), the most suited to differentiate species using authentic spectral signature in each waveband. In terms of vegetation, the top canopy reflectance largely used to detect the separability at species-specific level. Although a number of factors are affecting these reflectance values, including biophysical and biochemical characteristics of vegetation (Asner, 1998), soil or litter background, phenological dynamism, sun-view geometry (Jafarbiglu & Pourreza, 2023), and—in case of airborne instrument carrier—atmospheric conditions and bidirectional reflectance, the top canopy reflectance eventually gives better accuracy while discriminating species using hyperspectral remote approach than leaf reflectance of the vegetation (Cho et al., 2008). Mangrove species were identified and mapped using multispectral, hyperspectral, SAR, LiDAR, and aerial images by Maurya et al. (2021). Out of these techniques, the combination of spectral and spatial attributes in hyperspectral imageries always showed superior classification accuracy with or without different nonparametric machine learning algorithms (Maurya et al., 2021). It can be stated that HRS is one of the most robust approaches to detect mangrove species with higher precision.

2 Principles

Hyperspectral sensors measure the amount of radiance reflected by different earth features in the optical region. The optical region covers visible to infrared wavelength portion (400–2500 nm) in electromagnetic spectrum. Radiance measurements are then converted into reflectance values which are primarily used to distinguish one feature from others. The characteristic attribute of hyperspectral imaging is to capture these reflectance values in more than 200 narrow spectral channels to procure contiguous spectral signature of target earth features. Mangrove species are generally identified based on differences in their spectral signatures (Fig. 2) captured using HRS techniques.

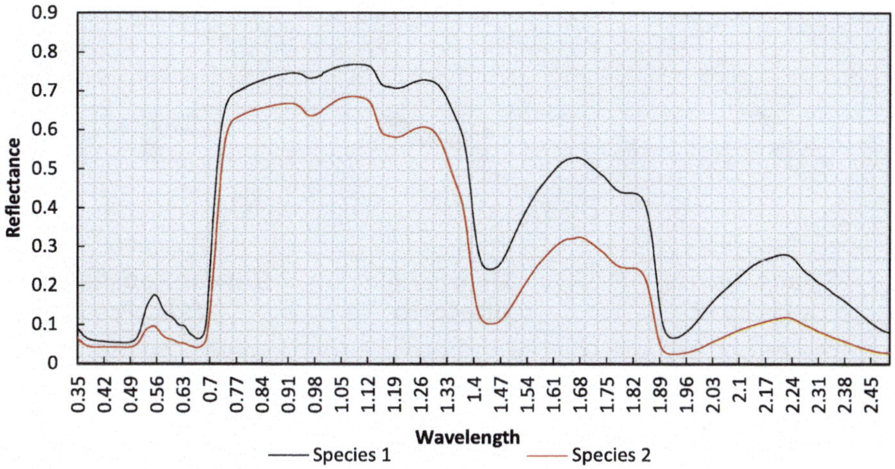

Fig. 2 Spectral signature of two different species

3 Methodologies

The sequential steps to identify a mangrove species begin from basic acquisition of field and hyperspectral data to process the suitable classification algorithm and perform accuracy assessment. Basic working flow to classify multiple mangrove species is given in Fig. 4.

3.1 Data Acquisition

Various hyperspectral satellites and aerial missions were launched by multiple countries, such as PRISMA by Italy, EnMAP by Germany, Pixxel by India, AVIRIS by USA, and so on. Raw satellite images are acquired from various publicly accessed data portals of satellite agencies if captured using space-borne platform. High-resolution datasets are often made accessible only on demand or by stating the purpose and future scope of acquisition. Some of the available remote hyperspectral platforms for data acquisition are shown in Fig. 3. Along with remote data, field inventory of MFs are often made to validate the accuracy of classified images. Thus, spatial distribution of mangrove species is required to be observed and recorded in the fields.

3.1.1 Space-Borne Platforms

A significant number of satellites mounted with hyperspectral sensors are commissioned by many global agencies till now. One of recently launched Indian origin constellation of hyperspectral satellites by Pixxel, a space data company, is called

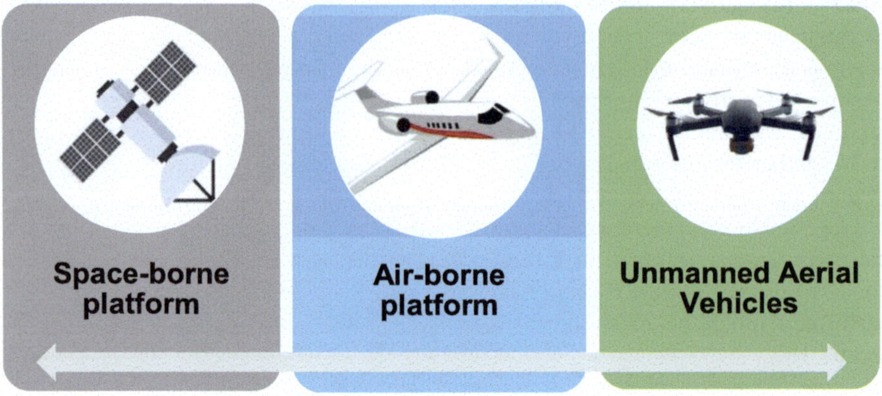

Fig. 3 Available platforms for hyperspectral instruments

Firefly. The satellites were begun to commission in January 2025. It is properly equipped to give the highest commercially available spatial resolution of 5 m in 250 narrow wavebands extending from Visible to SWIR regions with global covering datasets at every 24 h. The output products of such satellites will prove to be beneficial in preventing complex spectral mixing of different vegetation (hyperspectral imagery, Pixxel). It will provide highly detailed and robust output for accurate species identification. Archived datasets of many decommissioned satellites are also available for past species identifications. Some active earth observation satellites with hyperspectral payload are listed in Table 1.

3.1.2 Airborne Platforms

Hyperspectral sensors mounted on aircrafts and drones are prevalent for species classification at local scales. Previous studies show that airborne hyperspectral images were successfully utilized to classify forest species covering a small stretch of few thousand hectares to large areas of forest stands (Jones et al., 2010; Heinzel & Koch, 2012; Modzelewska et al., 2020). The airborne platforms further categorized as aircrafts and unmanned aerial vehicles (UAVs).

Aircrafts

Hyperspectral instruments such as AVIRIS-Classic, AVIRIS-NG, HyMap, HySpex, and Prospectir are mounted on aircrafts to capture datasets on demand in target study sites. Flights over mangrove forest covers are requested in many instances to conduct species identification analysis.

Table 1 List of space-borne hyperspectral mission suitable for species identification and other vegetation analysis

S. no.	Mission	Launch date	Sensor/payload	Spectral range (nm)	Spatial resolution	Number of spectral bands	Revisit duration	Company
1.	Firefly Constellation	January 2025	Hyperspectral Imager (HSI)	400–2500	5 m	250 (VNIR, SWIR)	24 h	Pixxel, India
2.	Aleph-1 Constellation	May 2016 to January 2025	–	460–900	30 m	29 (VNIR) Up to 600	5 min	Satellogic, Argentina
3.	PACE	February 2024	Ocean Color Instrument (OCI)	315–895	1.2 km		2 days	NASA
4.	GHOSt Constellation	April 2023 to March 2024	HSI	400–2500	8 m	512 (VNIR, SWIR)	24 h	Orbital Sidekick Inc., USA
5.	EnMAP	April 2022	HSI	420–2450	30 m	224	27 days	DLR Space Agency, Germany
6.	PRISMA	March 2019	HYC	400–2505 400–1010 (VNIR) 920–2505 (SWIR)	30 m	239 (VNIR-66, SWIR-173)	~10 days	Agenzia Spaziale Italiana, Italy

UAVs

Datasets captured using hyperspectral camera fitted to drones. Mangrove species at local level were monitored and analyzed using this approach. HySpex Mjolnir, Pika L, Pika IR-L, and Pika IR-L+ are some of the hyperspectral cameras mounted on drones to capture images for various real-time applications using spectral properties of objects.

3.2 *Pre-processing Techniques for Hyperspectral Images*

Raw images captured by hyperspectral sensors encounter numerous interferences. Pre-processing techniques increase the accuracy and robustness of processed results by diminishing or omitting these interferences. Noise, clouds, irregular illumination (Cozzolino et al., 2023), no data pixels, and many other disturbances are included in this list. It is important to perform appropriate pre-processing methods on raw images for enhanced quality data for further analysis. Some necessary methods include radiometric, geometric, and atmospheric corrections. Additional processing, such as noise removal, feature extraction, resolution enhancement, etc. (Fig. 4), is required according to the purpose of further processing. Velasco-Forero and Manian (2009) in their study indicated the vital role of spatial smoothing while performing classification algorithms on hyperspectral image. This pre-processing technique comparatively showed better results in classified image. Similarly, multiple pre-processing techniques (Vidal & Amigo, 2012) are incorporated in many cases to increase the classification accuracy.

3.3 *Image Classification Techniques for Mangrove Species*

According to Maurya et al. (2021), species classification in mangrove ecosystem might follow different approaches as each technique considers spatial and spectral attributes of species differently. Species may classify on the basis of their spatial and spectral properties using pixel-based, sub-pixel-based, object-based, or knowledge-based approach.

3.3.1 Pixel-Based Algorithm

Spectral signature of each pixel is analyzed, and classes are assigned after spectral patterns at pixel level are recognized (Weih & Riggan, 2010; Maurya et al., 2021). Combination of two bands at once is plotted for spectral values of all pixels in a scatter diagram for classification. This is a traditional approach as it aims to work at the basic unit, i.e., pixel, of satellite image.

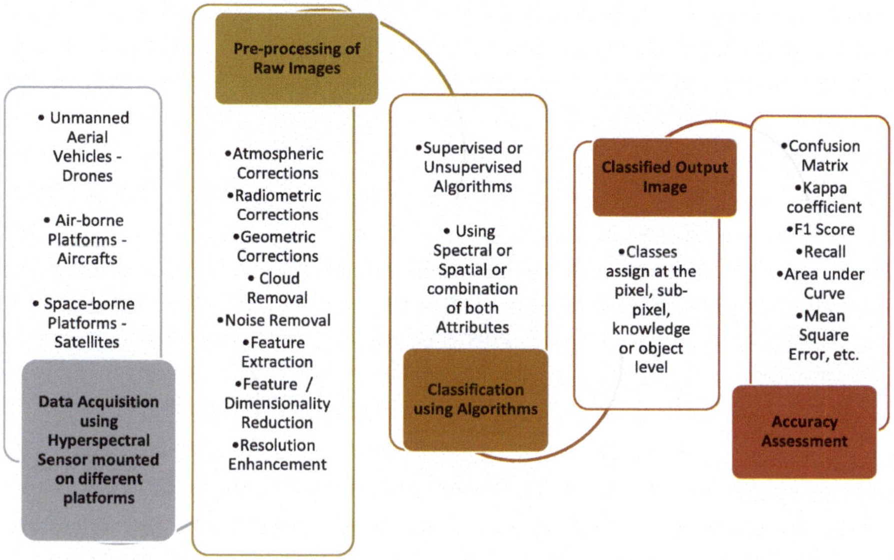

Fig. 4 Image classification using hyperspectral satellite imageries

3.3.2 Object-Based Algorithm

Pixels having identical spectral information are termed as objects. Object-based classifiers work on spectral as well as spatial features of objects. First step to perform this classification is segmentation. It breaks the image into multiple objects; these objects are then classified according to their features in various classes. The object segments are similar to real earth features as this algorithm aims to replicate the human-like visual interpretation (Liu & Xia, 2010; Maurya et al., 2021).

3.3.3 Sub-pixel-Based Algorithm

Frequent occurrence of mixed pixel in species identification is very usual. Area of single pixel in the ground mostly possesses patches of mixed species as homogenous patches of individual species in natural mangrove stand are less-likely to occur. In order to reduce this phenomenon, sub-pixel algorithms are adopted to classify different features of a pixel at sub-pixel level. Spectral unmixing techniques such as linear spectral unmixing (LSU), object-based image classifier (OBIA), Spectral Angle Mapper (SAM), fuzzy classification, simplex projection unmixing, and mixture tuned matched filtering (Kamal & Phinn, 2011; Chakravortty, 2013; Zhang et al., 2014; Kumar et al., 2022) are used on hyperspectral images in different scenarios and purposes. This approach gives good results in mangrove species identification but is often outperformed by other classification techniques.

3.3.4 Knowledge-Based Algorithm

Ancillary knowledge required to support the spectral properties in an image can be used in some cases to identify mangrove species (Kamal & Phinn, 2011; Chaube et al., 2019). This approach is preferred when it is uncertain to fully rely on spectral characteristics for feature identification. The ancillary knowledge include the elevation maps, thematic maps, slope information (Suchenwirth et al., 2012), object shape, texture, in-between distance, etc. In this algorithm, segments formed using object-based classification followed by pixel grouping. Classes are assigned to these segments on the basis of explicit rules made using ancillary knowledge. Combination of elevation model, surface model, gray level co-occurrence matrix, and a suitable spectral classification algorithm can be used in mangrove forest. Knowledge-based algorithm enhances the overall classification accuracy as it diminishes the probability of misclassification.

Another approach to classify mangrove species based on learning types includes the supervised and unsupervised classification approach. The pixel-based and object-based algorithms are often additionally classified using this approach. Supervised and unsupervised technique is opted based on learning technique involved to train the data.

3.3.5 Supervised Classification Algorithm (SCA)

Although SCA is further categorized into pixel-based supervised techniques and object-based supervised techniques, both approaches include similar machine learning and deep learning classification algorithms (Singh et al., 2025a, 2025b, 2025c). Various parametric-maximum likelihood classifier (MLC) (Satyanarayana et al., 2018) and nonparametric-artificial neural network (ANN) (Singh et al., 2025b), decision tree (Wong & Fung, 2014), k-nearest neighbor (k-NN) (Cao et al., 2018), support vector machine (SVM) (Ruwaimana et al., 2018; Manna & Raychaudhuri, 2020), Random Forest (RF), and Rotation of Forest (RoF) (Li et al., 2021; Paramanik et al., 2023) techniques can be utilized to detect mangrove species with high accuracy.

3.3.6 Unsupervised Classification Algorithm

It is further categorized in pixel- and object-based unsupervised classification algorithms. The pixel-oriented unsupervised classifiers include the vegetation index-based, K-mean cluster algorithm and Iterative Self-Organizing Data Analysis Technique Algorithm (ISODATA). Out of these algorithms, only K-mean algorithms are suitable for species detection and classification on hyperspectral datasets (Manjunath et al., 2013; Padma & Sanjeevi, 2014). The object-based unsupervised classifiers on hyperspectral dataset include object-oriented classification and k-NN classifier for mangrove species identification (Jia et al., 2014).

4 Case Studies

In the mid-2000s, airborne hyperspectral imaging product were utilized to identify nine mangrove species over Port Klang mangrove (Fig. 5a) stretch (Jusoff, 2006). The UPM-APSB's AISA Airborne Hyperspectral instrument mounted on SkyVan SC-7 aircraft to capture the datasets in 20 narrow spectral bands. This work mainly described the importance of NIR region in assessing spectral separability between different species. NIR region from 700 to 900 nm were analyzed in details to identify nine mangrove species using spectral signature of 19 randomly chosen mangrove species.

After almost a decade, mangrove species of Pichavaram Mangrove Ecosystem (Fig. 5b) in Tamil Nadu, India, identified using available eleven reference spectra in literatures to train different SCAs—Spectral Angle Mapper or SAM, Spectral Feature Fitting (SFF), and Spectral Information Diversion (SID). Hyperspectral images of Hyperion satellite incorporated for this purpose. Further, linear spectral unmixing (LSU) also performed to detect the precise location of species at sub-pixel level (Salghuna & Pillutla, 2017). In this experiment, SAM with LSU outperformed other algorithms in mapping mangrove species at higher accuracy. Ten separate species mentioned in Table 2 were included in this study to map estuarine mangrove of Pichavaram at species level. The progressive development in the HRS technology to discriminate species using various algorithms can be seen from previous works to this classification study.

A study conducted at mangrove stand of Bhitarkanika (Fig. 6a), India in 2019 where ten mangrove species (Table 2) were identified using hyperspectral output of Hyperion satellite. For this study, 196 bands out of 242 total bands remained for further processing after noisy, water-absorbing, and overlapping band removal. SCA SAM opted in this study to discriminate species with an overall accuracy of 84% and kappa coefficient of 0.81 (Pandey et al., 2019). In the similar year, the vicinity of Bhitarkanika along with dense patch of Sundarbans mangrove (Fig. 6b) was studied (Chaube et al., 2019) using AVIRIS-NG airborne data. AVIRIS-NG is a

Fig. 5 MFs of (**a**) Port Klang, Malaysia, (**b**) Pichavaram, Tamil Nadu, India

Table 2 List of species identified in the past studies conducted over MFs around the world

Classification	Pichavaram mangrove forest (Salghuna & Pillutla, 2017)	Sundarbans and Bhitarkanika mangrove forest (Chaube et al., 2019)	Bhitarkanika mangrove forest (Pandey et al., 2019)	Bhitarkanika mangrove forest (Paramanik et al., 2023)
Hyperspectral data	Hyperion	AVIRIS-NG	Hyperion	AVIRIS-NG
Calibrated bands (total bands)	198 (242)	(425)	196 (242)	(425)
Best classification algorithm	SAM	SAM	SAM	Random Forest
Overall accuracy	NA	84.5%	84%	88%
Kappa coefficient	NA	0.78	0.81	0.84
Number of species identified	10	15	10	3
1.	*Acanthus ilicifolius*	*Avicennia alba*	*Aegiceras corniculatum*	*Avicennia officinalis*
2.	*Aegiceras corniculatum*	*Aegiceras corniculatum*	*Cerbera odollam*	*Excoecaria agallocha*
3.	*Arthrocnemum indicum*	*Avicennia marina*	*Cynometra iripa*	*Heritiera fomes*
4.	*Avicennia marina*	*Avicennia officinalis*	*Excoecaria agallocha*	Combination of *Heritiera fomes-Excoecaria agallocha*
5.	*Avicennia officinalis*	*Aegialitis rotundifolia*	*Heritiera fomes*	Combination of *Avicennia officinalis-Excoecaria agallocha*
6.	*Bruguiera cylindrica*	*Bruguiera gymnorrhiza*	*Heritiera littoralis*	
7.	*Ceriops decandra*	*Bruguiera cylindrica*	*Intsia bijuga*	
8.	*Excoecaria agallocha*	*Heritiera fomes*	*Sonneratia apetala*	
9.	*Rhizophora apiculata*	*Ceriops decandra*	*Xylocarpus granatum*	
10.	*Rhizophora mucronate*	*Excoecaria agallocha*	*Xylocarpus mekongensis*	
11.		*Phoenix paludosa*		

Table 2 (continued)

Classification	Pichavaram mangrove forest (Salghuna & Pillutla, 2017)	Sundarbans and Bhitarkanika mangrove forest (Chaube et al., 2019)	Bhitarkanika mangrove forest (Pandey et al., 2019)	Bhitarkanika mangrove forest (Paramanik et al., 2023)
12.		*Casuarina equisetifolia*		
13.		*Sonneratia apetala*		
14.		*Sonneratia caseolaris*		
15.		*Xylocarpus mekongensis*		

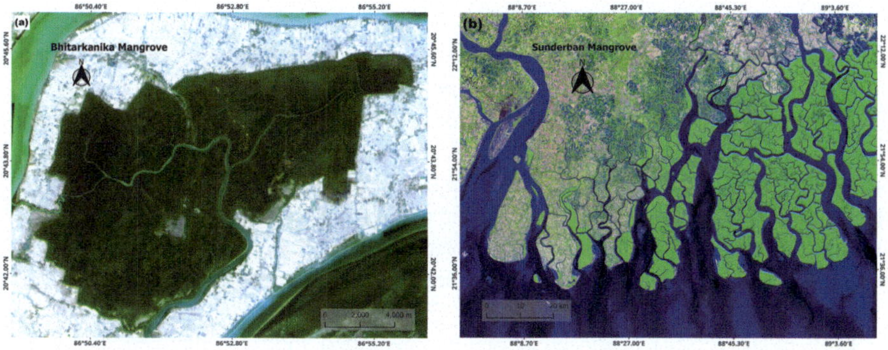

Fig. 6 MFs in (**a**) Bhitarkanika National Park and (**b**) Sundarbans, India

hyperspectral instrument covering the spectral range of 380–2510 nm in 425 contiguous bands. The study classified 15 mangrove species (listed in Table 2) using SAM classifier with an overall accuracy of 84.5% and kappa coefficient reaching 0.78 in values.

Integrated approach using Hyperspectral, LIDAR, and machine learning algorithms to identify mangrove species was reported by Cao et al. (2021). Mangrove Forest Nature Reserve in Qi'ao Island (Fig. 7a), China was analyzed using the Headwall Nano-hyperspec push-broom hyperspectral imager and Riegl VUX-1UAV laser scanner system (Cao et al., 2021) mounted on Scout B1–100 UAV helicopter. Seven mangrove species were identified in this study using ensemble learning model on the integrated hyperspectral and LiDAR data products. Rotation of Forest (RoF) approach identified these species with 96.88% accuracy using spectral and structural features, the combination of spectral (20 spectral bands) and LiDAR-derived canopy height model data. This approach regarded for monitoring the mangrove vegetation inventory with higher accuracy and promoted for restoration and management practices. Rapid advancement in machine learning approach largely contributes in increased algorithm accuracy. Impact of ML in mangrove species identification is evident in this study.

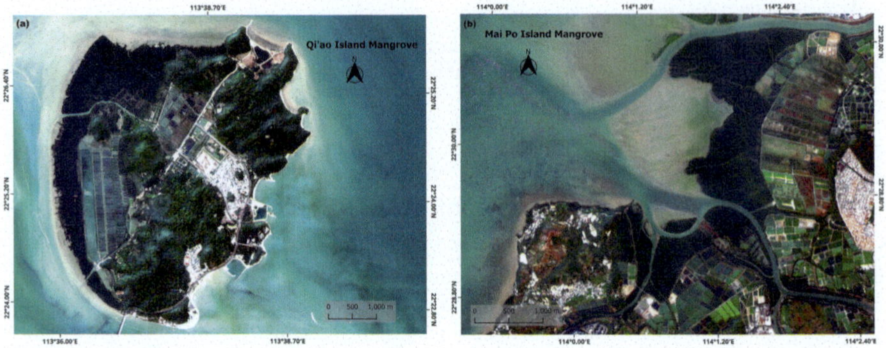

Fig. 7 Mangrove stretch of (**a**) Qi'ao Island, China, and (**b**) Mai Po Nature Reserve, China

Similar approach to classify mangrove species of Mai Po Nature Reserves (Fig. 7b) made by Li et al. (2021). Datasets of airborne hyperspectral instrument, multispectral WorldView-3, and LiDAR point cloud was incorporated to study vertical composition and species distribution of mangrove stand. ML algorithms—Random Forest (RF), support vector machine (SVM), and convolutional neural network (CNN) used to locate seven mangrove species. Combined input of hyperspectral and LiDAR gave the most accurate results. An overall accuracy of 0.86–0.88 attained in the three ML classifier, conforming the collaborative approach of HRS and LiDAR technology for highly reliable species identification results (Li et al., 2021).

Yang et al. (2022) proposed a newly developed index, Enhanced Mangrove Vegetation Index (EMVI), using one narrow band in green and two in SWIR region of ZY1-02D, GF-5, Hyperion, and PRISMA hyperspectral dataset. ZY1-02D and GF-5 satellites are equipped with AHSI payload which provides 30 m spatial resolution in 166 and 330 spectral bands, respectively. EO-1 of Hyperion and HSI of PRISMA also captures imageries in 30 m spatial resolution but have 242 and 239 narrow bands, respectively. EMVI generated in such a way that it does not get affected by high and low-tide situations. The EMVI mapped mangrove cover with an overall accuracy of more than 90%. Comparative analysis between multispectral-based Mangrove Vegetation Index (MVI) on Sentinel-2 output and hyperspectral-based EMVI showed that EMVI on hyperspectral images has higher accuracy than MVI or EMVI on multispectral images (Yang et al., 2022).

Distribution of three major species, along with two of their combinations (Table 2), in Bhitarkanika Mangrove studied by Paramanik et al. (2023) using AVIRIS-NG imageries, an airborne sensor that captures earth features in high spatial and spectral resolution. Spectral signatures of some individuals (Fig. 8) from each of the three species were extracted from the output imageries. Species were mapped using Random Forest algorithm after applying continuum removal and absorption band depth techniques. Overall accuracy of 88% and kappa coefficient of 0.84 was achieved in resultant species distribution map which is generated using absorption band depth technique. The study also concluded the water absorption

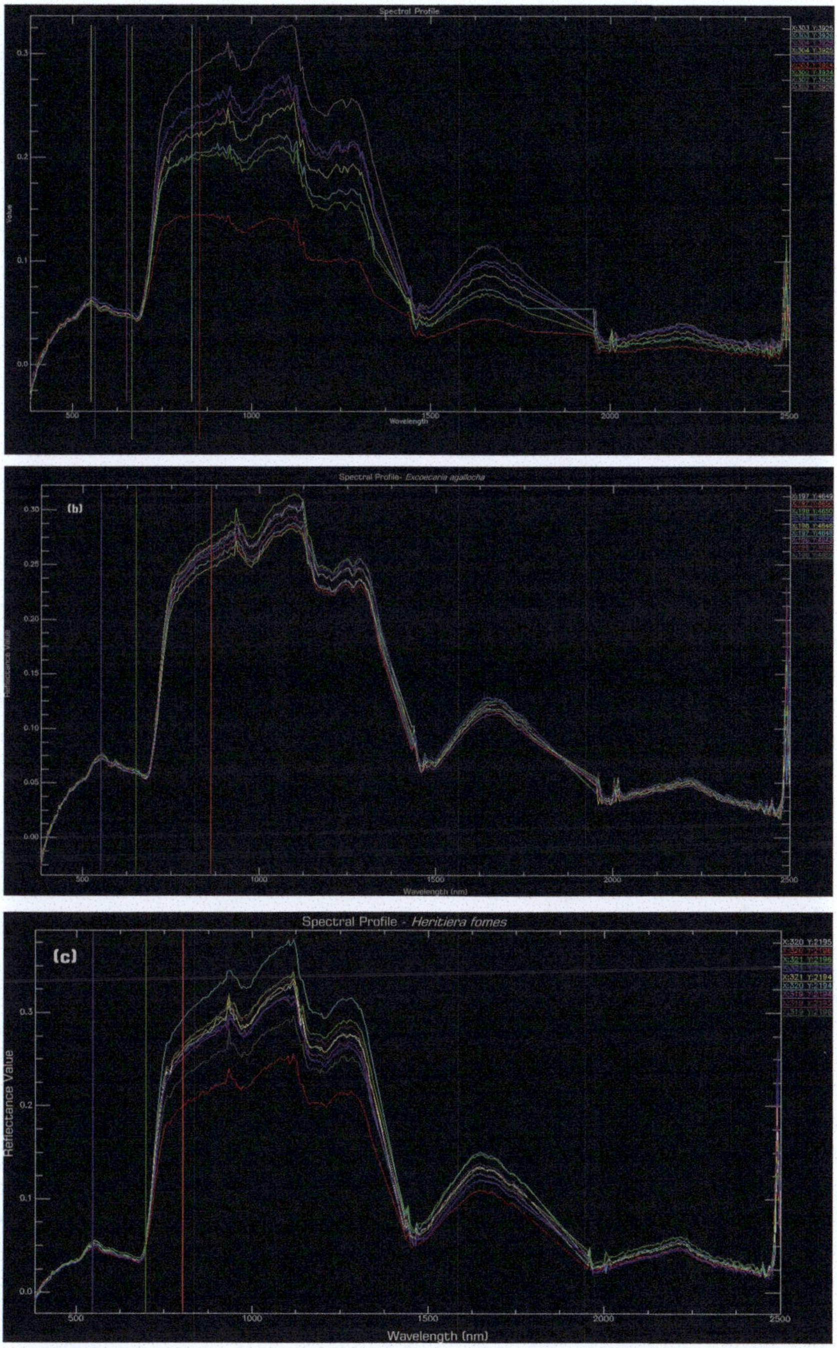

Fig. 8 Spectra generated from nine individuals of (**a**). Avicennia officinalis, (**b**) Excoecaria agallocha, and (**c**) Heritiera fomes, each species from distinct plots in Bhitarkanika Mangrove Forest. Here, *X*-axis represents the wavelength region, and *Y*-axis represents the reflectance values (Paramanik et al., 2023)

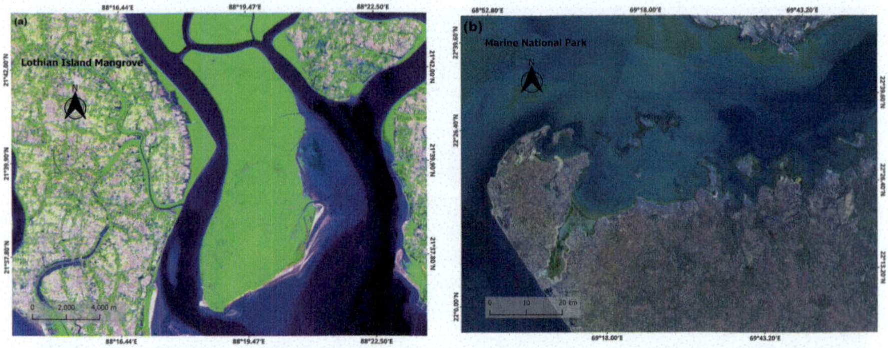

Fig. 9 MFs in (**a**) Lothian Island, Sundarbans, India, and (**b**) Marine National Park, Gujarat, India

bands, 972, 1172 and 1177 nm, that are crucial for mangrove species identification (Paramanik et al., 2023).

In another study, MFs of Marine National Park, Gujarat, and Lothian Island of West Bengal were classified at species level using Modified Spectral Angle Mapper (SAM) algorithm on hyperspectral datasets of AVIRIS-NG (Fig. 9). This adaptation of SAM classifier integrated change in spectral shape of mangrove species and angular change caused due to directional change as well as magnitude change in spectra of mangrove species as distance and spectral similarity measures. The modified SAM produced an overall accuracy of 93.48% in classified product of Lothian Island and 94.32% in classified image of Marine National Park (Maurya & Mahajan, 2024). This approach demonstrated enhanced results particularly in mixed-mangrove pixel regions.

5 Conclusion

HRS is an effective tool to identify mangrove species with high accuracy and reliability. Hyperspectral datasets are often incorporated in precise detection of individual species in natural forest habitat. The high spatial and spectral resolutions of hyperspectral datasets have many advantages in mangrove species identification. Spectral properties of species captured in more than 200 narrow bands is rigorously analyzed to discriminate different species. Multiple algorithms—supervised, unsupervised, pixel-based, object-based, etc.—are applied individually or in combination to identify mangrove species in different countries. The enhanced accuracy achieved using hyperspectral technique signifies its importance and future scope. In conclusion, HRS is playing a vital role in mangrove biodiversity assessment. It is evidently contributing in management and restoration of this important ecosystem. Progressive development of this technology will prove to be beneficial in mangrove monitoring, periodic evaluation, and detailed condition assessments.

References

Adame, M. F., Connolly, R. M., Turschwell, M. P., Lovelock, C. E., Fatoyinbo, T., Lagomasino, D., Goldberg, L. A., Holdorf, J., Friess, D. A., & Sasmito, S. D. (2021). Future carbon emissions from global mangrove forest loss. *Global Change Biology, 27*, 2856–2866.

Asner, G. P. (1998). Biophysical and biochemical sources of variability in canopy reflectance. *Remote Sensing of Environment, 64*, 234–253.

Axelsson, C., Skidmore, A. K., Schlerf, M., Fauzi, A., & Verhoef, W. (2013). Hyperspectral analysis of mangrove foliar chemistry using PLSR and support vector regression. *International Journal of Remote Sensing, 34*, 1724–1743.

Cao, J., Leng, W., Liu, K., Liu, L., He, Z., & Zhu, Y. (2018). Object-based mangrove species classification using unmanned aerial vehicle hyperspectral images and digital surface models. *Remote Sensing, 10*, 89.

Cao, J., Liu, K., Zhuo, L., Liu, L., Zhu, Y., & Peng, L. (2021). Combining UAV-based hyperspectral and LiDAR data for mangrove species classification using the rotation forest algorithm. *International Journal of Applied Earth Observation and Geoinformation, 102*, 102414.

Chakravortty, S. (2013). Analysis of end member detection and subpixel classification algorithms on hyperspectral imagery for tropical mangrove species discrimination in the Sunderbans Delta, India. *Journal of Applied Remote Sensing, 7*, 073523.

Chaube, N. R., Lele, N., Misra, A., Murthy, T., Manna, S., Hazra, S., Panda, M., & Samal, R. (2019). Mangrove species discrimination and health assessment using AVIRIS-NG hyperspectral data. *Current Science, 116*, 1136–1142.

Chen, X., Yang, Y., Zhang, D., Li, X., Gao, Y., Zhang, L., Wang, D., Wang, J., Wang, J., & Huang, J. (2023). Response mechanism of leaf area index and main nutrient content in mangrove supported by hyperspectral data. *Forests, 14*, 754.

Cho, M. A., Sobhan, I., Skidmore, A. K., & De Leeuw, J. M. (2008). Discriminating species using hyperspectral indices at leaf and canopy scales. *The International Archives of the Photogrammetry, Remote Sensing and Spatial Information Sciences, 5*, 369–376.

Choudhary, B., Dhar, V., & Pawase, A. S. (2024). Blue carbon and the role of mangroves in carbon sequestration: Its mechanisms, estimation, human impacts and conservation strategies for economic incentives. *Journal of Sea Research, 199*, 102504.

Cozzolino, D., Williams, P., & Hoffman, L. (2023). An overview of pre-processing methods available for hyperspectral imaging applications. *Microchemical Journal, 193*, 109129.

Flores-de-Santiago, F., Kovacs, J. M., & Flores-Verdugo, F. (2013). The influence of seasonality in estimating mangrove leaf chlorophyll-a content from hyperspectral data. *Wetlands Ecology and Management, 21*, 193–207.

Folkman, M. A., Pearlman, J., Liao, L. B., & Jarecke, P. J. (2001). EO-1/Hyperion hyperspectral imager design, development, characterization, and calibration. *Hyperspectral Remote Sensing of the Land and Atmosphere, 4151*, 40–51.

Friess, D. A., Rogers, K., Lovelock, C. E., Krauss, K. W., Hamilton, S. E., Lee, S. Y., Lucas, R., Primavera, J., Rajkaran, A., & Shi, S. (2019). The state of the world's mangrove forests: Past, present, and future. *Annual Review of Environment and Resources, 44*, 89–115.

Goetz, A. F., Vane, G., Solomon, J. E., & Rock, B. N. (1985). Imaging spectrometry for earth remote sensing. *Science, 228*, 1147–1153.

Goldberg, L., Lagomasino, D., Thomas, N., & Fatoyinbo, T. (2020). Global declines in human-driven mangrove loss. *Global Change Biology, 26*, 5844–5855.

Hamilton, S. E., & Casey, D. (2016). Creation of a high spatio-temporal resolution global database of continuous mangrove forest cover for the 21st century (CGMFC-21). *Global Ecology and Biogeography, 25*, 729–738.

Heinzel, J., & Koch, B. (2012). Investigating multiple data sources for tree species classification in temperate forest and use for single tree delineation. *International Journal of Applied Earth Observation and Geoinformation, 18*, 101–110.

Held, A., Ticehurst, C., Lymburner, L., & Williams, N. (2003). High resolution mapping of tropical mangrove ecosystems using hyperspectral and radar remote sensing. *International Journal of Remote Sensing, 24*, 2739–2759.

Heumann, B. W. (2011). Satellite remote sensing of mangrove forests: Recent advances and future opportunities. *Progress in Physical Geography, Earth and Environment, 35*, 87–108.

Inoue, T. (2018). Carbon sequestration in mangroves. In *Blue carbon in shallow coastal ecosystems: Carbon dynamics, policy, and implementation* (pp. 73–99). Springer.

Jafarbiglu, H., & Pourreza, A. (2023). Impact of sun-view geometry on canopy spectral reflectance variability. *ISPRS Journal of Photogrammetry and Remote Sensing, 196*, 270–286.

Jia, M., Zhang, Y., Wang, Z., Song, K., & Ren, C. (2014). Mapping the distribution of mangrove species in the Core Zone of Mai Po Marshes Nature Reserve, Hong Kong, using hyperspectral data and high-resolution data. *International Journal of Applied Earth Observation and Geoinformation, 33*, 226–231.

Jiang, X., Zhen, J., Miao, J., Zhao, D., Shen, Z., Jiang, J., Gao, C., Wu, G., & Wang, J. J. E. I. (2022). Newly-developed three-band hyperspectral vegetation index for estimating leaf relative chlorophyll content of mangrove under different severities of pest and disease. *Ecological Indicators, 140*, 108978.

Jones, T. G., Coops, N. C., & Sharma, T. (2010). Assessing the utility of airborne hyperspectral and LiDAR data for species distribution mapping in the coastal Pacific Northwest, Canada. *Remote Sensing of Environment, 114*, 2841–2852.

Jusoff, K. (2006). Individual mangrove species identification and mapping in Port Klang using airborne hyperspectral imaging. *Journal of Sustainability Science and Management, 1*, 27–36.

Kamal, M., & Phinn, S. (2011). Hyperspectral data for mangrove species mapping: A comparison of pixel-based and object-based approach. *Remote Sensing, 3*, 2222–2242.

Kumar, T., Kaur, P., Chandrasekar, K., & Bandyopadhyay, S. (2020). Aviris-ng hyperspectral data for mapping mangrove forests and their health spatially. *Journal of Tropical Forest Science, 32*, 317–331.

Kumar, V., Pandey, K., Panda, C., Tiwari, V., & Agrawal, S. (2022). Assessment of different spectral unmixing techniques on space borne hyperspectral imagery. *Remote Sensing in Earth Systems Sciences, 5*, 129–140.

Li, Q., Wong, F. K. K., & Fung, T. (2021). Mapping multi-layered mangroves from multispectral, hyperspectral, and LiDAR data. *Remote Sensing of Environment, 258*, 112403.

Liu, D., & Xia, F. (2010). Assessing object-based classification: Advantages and limitations. *Remote Sensing Letters, 1*, 187–194.

Manjunath, K., Kumar, T., Kundu, N., & Panigrahy, S. (2013). Discrimination of mangrove species and mudflat classes using in situ hyperspectral data: A case study of Indian Sundarbans. *GIScience & Remote Sensing, 50*, 400–417.

Manna, S., & Raychaudhuri, B. (2020). Mapping distribution of Sundarban mangroves using Sentinel-2 data and new spectral metric for detecting their health condition. *Geocarto International, 35*, 434–452.

Maurya, K., & Mahajan, S. (2024). Mangrove species discrimination using modified spectral angle mapper algorithm on hyperspectral imagery: A study at Marine National Park (MNP) Jamnagar, Gulf of Kutch, Gujarat and Lothian Island, Sundarbans, West Bengal. *Expert Systems With Applications, 237*, 121455.

Maurya, K., Mahajan, S., & Chaube, N. (2021). Remote sensing techniques: Mapping and monitoring of mangrove ecosystem—A review. *Complex & Intelligent Systems, 7*, 2797–2818.

Mishra, G., Govil, H., & Srivastava, P. K. (2021). Identification of malachite and alteration minerals using airborne AVIRIS-NG hyperspectral data. *Quaternary Science Advances, 4*, 100036.

Modzelewska, A., Fassnacht, F. E., & Stereńczak, K. (2020). Tree species identification within an extensive forest area with diverse management regimes using airborne hyperspectral data. *International Journal of Applied Earth Observation and Geoinformation, 84*, 101960.

Padma, S., & Sanjeevi, S. (2014). Jeffries Matusita-Spectral Angle Mapper (JM-SAM) spectral matching for species level mapping at Bhitarkanika, Muthupet and Pichavaram mangroves.

International Archives of the Photogrammetry, Remote Sensing and Spatial Information Sciences, 40, 1403–1411.
Pandey, P. C., Anand, A., & Srivastava, P. K. (2019). Spatial distribution of mangrove forest species and biomass assessment using field inventory and earth observation hyperspectral data. *Biodiversity and Conservation, 28*, 2143–2162.
Paramanik, S., Deep, N. R., Behera, M. D., Bhattacharya, B. K., & Dash, J. (2023). Species-level classification of mangrove forest using AVIRIS-NG hyperspectral imagery. *Remote Sensing Letters, 14*, 522–533.
Pham, T. D., Yokoya, N., Bui, D. T., Yoshino, K., & Friess, D. A. (2019). Remote sensing approaches for monitoring mangrove species, structure, and biomass: Opportunities and challenges. *Remote Sensing, 11*, 230.
Rayhana, R., Ma, Z., Liu, Z., Xiao, G., Ruan, Y., & Sangha, J. S. (2023). A review on plant disease detection using hyperspectral imaging. *IEEE Transactions on AgriFood Electronics, 1*, 108–134.
Ruwaimana, M., Satyanarayana, B., Otero, V. M., Muslim, A., Syafiq A, M., Ibrahim, S., Raymaekers, D., Koedam, N., & Dahdouh-Guebas, F. (2018). The advantages of using drones over space-borne imagery in the mapping of mangrove forests. *PLoS One, 13*, e0200288.
Salghuna, N., & Pillutla, R. (2017). Mapping mangrove species using hyperspectral data: A case study of Pichavaram mangrove ecosystem, Tamil Nadu. *Earth Systems and Environment, 1*, 1–12.
Saoum, M. R., & Sarkar, S. K. (2024). Monitoring mangrove forest change and its impacts on the environment. *Ecological Indicators, 159*, 111666.
Satyanarayana, B., Muslim, A. M., Horsali, N. A. I., Zauki, N. A. M., Otero, V., Nadzri, M. I., Ibrahim, S., Husain, M.-L., & Dahdouh-Guebas, F. (2018). Status of the undisturbed mangroves at Brunei Bay, East Malaysia: A preliminary assessment based on remote sensing and ground-truth observations. *PeerJ, 6*, e4397.
Shukla, A., & Kot, R. (2016). An overview of hyperspectral remote sensing and its applications in various disciplines. *IRA-International Journal of Applied Sciences (ISSN 2455-4499), 5*, 85–90.
Singh, P., Srivastava, P. K., Mall, R., Bhattacharya, B. K., & Prasad, R. (2022). A hyperspectral R based leaf area index estimator: Model development and implementation using AVIRIS-NG. *Geocarto International, 37*, 12792–12809.
Singh, P., Srivastava, P. K., Shah, D., Pandey, M. K., Anand, A., Prasad, R., Dave, R., Verrelst, J., Bhattacharya, B. K., & Raghubanshi, A. (2024). Crop type discrimination using Geo-Stat Endmember extraction and machine learning algorithms. *Advances in Space Research, 73*, 1331–1348.
Singh, H., Ang, L.-M., Paudyal, D., Acuna, M., Srivastava, P. K., & Srivastava, S. K. (2025a). A comprehensive review of empirical and dynamic wildfire simulators and machine learning techniques used for the prediction of wildfire in Australia. *Technology, Knowledge and Learning, 30*, 935–968. https://doi.org/10.1007/s10758-025-09839-5
Singh, H., Ang, L.-M., & Srivastava, S. K. (2025b). Benchmarking artificial neural networks and U-net convolutional architectures for wildfire susceptibility prediction: Innovations in geospatial intelligence. *IEEE Transactions on Geoscience and Remote Sensing, 63*, 1–15. https://doi.org/10.1109/TGRS.2025.3529134
Singh, H., Ang, L.-M., & Srivastava, S. K. (2025c). Active wildfire detection via satellite imagery and machine learning: An empirical investigation of Australian wildfires. *Natural Hazards, 121*, 9777–9800. https://doi.org/10.1007/s11069-025-07163-w
Srivastava, P. K., & Rawat, R. (2023). Retrieval of leaf chlorophyll content using drone imagery and fusion with Sentinel-2 data. *Smart Agricultural Technology, 6*, 100353.
Suchenwirth, L., Förster, M., Cierjacks, A., Lang, F., & Kleinschmit, B. (2012). Knowledge-based classification of remote sensing data for the estimation of below-and above-ground organic carbon stocks in riparian forests. *Wetlands Ecology and Management, 20*, 151–163.

Vaiphasa, C., Ongsomwang, S., Vaiphasa, T., & Skidmore, A. K. (2005). Tropical mangrove species discrimination using hyperspectral data: A laboratory study. *Estuarine, Coastal and Shelf Science, 65*, 371–379.

Velasco-Forero, S., & Manian, V. (2009). Improving hyperspectral image classification using spatial preprocessing. *IEEE Geoscience and Remote Sensing Letters, 6*, 297–301.

Vidal, M., & Amigo, J. M. (2012). Pre-processing of hyperspectral images. Essential steps before image analysis. *Chemometrics and Intelligent Laboratory Systems, 117*, 138–148.

Vidhya, R., Vijayasekaran, D., Ahamed Farook, M., Jai, S., Rohini, M., & Sinduja, A. (2014). Improved classification of mangroves health status using hyperspectral remote sensing data. *The International Archives of the Photogrammetry, Remote Sensing and Spatial Information, 40*, 667–670.

Wang, D., Qiu, P., Wan, B., Cao, Z., & Zhang, Q. (2022). Mapping α-and β-diversity of mangrove forests with multispectral and hyperspectral images. *Remote Sensing of Environment, 275*, 113021.

Weih, R. C., & Riggan, N. D. (2010). Object-based classification vs. pixel-based classification: Comparative importance of multi-resolution imagery. *The International Archives of the Photogrammetry, Remote Sensing and Spatial Information Sciences, 38*, C7.

Wong, F. K., & Fung, T. (2014). Combining eo-1 hyperion and envisat asar data for mangrove species classification in Mai Po Ramsar Site, Hong Kong. *International Journal of Remote Sensing, 35*, 7828–7856.

Yang, G., Huang, K., Sun, W., Meng, X., Mao, D., & Ge, Y. (2022). Enhanced mangrove vegetation index based on hyperspectral images for mapping mangrove. *ISPRS Journal of Photogrammetry and Remote Sensing, 189*, 236–254.

Zhang, C., Kovacs, J. M., Liu, Y., Flores-Verdugo, F., & Flores-de-Santiago, F. (2014). Separating mangrove species and conditions using laboratory hyperspectral data: A case study of a degraded mangrove forest of the Mexican Pacific. *Remote Sensing, 6*, 11673–11688.

Web Sources

EnMAP HSI Instrument Specification. https://www.enmap.org/data/doc/EnMAP_Specs.pdf

NOAA. (2024). *What is a mangrove forest?* National Ocean Service website. Retrieved June 16, 2024, from https://oceanservice.noaa.gov/facts/mangroves.html

Pixxel Hyperspectral Imagery. https://www.pixxel.space/hyperspectral-imagery

A Comparative Assessment of Global Land Cover Products and Mangrove Specific Datasets with High Spatial Resolution Imagery, with Emphasis on Mangrove Mapping

Mostafa Mahdavifard and Sadra Karimzadeh

Abstract Mangrove forests are vital and inseparable components of coastal ecosystems, primarily thriving in tropical regions. This study conducts a comparative assessment of a dedicated mangrove product, land cover products, spectral indices, and the Support Vector Machine (SVM) algorithm, with a particular focus on the identification of the mangrove class. For this purpose, 10-m spatial resolution products including High-resolution Global Mangrove Forests (HGMF), ESA WorldCover, ESRI Landcover, the Mangrove Vegetation Index (MVI), and the SVM classification algorithm were applied to two study areas: northwestern Australia and southern Iran (Sirik region). The results indicate that the HGMF and ESRI products in the Sirik region yielded lower Kappa coefficients (0.59 and 0.32, respectively) compared to their counterparts in Australia (0.93 and 0.61, respectively). In contrast, the ESA product, along with the remote sensing derivatives SVM and MVI, achieved higher Kappa values in Sirik (0.75, 0.92, and 0.82, respectively). Meanwhile, in northwestern Australia, the same remote sensing derivatives reached Kappa values of 0.86, 0.98, and 0.73, demonstrating significantly higher accuracy. In general, the findings suggest that these products exhibit greater sensitivity to extensive and close mangrove formations. The high density of such mangroves in the Australian study area contributed to higher classification accuracy. In contrast, the presence of both dense and sparse (open) mangroves in the Iranian region reduced classification performance, as sparse mangroves were less accurately

M. Mahdavifard
Remote Sensing Laboratory, Department of Remote Sensing and GIS, University of Tabriz, Tabriz, Iran

Department of Geographical Sciences, University of Hormozgan, Bandar Abbas, Iran

S. Karimzadeh (✉)
Remote Sensing Laboratory, Department of Remote Sensing and GIS, University of Tabriz, Tabriz, Iran

Department of Architecture and Building Engineering, Institute of Science Tokyo, Tokyo, Japan
e-mail: karimzadeh.s.aa@m.titech.ac.jp; sa.karimzadeh@tabrizu.ac.ir

S. K. Srivastava et al. (eds.), *Earth Observation for Monitoring Mangrove Ecosystems*, https://doi.org/10.1007/978-3-032-16668-5_9

detected by the products. On the other hand, machine learning algorithms effectively identified both close and open mangroves, likely due to the supervised nature of their classification. However, the MVI index caused pixel mixing in the study areas, resulting in the misclassification of some non-mangrove regions as mangrove areas.

Keywords Mangrove mapping · Products · HGMF · ESA Worldcover · ESRI Landcover · Machine learning · MVI · Remote sensing

1 Introduction

Forests are considered ecosystems or plant communities that support unique vegetation assemblages (Chazdon et al., 2016). Mangrove forests are tropical and subtropical ecosystems that grow along the interface between land and sea (Saleh, 2007; Atazadeh & Mahdavifard, 2021). These forests provide a wide range of essential functions, including mitigating the impacts of tropical storms and tsunamis, reducing coastal erosion, and storing large amounts of blue carbon (Brander et al., 2012; Pham et al., 2020). In addition, mangrove forests serve as critical habitats for fishery resources, functioning as nursery grounds, food sources, and breeding areas (Carrasquilla-Henao et al., 2019).

Mangroves are an integral part of coastal ecosystems, which are vital in protecting shorelines from extreme weather events and erosion. Despite their resilience to adverse anthropogenic and environmental impacts (Raihan et al., 2024), mangroves face significant threats such as climate change and unsustainable exploitation, which may lead to widespread degradation (Xu et al., 2024). Projections suggest that the loss of mangrove habitats due to the dual stressors of sea-level rise (SLR) and land-use change could range from 16.6 to 56.2% (Liang et al., 2023). Therefore, monitoring the spatial distribution of mangrove forests is crucial for natural resource agencies managing these ecosystems (Giri, 2023), as observing and analyzing their spatial patterns and dynamics is a fundamental step toward understanding the impacts of environmental and human-induced changes and developing effective management strategies (Cárdenas et al., 2017).

Analyzing the spatial and temporal patterns of mangroves enables responsible organizations and managers to plan effectively for mangrove conservation and development by selecting appropriate adaptive options and solutions (Feyzolahpour et al., 2023). In such remote areas, remote sensing technology can be of particular interest to mangrove conservation managers, as it allows for the monitoring and measurement of a wide range of mangrove forest characteristics at various spatial and temporal scales (Rondon et al., 2023). However, generating accurate maps using remote sensing data is a complex and specialized task that requires skill and expertise. For many non-specialist managers or researchers, directly processing these data may be challenging. Therefore, a variety of land use/land cover and forest map products with different spatial resolutions have been developed at national and global scales, which, by relying on remote sensing technology, have made access to land cover information more straightforward and have partially met research and management needs (Hailu et al., 2018). These land cover (LC) products provide essential data for a wide range of environmental studies at regional and global scales, including research on biodiversity, climate, and human health (Grekousis et al., 2015). Although land cover and forest products are generated at various scales, all remote sensing-derived products used for scientific and policy-making purposes require quantitative accuracy assessments to enhance confidence in the generated information (Olofsson et al., 2014; Nicolau et al., 2023). Therefore, it is essential to evaluate the accuracy of different remote sensing products to identify and quantify classification errors in the imagery (Zhu et al., 2023; Yi et al., 2024).

In this context, several studies have been conducted in the past. For example, Ximenes et al. (2023) compared four global mangrove datasets with a spatial resolution of 30 m and underscored the significance of precise delineation of mangrove extents in predicting their responses to climate change. Jia et al. (2023) introduced the HGMF 2020 dataset as the first global mangrove product at a 10-m resolution. With an overall accuracy of 95.2%, this dataset demonstrated superior capability in detecting small mangrove patches along coastal zones compared to the 30-m GMW product. Xiao et al. (2024) evaluated two global mangrove maps from 2020 GMW and LREIS, reporting that LREIS, with an accuracy of 83.2%, outperformed GMW, which had an accuracy of 63.4%. Liu et al. (2023) assessed the accuracy of six remote sensing products with different spatial resolutions for forest mapping in Southeast Asia and recommended the ESRI land cover product as a reliable option for monitoring mountainous and dense forest regions. Furthermore, Venter et al. (2022), the developers of the ESRI Global Land Cover dataset, found that tree cover, especially in northern forest biomes, was one of the most accurately classified land cover types, achieving an overall accuracy of 81%. In a comparative study, Ji et al. (2024) found that the forest class in the ESA WorldCover product exhibited user and producer accuracies exceeding 85%, highlighting the potential of high-resolution products in global forest monitoring.

As mentioned in previous studies, although limited research has been conducted on global land cover and mangrove datasets, a comprehensive review focusing specifically on dedicated mangrove products and high-resolution land cover products, especially for mangrove forests, has yet to be undertaken. Therefore, the main objective of this study is to evaluate the accuracy of various land cover products, dedicated mangrove products, mangrove-related indices, and the application of machine learning algorithms on high-resolution imagery. The primary focus of this study is on monitoring and analyzing mangrove forests along the southern coasts of Iran and the northwestern coasts of Australia. Furthermore, this study compares the different derivatives derived from remote sensing data in extracting mangrove classes and aims to critically analyze the advantages and disadvantages of each of these methods in the process of mangrove identification and classification.

2 Study Area

In this study, two mangrove regions with different environmental conditions were selected as study areas: one in the northwest of Australia and the other in southern Iran (Fig. 1). The Iranian mangrove area is located in Sirik County, with geographic coordinates ranging from 26.15° to 26.25° North latitude and 57.00° to 57.10° East longitude. This area is home to two mangrove species, *Rhizophora* sp. and *Avicennia marina*, with *Avicennia marina* being the dominant species. The climate in this region is hot and humid, with an average annual temperature of around 29 °C. The

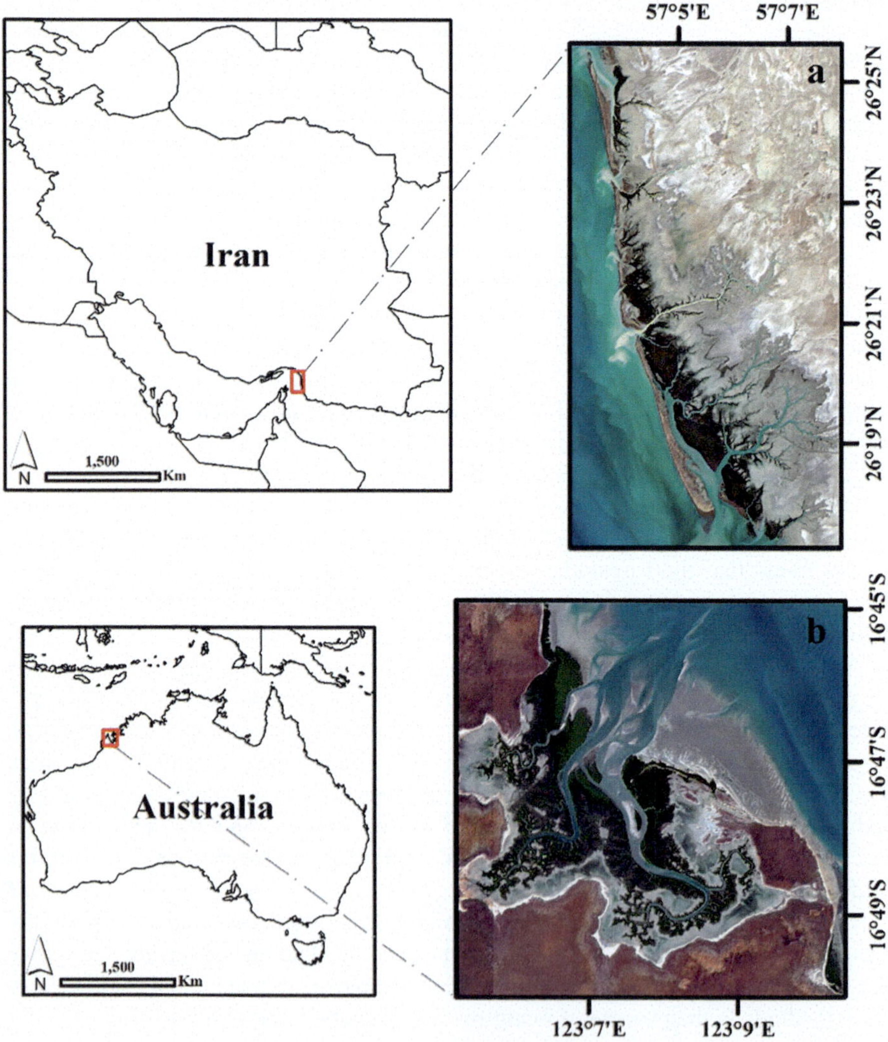

Fig. 1 Geographical location of the spatial distribution of mangroves within the study areas

second region is located in the northwest of Australia, near the Maddarr indigenous community, between the geographic coordinates of 16.45° to 16.50° South latitude and 123.05° to 123.10° East longitude. This area is situated in a tropical zone and is part of the Kimberley region, with an average annual temperature of approximately 27 °C, making it one of the hottest regions in Australia. The mangroves in this region have high biodiversity, including 41 different species. The dominant species is *Avicenna marina*, although resilient species such as *Nypa fruticans* and *Acrostichum speciosum* are also found in the area (Clancy, 2008).

3 Materials and Methods

In this study, the selected mangrove forest products include three land cover products: ESA WorldCover, ESRI LandCover, and the dedicated mangrove product HGMF. In addition to using these products, we utilized the time-series average images of the 2020–2021 period from Sentinel-2 L2A (BOA) satellite, which have undergone atmospheric and geometric correction (Karimzadeh et al., 2023), with a cloud cover percentage of 5% to implement the MVI index and perform classification using the SVM method. It is noteworthy that the main reason for choosing the specified time range in this study is the use of the HGMF product, as it was specifically designed and produced for monitoring mangrove forests in 2020. The general specifications of the products used are summarized in Table 1.

All of these products have a spatial resolution of 10 m and are considered high-resolution products extracted from Sentinel-2 images, which are freely available to users. The proposed method in this study is illustrated in Fig. 2. In this research, all processing tasks and land cover product extraction were carried out on satellite

Table 1 Characteristics of remote sensing derivatives used in the study

Products	Mangrove definition	Spatial resolution	Data source	Algorithm	Product accuracy
HGMF 2020	Mangroves are coastal trees that deliver vital ecosystem goods and services and significantly contribute to sustainable development	10 m	Sentinel-2	Object based and random forest	95.2%
ESA 2020	Woody vegetation communities that thrive in tidal zones of tropical and subtropical regions, especially within estuaries and river deltas	10 m	Sentinel-1 and Sentinel-2	Decision tree and machine learning	75.0%
ESRI 2020	Dense vegetation with a closed canopy, typically intermixed with water throughout the year, and approximately 15 feet or taller	10 m	Sentinel-2	Deep learning	85%

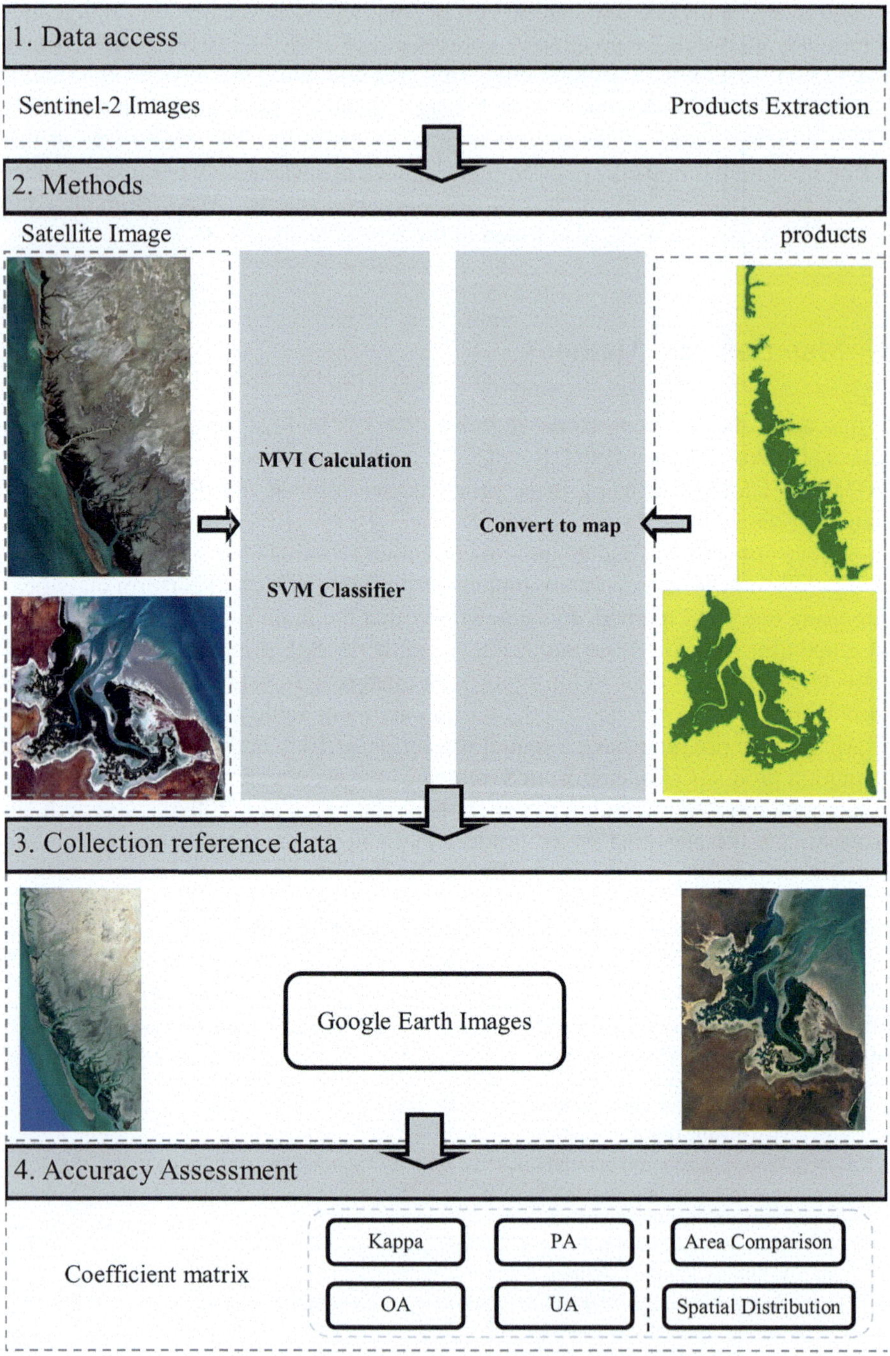

Fig. 2 Workflow of the methods for validation of remote sensing derivatives

images and the land cover products on the GEE platform. The entire global land cover products from ESA and ESRI are comprehensively and fully available on GEE, and access to them is free of charge.

3.1 HGMF Product

The HGMF dataset (Jia et al., 2023) is a global, dedicated mangrove forest map with a high spatial resolution of 10 m, produced based on Sentinel-2 images from 2020. This high-quality product is generated using object-based image analysis (OBIA) methods.

3.2 ESA Product

The land cover product provided by the European Space Agency (Zanaga et al., 2023) was developed within the framework of one of the agency's projects (access: https://worldcover2020.esa.int/). This product is based on 10-m spatial resolution data derived from Sentinel-1 radar images and Sentinel-2 optical images. The land cover map includes 11 different classes, with the "Mangrove" class identified by code 95. Mangroves are salt-tolerant trees that primarily grow in tidal coastal areas.

3.3 ESRI Product

The ESRI land cover product is a global land cover map (access: https://livingatlas.arcgis.com/landcover/). This product is created solely from 10-m optical Sentinel-2 images using deep learning techniques. It includes nine different classes. In this product, the mangrove class is not explicitly defined but is considered a subset of the Flooded Vegetation class (code 3) and the Trees class (code 2). These classes generally encompass areas such as seasonal floodplain grasslands, trees, dense vegetation, rice paddies, swamps, and mangrove forests (Venter et al., 2022).

3.4 MVI Index

This index is designed using the green, near-infrared (NIR), and short-wave infrared (SWIR) bands from Landsat-8 and Sentinel-2 satellite images. It takes into account the specific physiological characteristics of mangroves in coastal areas to identify significant differences in chlorophyll content between mangrove forests and other typical vegetation types (Baloloy et al., 2020).

3.5 SVM Classifier

In previous studies, the Support Vector Machine (SVM) classifier has been specifically used for mangrove mapping in Sentinel-2 multispectral images (Soffianian et al., 2023; Mahdavifard et al., 2023). SVM is one of the most common binary classification models based on machine learning, and it has been extended for multi-class classification (Jiang et al., 2021). This algorithm is also used for analyzing images and interpreting vegetation cover patterns (Lemenkova, 2024; Puertas et al., 2013). In this study, after extracting the classes, we classified the two classes: mangrove and non-mangrove using the SVM algorithm.

4 Accuracy Assessment

Maps derived from remote sensing data always contain pixels that are misclassified, so accuracy assessment is an essential step in any remote sensing process (Pour et al., 2023). In this study, we used high-resolution Google Earth imagery to collect random reference points in two classes: mangrove and non-mangrove, in the study areas. After collecting the points from Google Earth, an error matrix was used for the final evaluation of the classes. The error matrix is a quantitative method for describing the accuracy and performance of the classifier, which includes the Kappa coefficient, overall accuracy, producer's accuracy, and user's accuracy.

5 Results

In this section, we discuss the findings of our study aimed at evaluating and comparing remote sensing-derived products, including global mangrove datasets, land cover maps, the Mangrove Vegetation Index (MVI), and the Support Vector Machine (SVM) algorithm, with a particular emphasis on the mangrove class.

5.1 Validation of Remote Sensing-Derived Products

Table 2 presents the validation results of three global products, the MVI index, and the SVM algorithm across two study areas. The results include overall accuracy, user accuracy, and producer accuracy for the mangrove class. Overall, most

Table 2 Validation results of 5 remote sensing-based products

Sirik						
Accuracy (%)	Class	SVM	MVI	HGMF	ESA	ESRI
UA	Mangrove	94.44	91.18	91.49	89.23	79.41
	Non-mangrove	97.75	91.40	76.32	87.50	66.14
PA	Mangrove	97.14	88.57	61.43	82.86	38.57
	Non-mangrove	95.60	93.41	95.60	92.31	92.31
OA		96.27	91.30	80.74	88.19	68.94
KC		0.92	0.82	0.59	0.75	0.32

Australia						
Accuracy (%)	Class	SVM	MVI	HGMF	ESA	ESRI
UA	Mangrove	98.61	83.33	95.83	88.61	75.61
	Non-mangrove	100	91.55	97.44	98.59	86.76
PA	Mangrove	100	91.55	97.18	98.59	87.32
	Non-mangrove	98.73	82.28	96.20	88.61	74.68
OA		99.33	86.66	96.66	93.33	80.66
KC		0.98	0.73	0.93	0.86	0.61

products, indices, and algorithms—except one—achieved high overall accuracy in both regions, with values of 80 and 86%. In the Sirik region, the ESRI product showed the lowest overall accuracy at 68.94%. In contrast, the European Space Agency (ESA), High-resolution Global Mangrove Forests (HGMF), MVI index, and SVM algorithm achieved overall accuracies of 88.19%, 80.74%, 91.30%, and 96.27%, respectively. In the same region, the mangrove class specifically achieved the highest producer accuracy with SVM, MVI, and ESA, each exceeding 82%, whereas ESRI and HGMF recorded lower producer accuracies of 38.57% and 61.43%, respectively. Similarly, in the mangrove area of northwestern Australia, the producer accuracy of the SVM algorithm for the mangrove class was estimated at 100%. Other remote sensing-derived products also showed high accuracies above 87% in this region: MVI at 91.55%, HGMF at 97.18%, ESA at 98.59%, and ESRI at 87.32%. Notably, there was a significant difference in producer accuracy between the two regions. In Iran's Sirik region, ESRI and HGMF showed the lowest producer accuracies at 38.57% and 61.43%, respectively, whereas the same products performed much better in northwestern Australia, with accuracies of 87.32% and 97.18%.

5.2 *Comparison of Mangrove Class Area*

The area of mangrove forests across all products, indices, and classification algorithms was calculated and estimated in two mangrove regions (Figs. 3 and 4). In both the Sirik region in Iran and northwest Australia, significant differences and inconsistencies were observed in the area estimates across products. For instance,

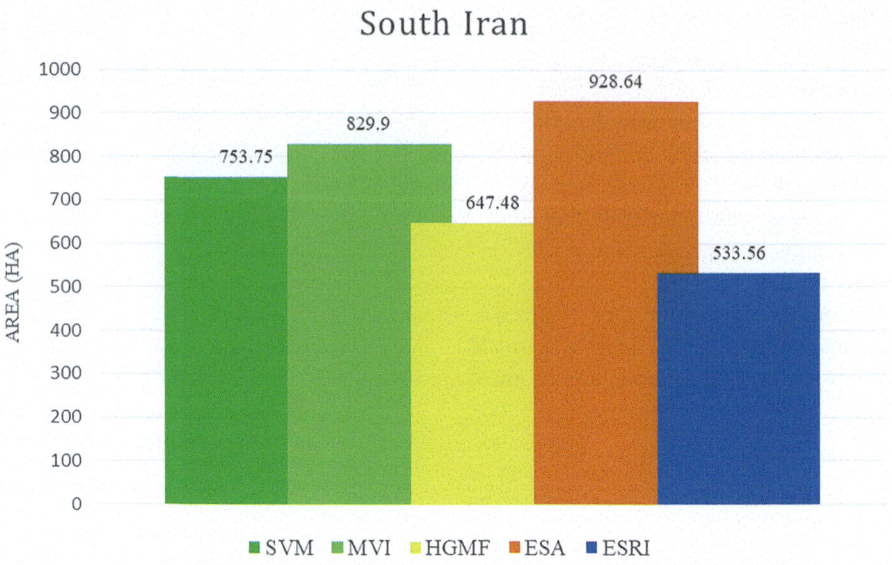

Fig. 3 Mangrove forest area of Sirik, Iran

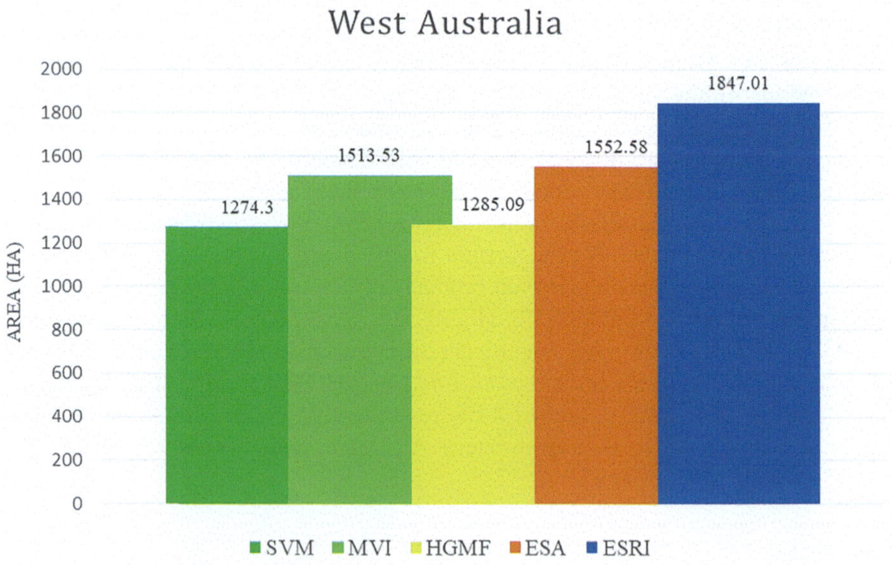

Fig. 4 Mangrove forest area of northwest Australia

ESRI and HGMF estimated the lowest mangrove coverage area in the Sirik region, which was below the natural threshold, while SVM, MVI, and ESA showed relatively smaller differences. In the northwest of Australia, the SVM algorithm effectively extracted mangrove coverage and provided a reasonably accurate estimate of the area compared to other remote sensing derivatives. However, the ESRI product showed the largest coverage area for mangroves, which appeared to be beyond the actual extent, indicating an overestimate. Overall, the indices MVI, HGMF, and the ESA product provided relatively similar estimates of the mangrove area. It is noteworthy that there was only about an 18% discrepancy in mangrove area estimates between ESA and MVI compared to HGMF.

5.3 Comparison of Mangrove Spatial Distribution

Figure 5 shows the level of spatial variation and alignment of mangroves across products, indices, and classification algorithms in the Sirik region of Iran and northwest Australia. Overall, in the Sirik region of Iran, three maps derived from remote sensing data, including SVM, MVI, and to some extent ESA, show relatively good spatial consistency with each other, while the HGMF and ESRI products have significantly lower spatial alignment. In the northwest Australia region, which has the largest mangrove coverage, considerable spatial consistency and similarity are observed between the products SVM, ESA, HGMF, and to some extent MVI. However, the ESRI product considers a larger area of mangrove coverage in the region.

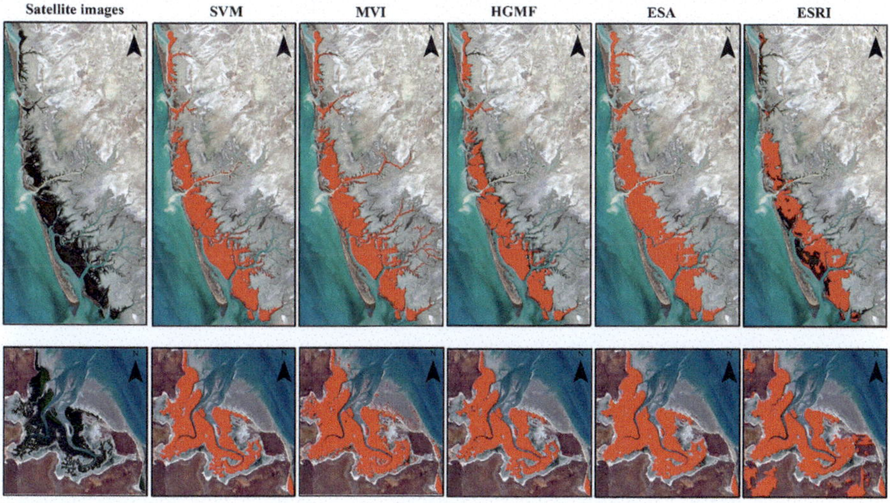

Fig. 5 Comparison of the spatial distribution of mangroves in maps derived from remote sensing data

6 Discussion

In this study, an attempt was made to briefly and concisely evaluate the latest global mangrove maps with high spatial resolution, as well as the emphasis on indices and classification algorithms for mangrove extraction and monitoring. In general, four products derived from global and regional remote sensing data in the studied areas showed suitable accuracy, with their overall accuracy (OA) exceeding 80%, indicating an acceptable evaluation of the accuracy. Overall, although the ESA product misclassified some mudflats as mangrove forests in certain parts of the study areas, it achieved the highest overall accuracy and Kappa coefficient among other land cover products with a 10-m spatial resolution. This performance aligns with the study by Ji et al. (2024), which stated that the ESA product has the highest accuracy among global land cover products, with the distinction that they focused on land cover. This good performance of the ESA land cover product, compared to ESRI in mangrove extraction, is due to the fact that the ESA global product, in addition to Sentinel-2 optical imagery, also utilizes Sentinel-1 radar data and machine learning and decision tree algorithms for class extraction (in this case, mangrove class), while the ESRI product only used Sentinel-2 optical imagery for class extraction (in this study, the mangrove class). The only advantage observed in this product is the use of deep learning techniques.

In general, three issues can be identified in the ESRI global product for the mangrove class. The first issue is that the mangrove class is not explicitly defined in this product. Additionally, the second problem, as mentioned earlier, is that mangroves are classified as part of the coastal floodplain tree and vegetation class. The third major issue is that part of the mangrove class was incorrectly classified as water, with no way to extract mangrove from this class and assign it to a separate mangrove class. This issue is more evident in the Sirik region. Since mangroves are classified as trees in the ESRI product, and adjacent to mangroves in northwest Australia, there are also non-mangrove forest trees, many of which were incorrectly classified as mangroves. The ESA product, despite its accuracy advantage, also has some drawbacks in mangrove class extraction. Given that the Sirik region has sparse mangrove forests or open mangroves, this product did not perform well in extracting sparse mangroves, though it did extract open mangroves in certain areas. Conversely, this issue was not observed in Australia due to the presence of dense mangrove cover, and the product performed well in extracting mangroves. Another issue observed in the Sirik region of Iran and northwest Australia is that, despite accurately extracting mangrove areas, the natural boundaries of mangroves were not respected, and some terrestrial areas were incorrectly classified as mangroves.

Bunting et al. (2022) emphasized the high accuracy of the GMW product, derived from a combination of Landsat data and the L-band ALOS Palsar satellite, for mangrove monitoring. In our study, the HGMF product, using 10-m Sentinel-2 resolution data and combining RF classification and object-based methods, also showed good accuracy in mangrove extraction without radar data. The HGMF-dedicated mangrove map, with its high accuracy in extracting Australia's dense mangroves, performed well in identifying the boundary of relatively large coastal mangroves, which aligns with the study by Jia et al. (2023), the developers of this global product, who also noted that this product efficiently identifies and extracts mangroves in Australia, Indonesia, and Brazil. In contrast, this product did not perform as well in the protected Sirik area in extracting open mangroves; however, dense mangroves were accurately identified. This result is consistent with the findings of Jia et al. (2023), where the impact of protective conditions and the isolated position of South Asia's mangroves were emphasized, confirming the superior performance of algorithms in extracting denser mangrove covers.

One issue with this product is that some muddy lands within the mangrove area were incorrectly identified as part of the mangrove class. However, overall, the product showed relatively high capability in mangrove identification. The study by Baloloy et al. (2020) indicated that the MVI effectively differentiates mangrove trees from other covers and suggested that mangrove maps extracted from Sentinel-2 perform logically and well compared to Landsat 8. This view slightly contrasts with our study, where mangrove maps produced using Sentinel-2 time series data in the Sirik region of Iran and northwest Australia struggled to differentiate from muddy covers and even secondary water channels. In the present study, the area of mangroves in Sirik, estimated by the SVM algorithm, was approximately 753.75 hectares, which closely matched the estimated mangrove area (773 hectares) in Parvaresh et al. (2011), indicating the high accuracy of the SVM classifier in both study areas. The algorithm was able to effectively distinguish between the classes related to open and dense mangroves, establishing a clear boundary between tidal, muddy, and mangrove areas. This research, by highlighting the differences in mangrove class extraction between mangrove and land cover products, the MVI index, and the SVM machine learning algorithm with 10-m spatial resolution, provides new insights into the advantages and disadvantages of these remote sensing data derivatives with an emphasis on mangroves.

What is clearly evident from the product results is that none of the products showed significant potential for extracting open mangroves, as most of the training data for mangrove mapping worldwide was collected from dense or compact mangrove areas. Overall, our findings emphasize the superiority of certain global products and classification algorithms, which add significant value to mangrove conservation management for organizations and decision-makers.

7 Conclusion

In this study, aiming to accurately evaluate the capabilities of various remote sensing data for mangrove forest extraction, the performance of several global land cover products, dedicated mangrove maps, the MVI index, and the SVM machine learning algorithm was assessed across two mangrove regions in Iran and Australia. The results showed that, overall, the ESA and HGMF products, the MVI index, and the SVM algorithm outperformed the ESRI product in accurately identifying and estimating the extent of mangroves. Among these, the SVM algorithm demonstrated superior performance due to its high accuracy and ability to distinguish between open and dense mangroves, while the MVI index also showed acceptable effectiveness, particularly in Australia. In contrast, the ESRI product performed poorly, especially in the Sirik region of Iran, due to the lack of a dedicated mangrove class, class mixing, and reliance solely on optical data.

At a broader level, the findings of this study indicate that using combined datasets (optical and radar) and advanced classification algorithms such as SVM and RF enhances the accuracy of monitoring sensitive mangrove ecosystems. Furthermore, the differences in product performance across the two study areas highlight the importance of considering local conditions such as canopy density and geomorphological features when selecting suitable tools and ready-to-use products for mangrove monitoring. The insights from this research can serve as a practical guide for environmental organizations and natural resource managers in choosing the most accurate and appropriate monitoring tools for mangrove forests in diverse geographic regions.

References

Atazadeh, E., & Mahdavifard, M. (2021). Application of remote sensing in natural sciences. In *Remote sensing*. IntechOpen.

Baloloy, A. B., Blanco, A. C., Ana, R. R. C. S., & Nadaoka, K. (2020). Development and application of a new mangrove vegetation index (MVI) for rapid and accurate mangrove mapping. *ISPRS Journal of Photogrammetry and Remote Sensing, 166*, 95–117.

Brander, L. M., Wagtendonk, A. J., Hussain, S. S., McVittie, A., Verburg, P. H., de Groot, R. S., & van der Ploeg, S. (2012). Ecosystem service values for mangroves in Southeast Asia: A meta-analysis and value transfer application. *Ecosystem Services, 1*(1), 62–69.

Bunting, P., Rosenqvist, A., Hilarides, L., Lucas, R. M., Thomas, N., Tadono, T., et al. (2022). Global mangrove extent change 1996–2020: Global mangrove watch version 3.0. *Remote Sensing, 14*(15), 3657.

Cárdenas, N. Y., Joyce, K. E., & Maier, S. W. (2017). Monitoring mangrove forests: Are we taking full advantage of technology? *International Journal of Applied Earth Observation and Geoinformation, 63*, 1–14.

Carrasquilla-Henao, M., Ban, N., Rueda, M., & Juanes, F. (2019). The mangrove-fishery relationship: A local ecological knowledge perspective. *Marine Policy, 108*, 103656.

Chazdon, R. L., Brancalion, P. H., Laestadius, L., Bennett-Curry, A., Buckingham, K., Kumar, C., et al. (2016). When is a forest a forest? Forest concepts and definitions in the era of forest and landscape restoration. *Ambio, 45*(5), 538–550.

Clancy, T. (2008). Australian forest profiles: Eucalypt. Bureau of Rural Sciences, Australian Government Department of Agriculture, Fisheries and Forestry.

Feyzolahpour, M., Ghasemlu, H., & Fard, M. M. (2023). The impact of human activities on the mangrove forests of the Qeshm Island, Iran. *AUC Geographica, 58*(1), 96–112.

Giri, C. (2023). Frontiers in global mangrove forest monitoring. *Remote Sensing, 15*(15), 3852.

Grekousis, G., Mountrakis, G., & Kavouras, M. (2015). An overview of 21 global and 43 regional land-cover mapping products. *International Journal of Remote Sensing, 36*(21), 5309–5335.

Hailu, B. T., Fekadu, M., & Nauss, T. (2018). Availability of global and national scale land cover products and their accuracy in mountainous areas of Ethiopia: A review. *Journal of Applied Remote Sensing, 12*(4), 041502.

Ji, X., Han, X., Zhu, X., Huang, Y., Song, Z., Wang, J., et al. (2024). Comparison and validation of multiple medium-and high-resolution land cover products in Southwest China. *Remote Sensing, 16*(6), 1111.

Jia, M., Wang, Z., Mao, D., Ren, C., Song, K., Zhao, C., et al. (2023). Mapping global distribution of mangrove forests at 10-m resolution. *Science Bulletin, 68*(12), 1306–1316.

Jiang, Y., Zhang, L., Yan, M., Qi, J., Fu, T., Fan, S., & Chen, B. (2021). High-resolution mangrove forests classification with machine learning using worldview and UAV hyperspectral data. *Remote Sensing, 13*(8), 1529.

Karimzadeh, S., Kamran, K. V., & Mahdavifard, M. (2023). A multisensor satellite image classification for the detection of mangrove forests in Qeshm Island (Southern Iran). *Applied Geomatics, 15*(1), 177–188.

Lemenkova, P. (2024). Support vector machine algorithm for mapping land cover dynamics in Senegal, west africa, using earth observation data. *Earth, 5*(3), 420–462.

Liang, S., Hu, W., Liu, J., Su, S., Chen, G., Chen, S., et al. (2023). Mapping mangrove sustainability in the face of sea level rise and land use: A case study on Leizhou Peninsula, China. *Journal of Environmental Management, 325*, 116554.

Liu, B., Yang, X., Wang, Z., Ding, Y., Zhang, J., & Meng, D. (2023). A comparison of six forest mapping products in Southeast Asia, aided by field validation data. *Remote Sensing, 15*(18), 4584.

Mahdavifard, M., Ahangar, S. K., Feizizadeh, B., Kamran, K. V., & Karimzadeh, S. (2023). Spatio-Temporal monitoring of Qeshm mangrove forests through machine learning classification of SAR and optical images on Google Earth Engine. *International Journal of Engineering and Geosciences, 8*(3), 239–250.

Nicolau, A. P., Dyson, K., Saah, D., & Clinton, N. (2023). Accuracy assessment: Quantifying classification quality. In *Cloud-based remote sensing with Google earth engine: Fundamentals and applications* (pp. 135–145). Springer International Publishing.

Olofsson, P., Foody, G. M., Herold, M., et al. (2014). Good practices for estimating area and assessing accuracy of land change. *Remote Sensing of Environment, 148*, 42–57.

Parvaresh, H., Abedi, Z., Farshchi, P., Karami, M., Khorasani, N., & Karbassi, A. (2011). Bioavailability and concentration of heavy metals in the sediments and leaves of grey mangrove, Avicennia marina (Forsk.) Vierh, in Sirik Azini Creek, Iran. *Biological Trace Element Research, 143*, 1121–1130.

Pham, T. D., Le, N. N., Ha, N. T., Nguyen, L. V., Xia, J., Yokoya, N., et al. (2020). Estimating mangrove above-ground biomass using extreme gradient boosting decision trees algorithm with fused sentinel-2 and ALOS-2 PALSAR-2 data in can Gio biosphere reserve, Vietnam. *Remote Sensing, 12*(5), 777.

Pour, A. B., Ranjbar, H., Sekandari, M., Abd El-Wahed, M., Hossain, M. S., Hashim, M., et al. (2023). Remote sensing for mineral exploration. In *Geospatial analysis applied to mineral exploration* (pp. 17–149). Elsevier.

Puertas, O. L., Brenning, A., & Meza, F. J. (2013). Balancing misclassification errors of land cover classification maps using support vector machines and Landsat imagery in the Maipo river basin (Central Chile, 1975–2010). *Remote Sensing of Environment, 137*, 112–123.

Raihan, A., Ali, T., & Mortula, M. (2024). Unveiling coastal ecosystem dependencies: Multifactorial analysis of climate change influencing mangrove system in the United Arab Emirates over the period 1990–2020. *Discover Sustainability, 5*(1), 498.

Rondon, M., Ewane, E. B., Abdullah, M. M., Watt, M. S., Blanton, A., Abulibdeh, A., et al. (2023). Remote sensing-based assessment of mangrove ecosystems in the Gulf Cooperation Council countries: A systematic review. *Frontiers in Marine Science, 10*, 1241928.

Saleh, M. A. (2007). Assessment of mangrove vegetation on Abu Minqar Island of the Red Sea. *Journal of Arid Environments, 68*(2), 331–336.

Soffianian, A. R., Toosi, N. B., Asgarian, A., Regnauld, H., Fakheran, S., & Waser, L. T. (2023). Evaluating resampled and fused Sentinel-2 data and machine-learning algorithms for mangrove mapping in the northern coast of Qeshm Island, Iran. *Nature Conservation, 52*, 1–22.
Venter, Z. S., Barton, D. N., Chakraborty, T., Simensen, T., & Singh, G. (2022). Global 10 m land use land cover datasets: A comparison of dynamic world, world cover and esri land cover. *Remote Sensing, 14*(16), 4101.
Xiao, Z., Jiang, W., Wu, Z., Ling, Z., Deng, Y., Zhang, Z., & Peng, K. (2024). Agreement analysis and accuracy assessment of multiple mangrove datasets in Guangxi Beibu Gulf and Guangdong-Hong Kong-Macau Greater Bay, China, for 2000–2020. *IEEE Journal of Selected Topics in Applied Earth Observations and Remote Sensing, 17*, 3438–3451.
Ximenes, A. C., Cavanaugh, K. C., Arvor, D., Murdiyarso, D., Thomas, N., Arcoverde, G. F., et al. (2023). A comparison of global mangrove maps: Assessing spatial and bioclimatic discrepancies at poleward range limits. *Science of the Total Environment, 860*, 160380.
Xu, X., Fu, D., Su, F., Lyne, V., Yu, H., Tang, J., et al. (2024). Global distribution and decline of mangrove coastal protection extends far beyond area loss. *Nature Communications, 15*(1), 10267.
Yi, C., Zhao, X., Sun, Q., & Wang, Z. (2024). Assessing the accuracy of remote sensing data products: A multi-granular spatial sampling method. *Future Generation Computer Systems, 159*, 151–160.
Zanaga, D., Van De Kerchove, R., De Keersmaecker, W., Souverijns, N., Brockmann, C., Quast, R., Wevers, J., Grosu, A., Paccini, A., Vergnaud, S., et al. (2023). *ESA2020 WorldCover 10 m 2020 v100. 2021*. Retrieved September 14, 2023, from https://zenodo.org/record/5571936
Zhu, L., Wang, H., Zhao, T., Li, W., Li, Y., Tong, C., et al. (2023). Disaggregation of remote sensing and model-based data for 1 km daily seamless soil moisture. *International Journal of Applied Earth Observation and Geoinformation, 125*, 103572.

Multi-temporal Data Usage for Aboveground Biomass Estimation in Bhitarkanika Mangrove Forests, Eastern India

Sujit M. Ghosh, Mukunda Dev Behera, and Bimal K. Bhattyacharya

Abstract Measuring biophysical indicators of carbon storage by forest vegetation, e.g., tree canopy height, growing stock volume, and aboveground biomass (AGB) is useful in understanding the carbon balance in terrestrial environment. Estimating the carbon content of forests accurately provides critical inputs for United Nations Framework Convention on Climate Change (UNFCCC), as this carbon content estimate forms the basis of reporting for initiatives developed under the UNFCCC, such as Clean Development Mechanisms (CDM) and the Reducing Emissions from Deforestation and forest Degradation (REDD) in developing countries. Forest biomass can be estimated in mainly two ways, field-based measurements and remotely sensed methods. Despite its exceptional accuracy, it is not always feasible to measure biomass based on field-based methods only. Conducting field measurements over a regional scale demands significant time and resources and is often impractical in inaccessible locations. Remote sensing, however, offers an efficient approach for assessing and monitoring biomass at broader spatial scales. This study presents how accurately the AGB of a mangrove forest can be estimated, using C band SAR backscatter alongside with textural parameters and optical data-derived vegetation indices. This study demonstrated that machine learning methods can establish a complex non-linear relationship between the predictor and predicted variables and thereby offers improved accuracy in AGB estimates.

Keywords Aboveground biomass · Reducing emissions from deforestation and forest degradation · Synthetic Aperture Radar · Biophysical variables

S. M. Ghosh · M. D. Behera (✉)
Centre for Ocean, River, Atmosphere and Land Sciences (CORAL), Indian Institute of Technology Kharagpur, Kharagpur, India
e-mail: mdbehera@coral.iitkgp.ac.in

B. K. Bhattyacharya
Space Applications Centre, ISRO, Ahmedabad, India

S. K. Srivastava et al. (eds.), *Earth Observation for Monitoring Mangrove Ecosystems*, https://doi.org/10.1007/978-3-032-16668-5_10

1 Introduction

The tropical forests are expected to be affected due to rapid climate changes over the next decades (Biswas et al., 2007; Ravindranath et al., 2006). Approximately one fifth of the global carbon stock is stored in tropical forests, (Donato et al., 2011; Hunter et al., 2013), while tropical forests are understudied in comparison to temperate or alpine forests (FSI, 1996, Vaglio Laurin et al., 2014). The appraised global coverage of mangrove forest is 137,760 km^2 (Giri et al., 2014; Marchio et al., 2016); this equates to 0.1% of the earth's surface (Cornforth et al., 2013). These mangroves are declining worldwide at an alarming rate (Datta & Deb, 2012). Globally, 36,000 km^2 of mangroves have been lost since 1980, primarily due to conversion to agriculture (Chitale & Behera, 2014) and aquaculture, urbanization, and timber extraction (Asbridge et al., 2016). Forest biomass can be estimated using field-based measurements and remotely sensed methods (Connette et al., 2016; Su et al., 2016; Ali et al., 2015; Babcock et al., 2015; KaaSalainen et al., 2015). In this study we tried to find out how accurately the aboveground biomass (AGB) of an Indian tropical forest can be estimated, using C band Synthetic Aperture Radar (SAR) backscatter alongside with textural parameters and optical data derived vegetation indices. Remote sensing systems like Radio Detecting and Ranging (RaDAR) and Light Detecting and Ranging (LiDAR) have the capability to measure the height dimension of forest structure (Duncanson et al., 2010). Though both the systems are sensitive to the vegetation structure and aboveground biomass, Synthetic Aperture Radar (SAR) offers wall-to-wall coverage unlike LiDAR systems (Su et al., 2016). Several works (Cartus et al., 2012; Carreiras et al., 2012; Deng et al., 2014) have shown the ability of SAR images in estimating forest AGB.

SAR data are acquired generally in X, C, L bands and sometimes in S and P bands also. The longer-wavelength L and P band SAR backscatter (Sun et al., 2011) is more related to the biophysical parameters like AGB as it penetrates through the surface layers of vegetation and is scattered by the trunk and main branches (Ghasemi et al., 2011). The number of long-term spaceborne L band missions is only three, Japanese Earth Resources Satellite 1 (JERS-1), Advanced Land Observing Satellite (ALOS), and ALOS-2. And there are huge time gaps in between them. So for preparing a time series biomass map, we can't rely only on L band SAR data. Therefore smaller band, i.e., X and C band SAR data must be utilized in proper way so that it can replicate almost the same result as L band data. Because of their own attributes in rendering forest stand structures, combining radar and optical sensor data has the potential to improve the estimation of forest AGB (Zhao et al., 2016; Pandey et al., 2019a). Several earlier studies (Maynard et al., 2007; Gasparri et al., 2010; Das & Singh, 2016; Günlü et al., 2014) showed that vegetation indices (VIs) obtained from transformation of optical bands reflectance are very useful for correlating biomass with optical data. Kuplich et al. (2003) found that for a tropical forest, SAR image texture is strongly correlated with its biomass. Texture can be described as the spatial distribution of grey levels of an image represented by a set of statistical measurements (Kuplich et al., 2005). Sarker et al. (2013) in their study showed that rather than using only backscatter values, if texture parameters are used as predictor variable, then the biomass estimation accuracy could be improved drastically, even for dense forests. So in our study, to improve the biomass estimation accuracy, apart from using only SAR backscatter as predictor variable, VIs from optical data and SAR texture images are also used in synergy (Haralick et al., 1973; Ouchi, 2013).

The retrieval of biophysical parameters, like biomass, using remotely sensed imagery has usually been done by the derivation of some standard relationships between several associated datasets (Malhi et al., 2022). This technique depends on a set of field measured referencc samples of the desired target variable related with the corresponding remote sensor measurements. Statistical regression methods along with parametric (linear, logarithmic, or polynomial) functions are usually applied to derive any relationship between measurements from earth observation and field (Ali et al., 2015). This forms the basis of extending to larger area on the basis of satellite data (Heiskanen, 2006). This kind of relationship is extremely interesting due to their agility. Besides, they represent complex physical phenomena to a fairly simpler level, which can be easily addressed by people without a certain knowledge in the field (Ali et al., 2015). Despite having its fair share of positive attributes, traditional parametric function-based regression will not be reliable for the high complexity and non-linearity of retrieval problems.

Nonparametric machine learning techniques are powerful non-linear regression methods that can be used as an alternative to traditional parametric methods to handle complex and non-linear problems (Gwal et al., 2026) . Machine learning algorithms have mainly two advantages, first is advanced learning approach, which results in establishing complex non-linear relationships using information held in a set of reference sample (Ali et al., 2015). Another advantage is the fact that non-linear machine learning methods are distribution-free, i.e., no assumptions are made about the data distribution. As a result, data coming from different sources with poorly defined (or unknown) probability density functions and relating well to the target variable can be integrated in the retrieval process (Ali et al., 2015). Joshi and Ghose (2014) found that in general the forest was dominated by small sized trees; only 2.7% of the trees exceeded 10 cm dbh and 2.6% exceeded 6 m height. Tree density ranged from 4723 trees ha^{-1} to 23,751 trees ha^{-1}. Aboveground biomass was low and ranged from 8.9 t ha^{-1} to 50.9 t ha^{-1} in different communities, depending on the structural characteristics. Hong et al. (2015) applied an object-oriented image analysis approach to classify vegetation types of a subtropical wetland system with the decomposition results of quad-pol TerraSAR-X data. Behera et al. (2015) investigated the potential use of a one time, Quad-pol Advanced Land Observing Satellite (ALOS) Phased Array L-band Synthetic Aperture Radar (PALSAR) 25 m data to estimate aboveground biomass (AGB) of a tropical forest using a water cloud model (WCM). Ali et al. (2015) done a review of advanced machine learning methods and concluded that machine learning methods have shown their versatility in different contexts by using optical and radar data, by fusing remotely-sensed data with ground data, as well as exploiting data derived from a UAV platform (Suman et al., 2026).

Using data from ALOS PALSAR, Carreiras et al. (2012) compared a semi-empirical and machine learning algorithm, with the latter based on bagging stochastic gradient boosting (BagSGB), for retrieving the AGB of tropical savanna woodland vegetation. They have shown that the BagSGB outperformed the semi-empirical algorithm. Dube et al. (2014) tested two machine learning algorithms, stochastic gradient boosting (SGB) and random forest (RF) regression trees to predict intra- and interspecies biomass using high resolution RapidEye reflectance bands as well as the derived vegetation indices in a commercial plantation (Hopkinson et al., 2016). The results showed that the SGB algorithm yielded the best performance for intra- and interspecies biomass prediction, using all the predictor variables as well as based on the most important selected variables. This study finds that the combination of PALSAR and WorldView-2 data is a promising approach to improve biomass estimation.

Biomass inventory data is limited from tropical ecosystems due to operational limitations and (Houghton et al., 2009). Therefore, there is a need for development of new methodologies using latest dataset to reduce the ambiguity between field-measured values and remote estimates. There is limited study about the usefulness of multi-sensor data merging biomass estimation of mangrove forests (Pandey et al., 2019b).

2 Study Area

Bhitarkanika Wildlife Sanctuary (BWS) is sprawled over an area of 672 sq. Km in Kendrapara district of Odisha, India (Reddy et al., 2002). Mangroves, consisting of 58 species (Reddy et al., 2006), cover an area of 130 sq. Km (Mishra et al., 2005) of BWS. BWS receives a high average annual rainfall of about 1642 mm, and most of it occurs from June to October (Raha et al., 2012). This area has a typical warm and humid tropical climate with temperature reaches a maximum in May and lowest in January (Pattanaik et al., 2008). The mangroves area of BWS is surrounded by 81 villages. As a result, anthropogenic activities pose a severe threat to this region.

2.1 Methodology

As the first step, field sampling survey was conducted in November–December 2018. Circumference at breast height (CBH) of each tree was measured for 20 m × 20 m plots. It was observed that mainly four species dominates the inland mangrove forest. They are *Heritiera fomes*, *Excoecaria agallocha*, *Avicennia officinalis*, and *Cynometra iripa*. Among these top canopy is formed by first three only. *Cynometra iripa* generally remains understory. A total of 140 plots were generated. Barring a few, all of them were a mixture of different species, and almost all the mixed plots have a dominant species. Location of the field plots is given in Fig. 1. CBH was converted to DBH and country, and species-specific allometric equations developed by Komiyama et al. (2005) were used to estimate the wood volume. This volume is converted to AGB using specific gravity information compiled by Chave et al. (2009).

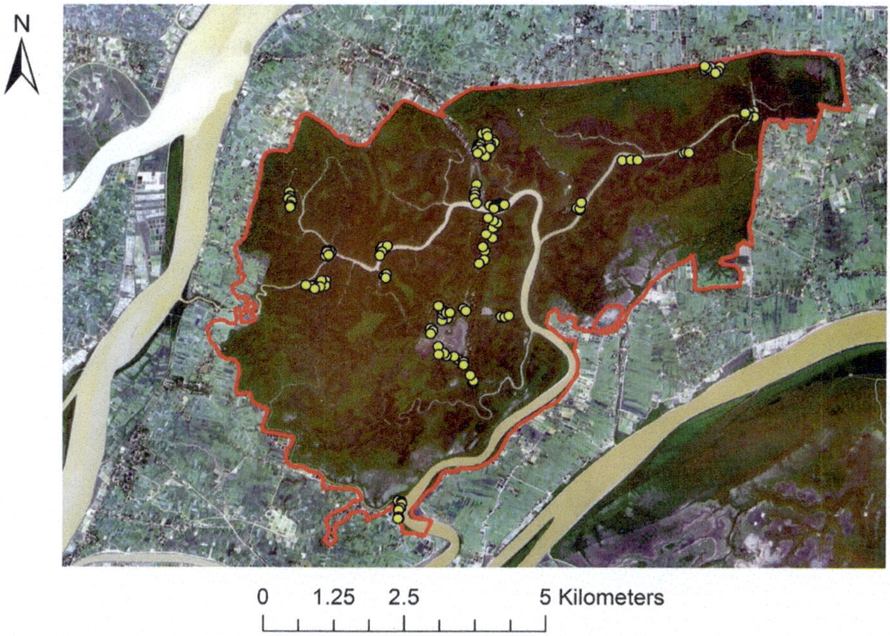

Fig. 1 Study area: a natural color composite of Bhitarkanika mangrove forest, India

The main objectives are to estimate the biomass and measuring the usefulness of multi-temporal data (Tian et al., 2016). For that purpose, five period data for Sentinel-1 and four time period data for Sentinel-2 are procured from Copernicus hub. Sentinel-1 data were dated from 12 October to 29 November 2018, with 12 days gap in between them. All tiles of Sentinel-1 data was calibrated, terrain corrected at first using Sentinel Application Platform (SNAP). Then the tiles were stacked together and multi-temporal speckle filtering was applied. Multi-temporal speckle filtering has the advantage of obtaining a better result than single date filtering while preserving the edge information and major details. Acquisition dates of Sentinel-2 images were 16 and 21 October and 17 and 22 November 2018. Atmospheric correction of Sentinel-2 data was done using sen2Cor processor. Vegetation indices were generated using the surface reflectance data. Variation of VIs over the 1 month time period was checked as it is a major drawback while correlating VIs to biomass.

Location of the center of the plots was measured with handheld GPS. Without differential corrections GPS readings are prone to errors. Due to multipath propagation and blockage of satellite signal by canopy, an error of 8–10 m is quite common in ordinary GPS measurement. The common procedure followed to eradicate this error is to create a buffer around the plot center locations. Buffer radius should be restricted to such a value that it did not encroach non-homogeneous areas. A 40 m buffer was drawn here for the purpose.

After obtaining the remote sensing parameters for each plot, it was correlated with plot biomass values estimated with the help of ground data. For retrieval of biophysical parameters like biomass in most remote sensing-based studies, linear regression is used to establish the equation. It assumes that a biophysical parameter and its electromagnetic response follow a linear relationship (Srivastava et al., 2026). However, this may not be the case most of the time. Biophysical parameters and electromagnetic responses captured in remote sensing images tend to be more complex and non-linear (Ali et al., 2015). To handle the problem of non-linearity and complexity of retrieval problems, use of machine learning algorithms is a common practice nowadays. Several studies (Dube et al., 2014; Gleason & Im, 2012; Jachowski et al., 2013; Mutanga et al., 2012; Adam et al., 2014) were done to check the effectiveness of different algorithms. Out of all the machine learning algorithms, decision tree-based models (Baker et al., 2006; Lawrence et al., 2004) such as random forest (Breiman, 2001) and stochastic gradient boosting (Friedman, 2002) have shown better result in forest biomass estimation. Here we have used both to compare their effectiveness.

3 Results

3.1 Field-Plot Data

Figure 2 shows the name of the major species and their frequency. *Heritiera fomes* is the most dominant species followed by *Excoecaria agallocha*. They together comprise almost 90% of the total inland trees. There is some occurrence of *Avicennia officinalis*, *Ceriops decandra*, and *Cynometra iripa*, but rest is negligible. *Sonneratia* is the most abandoned species in river banks. Although the number of *Avicennia officinalis* trees is less, due to its large structure, it covers a significant portion of the top canopy. Out of the four dominant species *Avicennia officinalis* has highest average diameter followed by *Heritiera fomes*, *Excoecaria agallocha*, and *Cynometra iripa*. Diameter of all the species varies widely (Fig. 3). For *Avicennia officinalis*, it ranges from 5 to 147 cm, 3 to 100 cm for *Heritiera fomes*, 3 to 104 cm for *Excoecaria agallocha*, and 3 to 32 cm for *Cynometra iripa*.

Tree density per plot varies from 26 to 140. In average field plots showed high tree density (Fig. 4). Seventy-seven plots contain somewhere between 60 and 100 trees per plot. Twenty plots contain more than 100 trees, and 43 plots contain 20–60 trees. Biomass density of plots is generally low (Fig. 5). Majority of the plots have a biomass value of 12 t or less. And a total of 110 plots have a biomass value of 21.5 t or less. The number of trees in a plot and plot biomass follows an inverse relationship (Fig. 6). With the increase in the number of trees, total biomass of the plot generally goes down. This happens because in high tree density plots, most trees are young ones with less diameter but in less dense plots has more mature and bigger trees. Species composition also matters in this case. A single *Avicennia officinalis* tree can weigh up to 10 tonnes, while a mature *Excoecaria agallocha* weighs a couple of hundred kilos. If a plot contains bigger trees like *Avicennia officinalis*, which generally occurs in the plots with a lower number of trees, it increases the overall biomass drastically (Johnsingh et al., 2004). High tree density plots contain more *Excoecaria agallocha* which results in lower biomass.

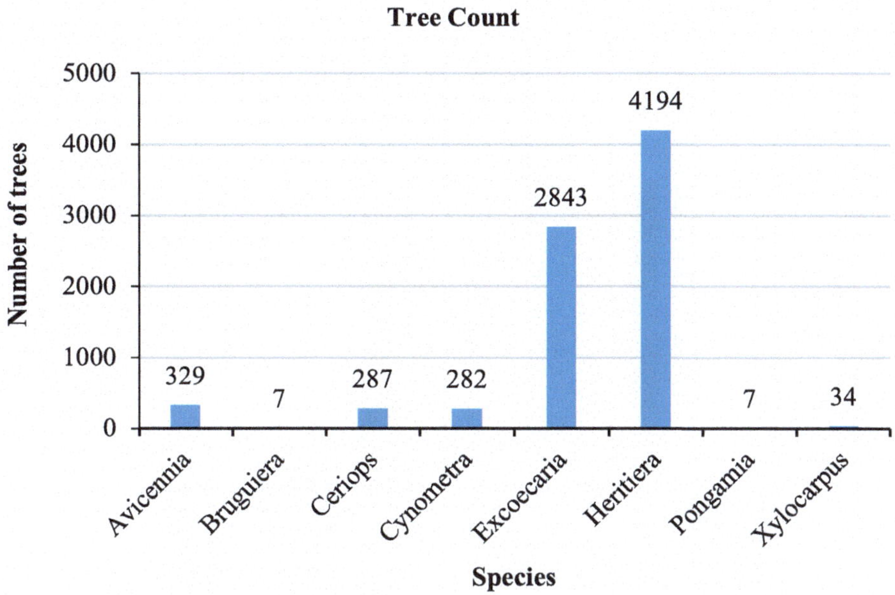

Fig. 2 Dominant species and their frequency

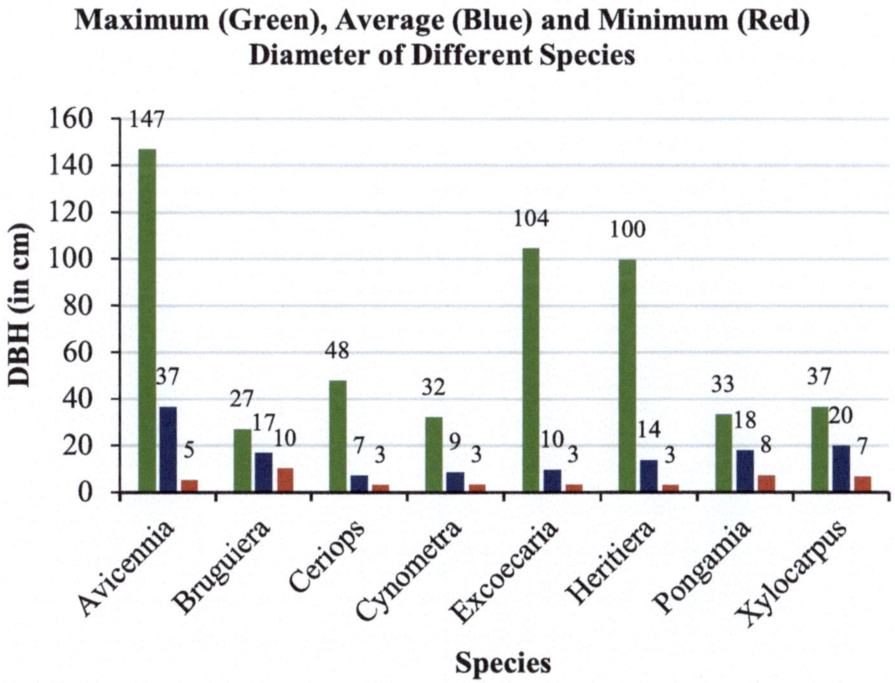

Fig. 3 DBH statistics of different species

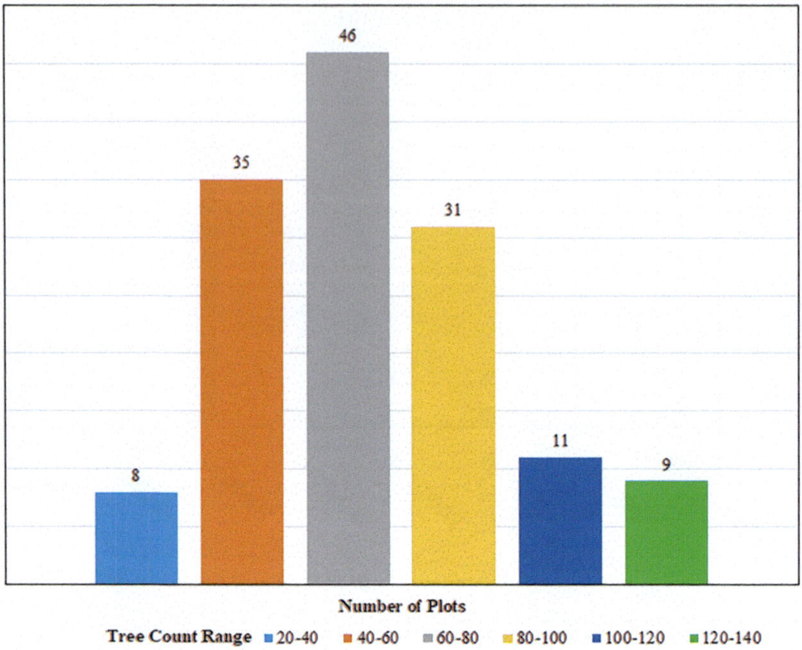

Fig. 4 Tree count range and associated number of plots

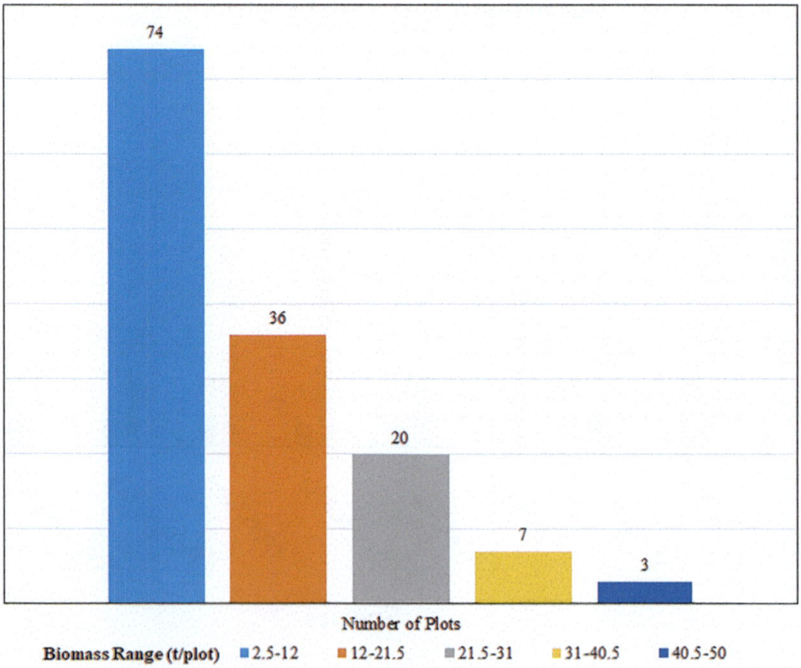

Fig. 5 Biomass range and associated number of plots

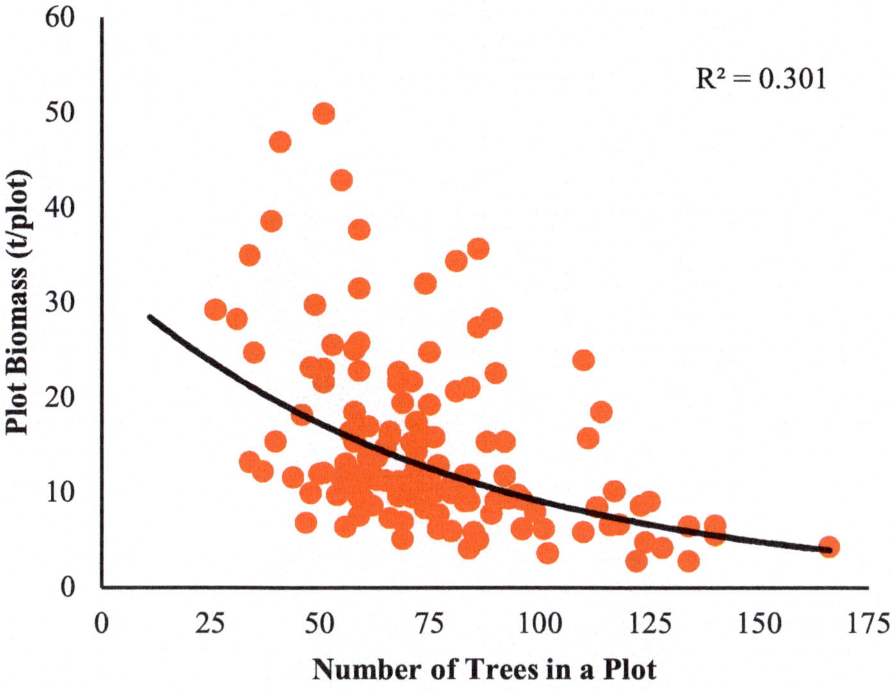

Fig. 6 The relationship between plot biomass and number of trees in the plot

4 Temporal Variation of Vegetation Indices

Field plot values are correlated with remote sensing parameters as the next step. Four vegetation indices are prepared using the Sentinel-2 data. They are Normalized Difference Vegetation Index (NDVI), Green NDVI, Red-Edge NDVI, Soil Adjusted Vegetation Index (SAVI), Enhanced Vegetation Index (EVI), and EVI2 (Huete, 1988). First, we checked its variation over the acquisition period. Figure 7 shows the variation of all vegetation indices for all 140 field plots. Even for the cloud-free, atmospherically corrected data from the same sensor, there are considerable variations in the values of indices within a time of 40 days. Almost all the field plots show a declining trend. EVI values are less affected due to temporal variation than the other indices, while changes in NDVI and GNDVI values are much bigger (Liu & Huete, 1995). As a result, we didn't consider them while establishing the biomass estimation model. Some of the field plots are showing more erratic trend than the others. To understand if this behavior is in any way related to plot biomass, we analyzed it carefully. To do that all plots are divided into four parts based on plot-wise biomass range and check the variation again. We can observe from the resulting graphs (Fig. 8) that those plots are almost equally distributed among all the *ranges*. As the number of these plots is very low compared to the total number of plots, we discard these plots for the time being for further analysis.

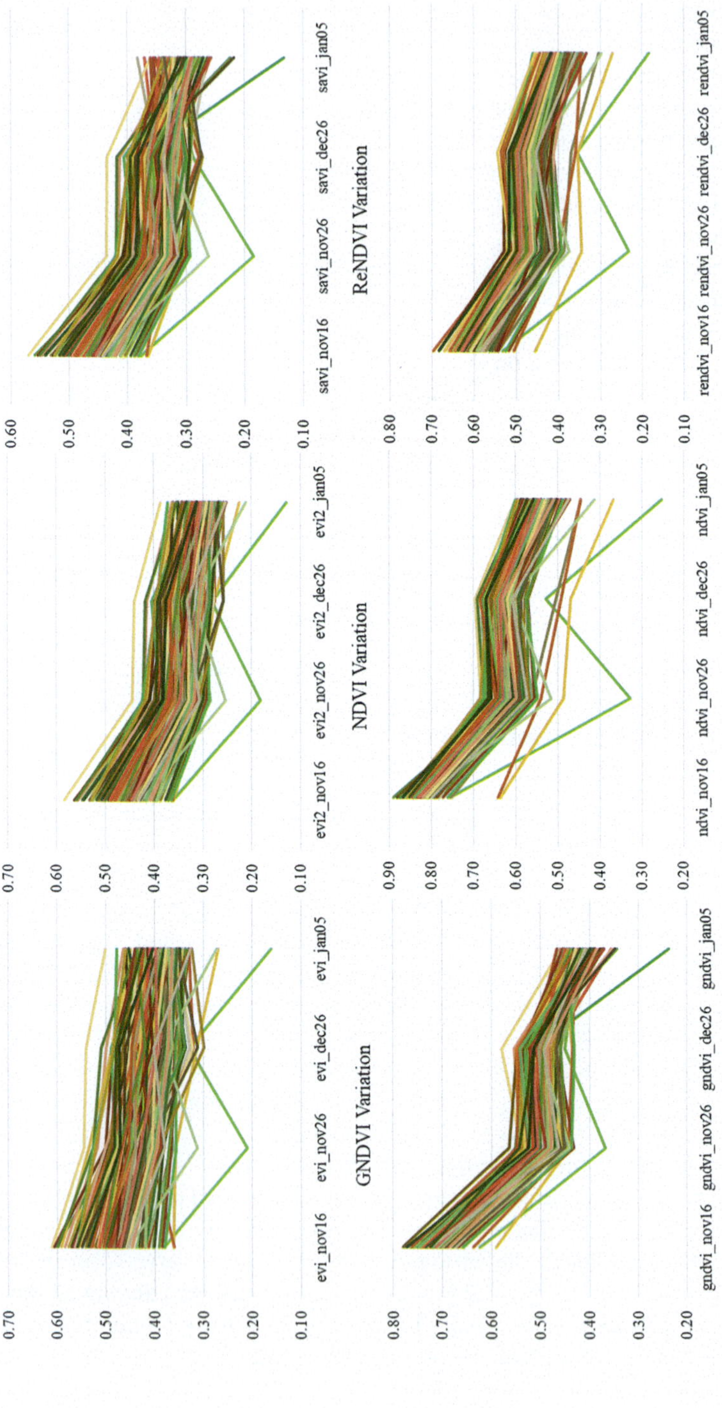

Fig. 7 Plot-wise variation of vegetation indices over time

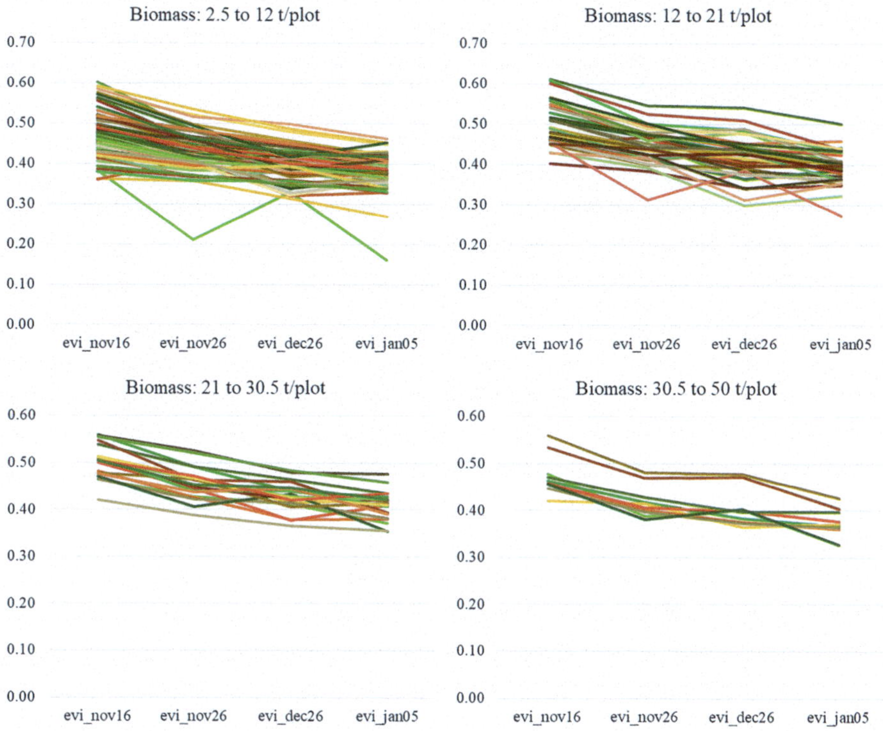

Fig. 8 Plot biomass range wise variation

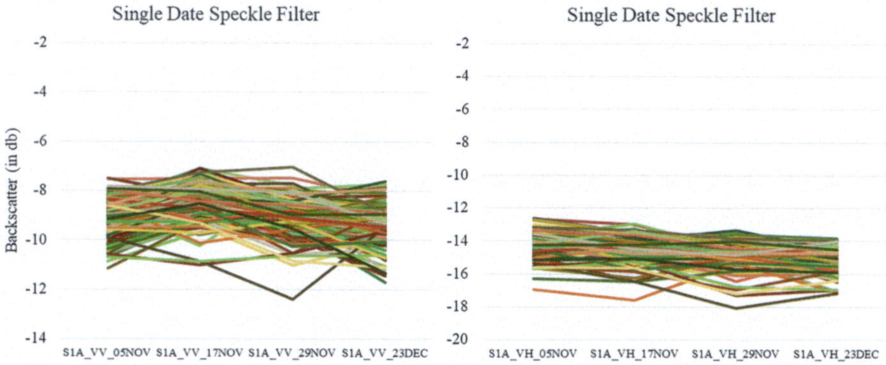

Fig. 9 Plot-wise variation of SAR backscatter over time

For SAR data also, when we see the graphs, the backscatter values in both VV and VH polarization varies widely (Fig. 9). It is evident from the result that to establish a relationship between biomass and remote sensing variables, correlating the biomass with a single time data is not the correct procedure.

5 Biomass Estimation

To estimate the biomass of the area, we used multi-time period data to reduce the effect of temporal variation. The vegetation indices were prepared for each data, and they were averaged to get the final image. Four Sentinel-1 images taken for the study were also averaged to generate images for both polarizations. Finally, an average of two polarizations and the square root of their product are also produced. For all these images maximum, minimum, average, and standard deviation values for a 50 m diameter buffer radius around plot centers are calculated. Buffer distance will vary depending on the accuracy of GPS. A total of 40 variables were generated to be used as a predictor. Variable values for 140 plot locations were extracted. Using this data, random forest and GBM model was run through CARET package of R software (Liaw & Wiener, 2002). Seventy percent of the points were used for the model establishment, and the rest were used for validation. At first variables from SAR dataset were used as a predictor. Then optical data-based variables were used, and finally, both were used in tandem. Figures 10 and 11 show the model result for RF and GBM model using vegetation indices as predictor variable.

Random forest gives best result while selecting five variables at each node for splitting. In GBM, lowest RMSE was found using interaction depth of 1 and a shrinkage ratio of 0.001. These models also generate the list of important variables. Figure 12 shows the list for RF model and Fig. 13 for GBM model.

The validation results of both random forest and Gradient Boosted Model (GBM) are shown in Table 1. In the random forest model, we can see SAR data is not correlated with field plot biomass values. It shows a correlation of 0.15 with a root mean square error of 6 t/ha. Vegetation indices show an excellent relationship with biomass with a correlation value of 0.69. RMSE also improves to 5.706 t/plot. Using SAR and optical data together produces the best RMSE of 5.303 t/plot, but

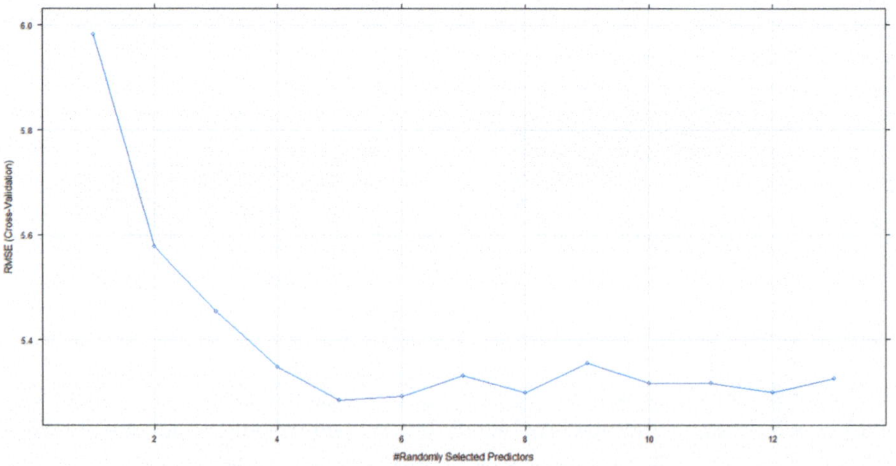

Fig. 10 Result of random forest model using vegetation indices as predictor

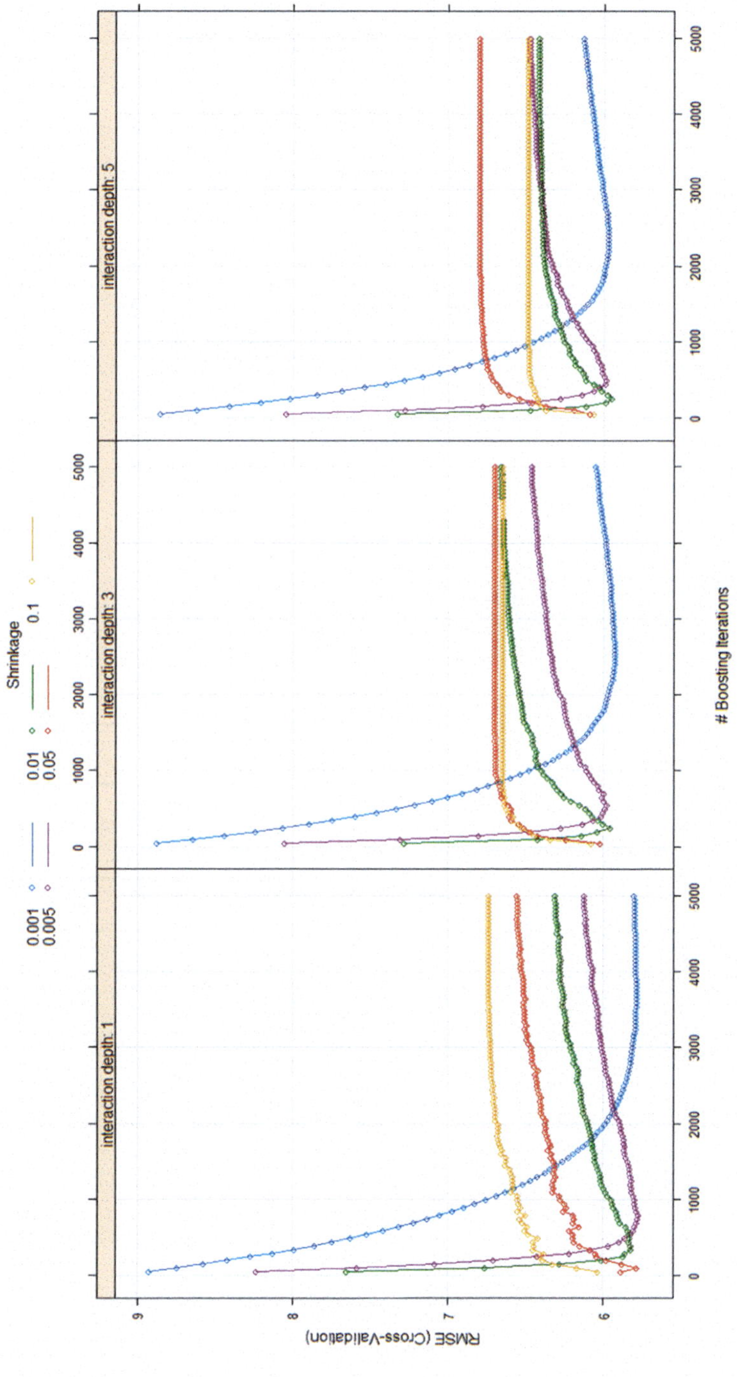

Fig. 11 Result of GBM model using vegetation indices as a predictor

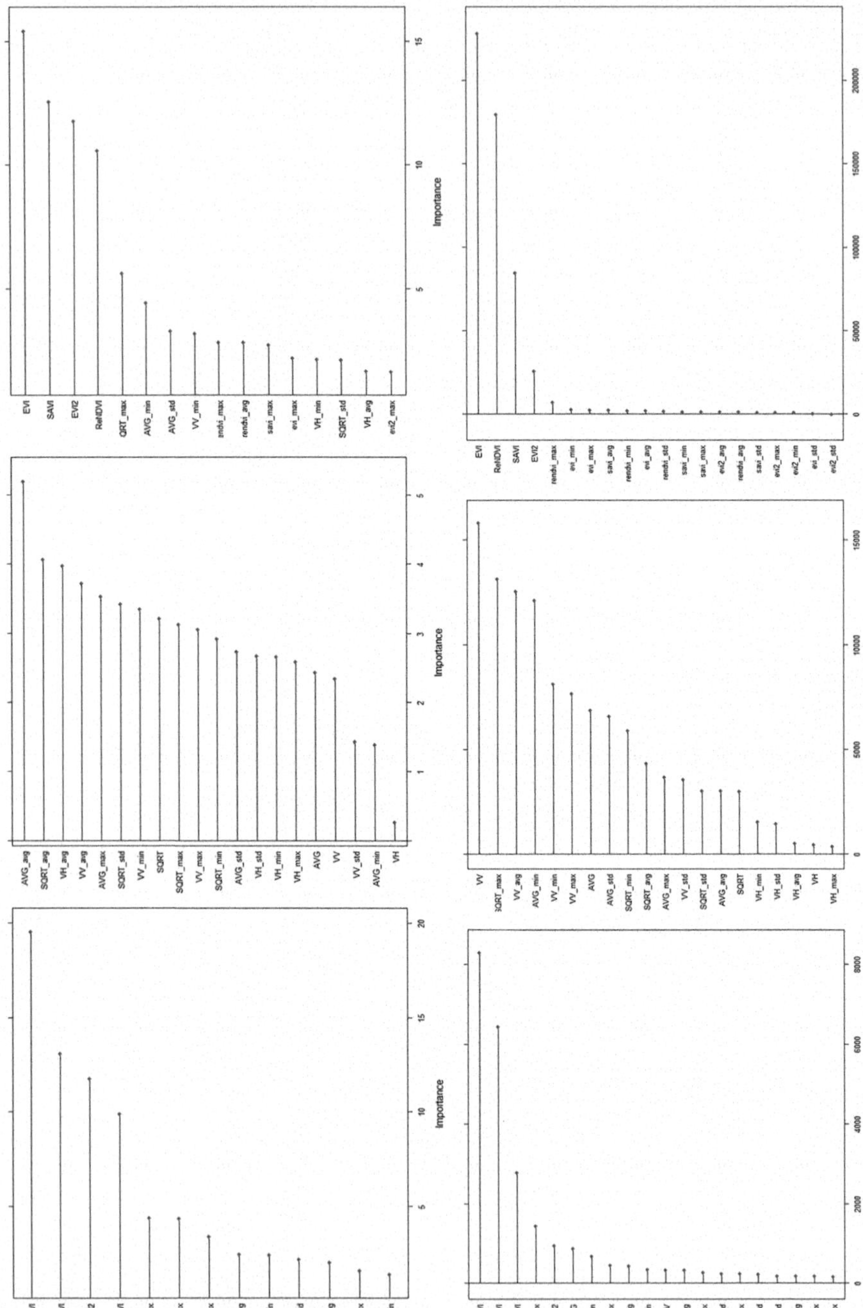

Fig. 12 Variable importance RF (left) and GBM (right)

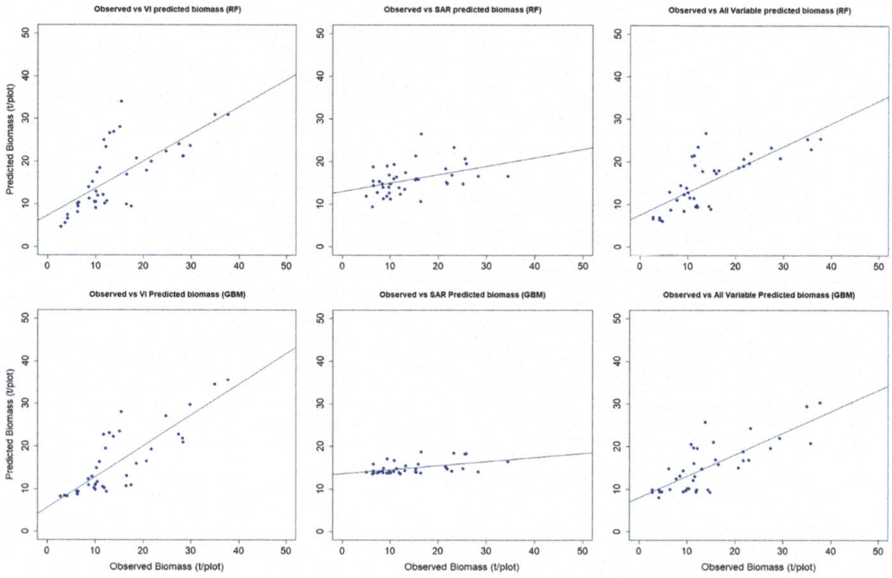

Fig. 13 Observed and predicted biomass correlation plot

Table 1 Biomass prediction results

Model	Variable	Correlation	RMSE
Random forest	SAR	0.15	6.691
Random forest	Optical	0.69	5.706
Random forest	All	0.61	5.303
Gradient Boosted Model	SAR	0.15	7.932
Gradient Boosted Model	Optical	0.79	4.878
Gradient Boosted Model	All	0.79	3.938

correlation slightly decreases to 0.61. GBM improves the result of SAR data-predicted biomass with a correlation of 0.15 and RMSE of 7.932 t/plot. For optical data, correlation value improves to 0.79, and RMSE decreases to 4.878 t/plot. However, we get the best result from the combination of both optical and SAR while using GBM. It shows a correlation of 0.79 and RMSE of 3.938 t/ha. A final biomass map was prepared based on the relationship which uses both SAR and optical data.

6 Discussion

In our study biomass density varies from 2.8 t/plot or 70 t ha^{-1} to 49.945 t/plot or 1248.625 t ha^{-1} (Fig. 14). Biomass density of Bhitarkanika mangroves is on the higher side if we compare it with the result of other studies done on other parts of

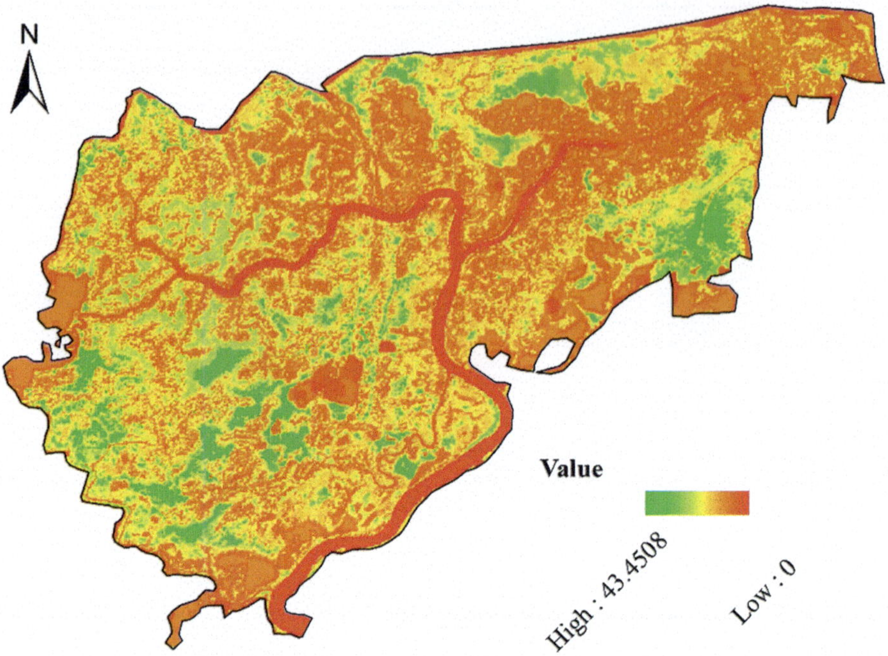

Fig. 14 Biomass map of the study area (t/plot)

the world and India. In India, most of the studies done on mangrove biomass are based on Sundarbans. Still, there is no consensus about the biomass density of Sundarbans. In one study the biomass range varies from 3 to 20.9 t ha^{-1} (Manna et al., 2014), while in another work, it changes from 8.9 to 50.9 t ha^{-1} (Joshi & Ghose, 2014), which is considerably lower than the biomass density of Bhitarkanika. Mangroves of Vietnam have a high biomass density which varies from 30 to 596 t ha^{-1} (Pham & Brabyn, 2017). In Philippines, biomass density of mangroves ranges from 1.1 to 210 t ha^{-1} (Castillo et al., 2017). High biomass density of Bhitarkanika can be attributed to its species composition. The study on Sundarbans was done on a young *Avicennia marina* plantation, while the Joshi and Ghose study was conducted on an area dominated by smaller trees (Manna et al., 2014), while mature and bigger trees dominate our study area. So, naturally, we got a much higher biomass density than other regions.

The reflectance of electromagnetic radiation by a vegetation cover is not a constant quantity. It varies depending on multiple factors like moisture content of the vegetation, type of the vegetation, and its orientation. As a result, the remote sensing parameters captured by sensors will also differ from time to time. In our study, we observed how the values of Sentinel-1 backscatter and Sentinel-2-derived vegetation indices changes over time. Although the observation period was only 50 days for Sentinel-2 and 48 days for Sentinel-1, variations we observed is rather high. Our observation of the area starts from early and mid-November. Examining the plots we can see that all indices begin with a high value. Then it reduces near the end of November. The index value increases meagerly during the end of December and then again reduces in early January. As the period is too low for occurring any significant shift in species composition and its orientation, much of the change can be traced to moisture content. According to Indian Meteorological Department data (source URL: http://www.imdpune.gov.in/Seasons/Pre_Monsoon/premonsoon.html), coastal regions of Odisha received slight excess to average rainfall during October. So, greenness of the plants in early November remains high, resulting in high vegetation index values. But November sees a less-than-usual rainfall, and as a consequence index values for late November take a plunge. Now again in December, the area gets an average rainfall which results in a minor increase in index values. But again in January, it gets reduced due to lack of rainfall. Sentinel-1 backscatter does not follow any specific trend.

Different properties of the tree canopy affect the values of different remote sensing parameters. Vegetation indices depend on the conditions of the leaves. Healthier leaves result in higher vegetation indices (Dong et al., 2003). As healthier leaves are also a good indicator of vegetation, vegetation indices work as a good predictor of productivity. Nonetheless, vegetation indices are produced from optical data which only gives us information about top of the canopy. Previous works have shown (Reddersen et al., 2014; Tilly et al., 2015) that vegetation indices generally saturate at lower biomass value. However, Sentinel-2 data is different and more advanced than other optical dataset. It is better than others in terms of spatial, spectral, and temporal resolutions. It also has the advantage of three well-placed red-edge bands (Han et al., 2017). Thus Sentinel-2 data is more suitable for monitoring vegetation health information. C-band SAR can't penetrate the canopy due to its shorter wavelength and backscattered from the top. So, primarily, moisture content and structural parameters like size, density, and orientation of leaves affect its backscatter (Castel et al., 2002). Longer-wavelength SAR has the ability to penetrate the canopy and interact with the trunk containing most of the biomass. So, when we combine longer wavelength SAR data with vegetation indices, it improves the predictions. But as C-band SAR can't penetrate the canopy, it cannot improve the estimation accuracy. Our results also prove these points. While Sentinel-2 vegetation indices act as the better predictor variable for biomass, Sentinel-1 data gives us a very poor prediction of biomass. Earlier studies (Carreiras et al., 2013; Dube et al., 2014; García-Gutiérrez et al., 2015; Gleason & Im, 2012) have shown that machine learning models outperform other models in biomass estimation.

7 Conclusions

We observed that carbon storage capacity (Spalding et al., 2010) of Indian mangroves is very high. We need to more studies in other major mangrove areas of India, especially Sundarbans, to find their true carbon sequestration potential. In the near future, NASA-ISRO Synthetic Aperture Radar (NISAR) will become the world's first dual frequency (L and S bands) spaceborne SAR system. As NISAR will operate in a higher frequency, its data will be more suitable for forest biomass measurement. ESAs' P band BIOMASS mission will also come into operation in next decade as a major addition to current global efforts to build a carbon data monitoring system (Scipal et al., 2010, Rodda et al., 2016). These missions will be able to produce a global map of aboveground biomass. Biomass map generated using Sentinel-1 and Sentinel-2 data can be used to validate the aboveground biomass maps generated from these future missions.

References

Adam, E., Mutanga, O., Abdel-Rahman, E. M., & Ismail, R. (2014). Estimating standing biomass in papyrus (Cyperus papyrus L.) swamp: Exploratory of in situ hyperspectral indices and random forest regression. *International Journal of Remote Sensing, 35*(2), 693–714. https://doi.org/10.1080/01431161.2013.870676

Ali, I., Greifeneder, F., Stamenkovic, J., Neumann, M., & Notarnicola, C. (2015). Review of machine learning approaches for biomass and soil moisture retrievals from remote sensing data. *Remote Sensing, 7*(12), 16398–16421. https://doi.org/10.3390/rs71215841

Asbridge, E., Lucas, R., Ticehurst, C., & Bunting, P. (2016). Mangrove response to environmental change in Australia's Gulf of Carpentaria. *Ecology and Evolution, 6*(11), 3523–3539. https://doi.org/10.1002/ece3.2140

Babcock, C., Finley, A. O., Bradford, J. B., Kolka, R., Birdsey, R., & Ryan, M. G. (2015). LiDAR based prediction of forest biomass using hierarchical models with spatially varying coefficients. *Remote Sensing of Environment, 169*, 113–127. https://doi.org/10.1016/j.rse.2015.07.028

Baker, C., Lawrence, R., Montagne, C., & Patten, D. (2006). Mapping wetlands and riparian areas using Landsat ETM+ imagery and decision-tree-based models. *Wetlands, 26*(2), 465–474. https://doi.org/10.1672/0277-5212(2006)26[465:MWARAU]2.0.CO;2

Behera, M., Tripathi, P., Mishra, B., Kumar, S., Chitale, V., & Behera, S. K. (2015). Above-ground biomass and carbon estimates of Shorea robusta and Tectona grandis forests using QuadPOL ALOS PALSAR data. *Advances in Space Research, 57*, 552–561. https://doi.org/10.1016/j.asr.2015.11.010

Biswas, S. R., Choudhury, J. K., Nishat, A., & Rahman, M. M. (2007). Do invasive plants threaten the Sundarbans mangrove forest of Bangladesh? *Forest Ecology and Management, 245*, 1–9.

Breiman, L. (2001). Random forests. *Machine Learning, 45*, 5–32.

Carreiras, J. M. B., Vasconcelos, M. J., & Lucas, R. M. (2012). Understanding the relationship between aboveground biomass and ALOS PALSAR data in the forests of Guinea-Bissau (West Africa). *Remote Sensing of Environment, 121*, 426–442. https://doi.org/10.1016/j.rse.2012.02.012

Carreiras, J. M. B., Melo, J. B., & Vasconcelos, M. J. (2013). Estimating the above-ground biomass in miombo savanna woodlands (Mozambique, East Africa) using L-band synthetic aperture radar data. *Remote Sensing, 5*(4), 1524–1548. https://doi.org/10.3390/rs5041524

Cartus, O., Santoro, M., & Kellndorfer, J. (2012). Mapping forest aboveground biomass in the Northeastern United States with ALOS PALSAR dual-polarization L-band. *Remote Sensing of Environment, 124*, 466–478. https://doi.org/10.1016/j.rse.2012.05.029

Chitale, V. S., & Behera, M. D. (2014). Analysing land and vegetation cover dynamics during last three decades in Katerniaghat wildlife sanctuary, India. *Journal of Earth System Science, 123*(7), 1467–1479.

Connette, G., Oswald, P., Songer, M., & Leimgruber, P. (2016). Mapping distinct forest types improves overall forest identification based on multi-spectral landsat imagery for Myanmar's Tanintharyi Region. *Remote Sensing, 8*, 882.

Cornforth, W. A., Fatoyinbo, T. E., Freemantle, T. P., & Pettorelli, N. (2013). Advanced land observing satellite phased array type L-Band SAR (ALOS PALSAR) to inform the conservation of mangroves: Sundarbans as a case study. *Remote Sensing, 5*(1), 224–237. https://doi.org/10.3390/rs5010224

Datta, D., & Deb, S. (2012). Analysis of coastal land use/land cover changes in the Indian Sunderbans using remotely sensed data. *Geo-Spatial Information Science, 5020*, 1–10. https://doi.org/10.1080/10095020.2012.714104

Deng, S., Katoh, M., Guan, Q., Yin, N., & Li, M. (2014). Estimating forest aboveground biomass by combining ALOS PALSAR and WorldView-2 data: A case study at Purple Mountain National Park, Nanjing, China. *Remote Sensing, 6*(9), 7878–7910. https://doi.org/10.3390/rs6097878

Donato, D. C., Kauffman, J. B., Murdiyarso, D., Kurnianto, S., Stidham, M., & Kanninen, M. (2011). Mangroves among the most carbon-rich forests in the tropics. *Nature Geoscience, 4*, 293–297.

Dube, T., Mutanga, O., Elhadi, A., & Ismail, R. (2014). Intra-and-inter species biomass prediction in a plantation forest: Testing the utility of high spatial resolution spaceborne multispectral RapidEye sensor and advanced machine learning algorithms. *Sensors (Basel, Switzerland), 14*(8), 15348–15370. https://doi.org/10.3390/s140815348

Duncanson, L. I., Niemann, K. O., & Wulder, M. A. (2010). Estimating forest canopy height and terrain relief from GLAS waveform metrics. *Remote Sensing of Environment, 114*(1), 138–154. https://doi.org/10.1016/j.rse.2009.08.018

Forest Survey of India (FSI). (1996). *Volume equations for forests of India, Nepal and Bhutan.* Ministry of Environment and Forests (MOEF), Govt. of India.

Giri, S., Mukhopadhyay, A., Hazra, S., Mukherjee, S., Roy, D., Ghosh, S., & Mitra, D. (2014). A study on abundance and distribution of mangrove species in Indian Sundarbans using remote sensing technique. *Journal of Coastal Conservation, 18*(4), 359–367. https://doi.org/10.1007/s11852-014-0322-3

Heiskanen, J. (2006). Estimating aboveground tree biomass and leaf area index in a mountain birch forest using ASTER satellite data. *International Journal of Remote Sensing, 27*(6), 1135–1158. https://doi.org/10.1080/01431160500353858

Hong, S. H., Kim, H. O., Wdowinski, S., & Feliciano, E. (2015). Evaluation of polarimetric SAR decomposition for classifying wetland vegetation types. *Remote Sensing, 7*(7), 8563–8585. https://doi.org/10.3390/rs70708563

Hopkinson, C., Chasmer, L., Gynan, C., Mahoney, C., & Sitar, M. (2016). Multisensor and Multispectral LiDAR Characterization and Classification of a Forest Environment. Canadian *Journal of Remote Sensing, 42*(5), 501–520.

Huete, A. R. (1988). A soil-adjusted vegetation index (SAVI). *Remote Sensing of Environment, 25*(3), 295–309. https://doi.org/10.1016/0034-4257(88)90106-X

Hunter, M. O., Keller, M., Victoria, D., & Morton, D. C. (2013). Tree height and tropical forest biomass estimation. *Biogeosciences, 10*, 8385–8399. https://doi.org/10.5194/bg-10-8385-2013

Johnsingh, A. J. T., Ramesh, K., Qureshi, Q., David, A., Goyal, S. P., Rawat, G. S., Rajapandian, K., & Prasad, S. (2004). *Conservation status of tiger and associated species in the Terai Arc Landscape*. Wildlife Institute of India.

Joshi, H. G., & Ghose, M. (2014). Community structure, species diversity, and aboveground biomass of the Sundarbans mangrove swamps. *Tropical Ecology, 55*(3), 283–303.

Kaasalainen, S., Holopainen, M., Karjalainen, M., Vastaranta, M., Kankare, V., Karila, K., & Osmanoglu, B. (2015). Combining lidar and synthetic aperture radar data to estimate forest biomass: Status and prospects. *Forests, 6*(1), 252–270. https://doi.org/10.3390/f6010252

Lawrence, R., Bunn, A., Powell, S., & Zambon, M. (2004). Classification of remotely sensed imagery using stochastic gradient boosting as a refinement of classification tree analysis. *Remote Sensing of Environment, 90*(3), 331–336. https://doi.org/10.1016/j.rse.2004.01.007

Liaw, A., & Wiener, M. (2002). Classification and regression by randomForest. *R News, 2*, 18–22. https://doi.org/10.1177/154405910408300516

Liu, H. Q., & Huete, A. R. (1995). A feedback based modification of the NDVI to minimize canopy background and atmospheric noise. *IEEE Transactions on Geoscience and Remote Sensing, 42*, 457–465.

Malhi, R. K. M., Anand, A., Srivastava, P. K., Chaudhary, S. K., Pandey, M. K., Behera, M. D., Kumar, A., Singh, P., & Kiran, G. S. (2022). Synergistic evaluation of Sentinel 1 and 2 for biomass estimation in a tropical forest of India. *Advances in Space Research, 69*(4), 1752–1767.

Manna, S., Nandy, S., Chanda, A., Akhand, A., Hazra, S., & Dadhwal, V. K. (2014). Estimating aboveground biomass in\n Avicennia marina\n plantation in Indian Sundarbans using high-resolution satellite data. *Journal of Applied Remote Sensing, 8*(1), 83638. https://doi.org/10.1117/1.JRS.8.083638

Marchio, D. A., Savarese, M., Bovard, B., & Mitsch, W. J. (2016). Carbon sequestration and sedimentation in mangrove swamps influenced by hydrogeomorphic conditions and urbanization in Southwest Florida. *Forests, 7*(6). https://doi.org/10.3390/f7060116

Ouchi, K. (2013). Recent trend and advance of synthetic aperture radar with selected topics. *Remote Sensing, 5*, 716–807. https://doi.org/10.3390/rs5020716

Pandey, P. C., Srivastava, P. K., Chetri, T., Choudhary, B. K., & Kumar, P. (2019a). Forest biomass estimation using remote sensing and field inventory: A case study of Tripura, India. *Environmental Monitoring and Assessment, 191*(9), 593.

Pandey, P. C., Anand, A., & Srivastava, P. K. (2019b). Spatial distribution of mangrove forest species and biomass assessment using field inventory and earth observation hyperspectral data. *Biodiversity and Conservation, 28*(8), 2143–2162.

Raha, A., Das, S., Banerjee, K., & Mitra, A. (2012). Climate change impacts on Indian Sunderbans: A time series analysis (1924–2008). *Biodiversity and Conservation, 21*(5), 1289–1307. https://doi.org/10.1007/s10531-012-0260-z

Ravindranath, N. H., Joshi, N. V., Sukumar, R., & Saxena, A. (2006). Impact of climate change on forests in India. *Current Science, 90*(3), 354–361.

Rodda, S. R., Thumaty, K. C., Jha, C. S., & Dadhwal, V. K. (2016). Seasonal variations of carbon dioxide, water vapor and energy fluxes in tropical indian mangroves. *Forests, 7*, 35. [CrossRef].

Spalding, M., Kainuma, M., & Collins, L. (2010). *World atlas of mangroves* (p. 319). Earthscan.

Su, Y., Guo, Q., Xue, B., Hu, T., Alvarez, O., Tao, S., & Fang, J. (2016). Spatial distribution of forest aboveground biomass in China: Estimation through combination of spaceborne lidar, optical imagery, and forest inventory data. *Remote Sensing of Environment, 173*, 187–199. https://doi.org/10.1016/j.rse.2015.12.002

Sun, G., Ranson, K. J., Guo, Z., Zhang, Z., Montesano, P., & Kimes, D. (2011). Forest biomass mapping from lidar and radar synergies. *Remote Sensing of Environment, 115*(11), 2906–2916. https://doi.org/10.1016/j.rse.2011.03.021

Tian, S., Zhang, X., Tian, J., & Sun, Q. (2016). Random forest classification of wetland landcovers from multi-sensor data in the arid region of Xinjiang, China. *Remote Sensing, 8*, 954.

Vaglio Laurin, G., Chen, Q., Lindsell, J. a., Coomes, D. a., Del Frate, F., Guerriero, L., & Valentini, R. (2014). Above ground biomass estimation in an African tropical forest with lidar and hyperspectral data. *ISPRS Journal of Photogrammetry and Remote Sensing, 89*, 49–58. https://doi.org/10.1016/j.isprsjprs.2014.01.001

Zhao, P., Lu, D., Wang, G., Liu, L., Li, D., Zhu, J., & Yu, S. (2016). Forest aboveground biomass estimation in Zhejiang Province using the integration of Landsat TM and ALOS PALSAR data. *International Journal of Applied Earth Observation and Geoinformation, 53*, 1–15. https://doi.org/10.1016/j.jag.2016.08.007

Gwal, S., Gupta, A., Singh, P., Srivastava, P. K., & Srivastava, S. K. (2026). AI-driven approaches for forest growth assessment and management. In Advanced Geospatial and Ground Based Techniques in Forest Monitoring (pp. 413-433). Elsevier.

Srivastava, P.K., Sharma, B., Gupta, A., Gwal, S., Pandey, P.C., Srivastava, S.K., 2026. New satellite missions and sensors for forest monitoring. Advanced Geospatial and Ground Based Techniques in Forest Monitoring. Elsevier, pp. 365-391.

Suman, S., Gupta, A., Srivastava, P.K., 2026. Role of airborne datasets in forest monitoring. Advanced Geospatial and Ground Based Techniques in Forest Monitoring. Elsevier, pp. 327-343.

Remote Sensing and Cloud-Based Web Application for Mangrove Visualisation: A Google Earth Engine Approach for Monitoring the Sundarbans

Harikesh Singh and Sanjeev Kumar Srivastava

Abstract The Sundarbans, the world's largest contiguous mangrove forest, is a critical ecosystem facing significant threats from climate change, sea-level rise, and human activities. This study introduces a cloud-based web application developed using Google Earth Engine (GEE) to monitor and analyse the Sundarbans' vegetation and hydrological dynamics. The application integrates remote sensing indices such as the Normalised Difference Vegetation Index (NDVI) and the Normalised Difference Water Index (NDWI), along with rainfall data, to provide an interactive platform for tracking forest health, water body changes, and precipitation patterns over time. Key features include time-series analyses, split-panel visualisations for temporal comparisons, and rainfall distribution mapping. This innovative tool empowers researchers and policymakers with data-driven insights for sustainable forest management and conservation planning. By utilising cloud computing and geospatial data, the study underscores the potential of integrating advanced technologies to address environmental challenges in fragile ecosystems like the Sundarbans.

Keywords Sundarbans · Google Earth Engine · NDVI · NDWI · Remote sensing · Forest monitoring · Rainfall analysis

H. Singh
Geospatial Analytics for Conservation and Management, School of Science Technology and Engineering, University of the Sunshine Coast, Sippy Downs, QLD, Australia

SmartSat Cooperative Research Centre, Adelaide, SA, Australia
e-mail: harikesh@research.usc.edu.au

S. K. Srivastava (✉)
Geospatial Analytics for Conservation and Management, School of Science Technology and Engineering, University of the Sunshine Coast, Sippy Downs, QLD, Australia
e-mail: Sanjeev.Srivastava@usc.edu.au

S. K. Srivastava et al. (eds.), *Earth Observation for Monitoring Mangrove Ecosystems*, https://doi.org/10.1007/978-3-032-16668-5_11

1 Introduction

Forests play a critical role in maintaining ecological balance, providing a wide array of ecosystem services, including carbon sequestration, biodiversity conservation, and the regulation of water cycles (Miura et al., 2015; Watson et al., 2018). Among the world's vital forest ecosystems, the Sundarbans stand out as the largest contiguous mangrove forest, stretching across India and Bangladesh (Mitra, 2013). This unique region, recognised as a UNESCO World Heritage Site, is characterised by complex tidal waterways, mudflats, and a dynamic ecosystem that supports a rich diversity of flora and fauna, including the iconic Bengal tiger (Ahmad et al., 2022; Sarkar, 2022; Zaman & Chowdhury, 2024). However, the Sundarbans face numerous threats, such as rising sea levels, climate change, cyclonic storms, and anthropogenic pressures, making it crucial to monitor and manage its health and extent (Das, 2022; Islam, 2019; Kumar & Singh, 2019a, 2019b; Mondal et al., 2022).

Remote sensing technologies have transformed forest monitoring, offering a powerful means to analyse changes over large areas with high temporal frequency and spatial detail (Singh et al., 2024, 2025b; Singh & Pandey, 2021; Singh & Srivastava, 2024, 2025; Sos et al., 2023). Among these advancements, cloud-based platforms, such as Google Earth Engine (GEE), have emerged as innovative tools that streamline the processing and analysis of vast satellite imagery datasets (Singh et al., 2022b). GEE enhances the scalability of cloud computing to provide users with on-demand access to preprocessed satellite data, algorithms, and interactive mapping capabilities (Singh et al., 2025c; Tamiminia et al., 2020). This enables researchers and policymakers to detect, assess, and respond to environmental changes more effectively.

This study presents a cloud-based web application developed using Google Earth Engine to visualise and analyse forest dynamics in the Sundarbans. The application integrates various remote sensing indices, including the Normalised Difference Vegetation Index (NDVI) and the Normalised Difference Water Index (NDWI), along with rainfall data visualisation, to offer an interactive and comprehensive view of forest changes over time. By enabling users to explore NDVI and NDWI split-panel comparisons, rainfall patterns, and probable forest cover change detection, this tool facilitates informed decision-making for forest management and conservation.

The objectives of this research are to (1) demonstrate the application of GEE in building a user-friendly web interface for forest visualisation, (2) analyse spatial and temporal changes in vegetation health and water content in the Sundarbans, and (3) identify potential areas of forest cover change. This work underscores the significance of integrating remote sensing data with cloud-based technologies to promote proactive forest management strategies and contributes to a better understanding of the challenges facing Sundarbans.

2 Study Area

The Sundarbans, the world's largest contiguous mangrove forest, spans approximately 10,000 km^2 across India and Bangladesh at the delta of the Ganges, Brahmaputra, and Meghna rivers (Samanta et al., 2021). It serves as a vital natural barrier against cyclones and storm surges while supporting rich biodiversity, including the Bengal tiger, spotted deer, and diverse aquatic life. Recognised as a UNESCO World Heritage Site, the Sundarbans play a crucial role in carbon sequestration and coastal protection (Neogi et al., 2017; Srivastava et al., 2025).

Despite its ecological significance, the Sundarbans face numerous threats, including sea-level rise, cyclones, deforestation, pollution, resource exploitation, and salinity intrusion (Gopal & Chauhan, 2006; Mandal et al., 2019). These challenges highlight the urgent need for robust monitoring and management strategies to preserve its unique ecosystem. Remote sensing tools, such as those developed using Google Earth Engine, provide an innovative approach to track and analyse environmental changes, supporting conservation and sustainable management efforts (Harikesh et al., 2020; Singh et al., 2022a; Singh & Pandey, 2021).

3 Materials and Methods

3.1 Data Collection

To develop the webpage for monitoring and analysing changes in the Sundarbans forest, data was collected and processed using Google Earth Engine (GEE). The primary focus was to analyse and visualise changes using various indices, such as the Normalised Difference Vegetation Index (NDVI) and Normalised Difference Water Index (NDWI). This allows for effective tracking of forest health, vegetation dynamics, and water body changes over time. Data was sourced from various satellite missions, providing high-resolution and multi-temporal imagery essential for accurate monitoring (Table 1).

3.2 Data Preprocessing

Data preprocessing is a critical step to ensure the reliability and accuracy of subsequent analyses. For this study, preprocessing steps were applied to all datasets used in monitoring the Sundarbans forest. Satellite imagery from Sentinel-2 and Landsat missions underwent initial preprocessing, including radiometric calibration and geometric correction, to enhance data consistency (Singh et al., 2025c). Cloud masking was applied using Google Earth Engine's built-in cloud mask algorithms to minimise noise caused by atmospheric interference (Singh et al., 2022b; Yang

Table 1 Overview of datasets used for Sundarbans forest monitoring

Data source	Dataset name	Description	Spatial resolution	Temporal coverage	Purpose
Sentinel-2 imagery	Copernicus Programme (ESA)	Multispectral satellite imagery with a revisit time of 5 days, capturing detailed information on vegetation, water bodies, and land cover	10 m	From 2015 to present	Used for generating NDVI and NDWI indices for vegetation health and water extent analysis
Landsat-8	USGS/NASA	Long-term satellite imagery with a revisit cycle of 16 days, providing historical data for change detection	30 m	From 2013 to present	Used for historical change detection and comparison of forest cover over time
Rainfall data	CHIRPS (Climate Hazards Group InfraRed Precipitation with Station data)	High-resolution precipitation data that combines satellite and station-based observations	~5 km	1981–present	Used for visualising rainfall patterns over the Sundarbans region, supporting analysis of hydrological influences on vegetation and water dynamics
MODIS	MODIS/006/MOD13Q1	Moderate-resolution imaging spectroradiometer dataset providing 16-day vegetation indices for monitoring large-scale changes	250 m	From 2000 to present	NDVI, Enhanced Vegetation Index (EVI)
Global Surface Water	JRC/GSW1_3/ GlobalSurfaceWater	Historical Global Surface Water dataset used for analysing water body changes in the Sundarbans	30 m	From 1984 to 2019	Water occurrence, change intensity
Digital Surface Model (DSM)	USGS/SRTMGL1_003	Shuttle Radar Topography Mission DEM providing elevation data of the region, used for terrain analysis	30 m	Static coverage	Elevation, slope

et al., 2022). This step ensured only cloud-free pixels were retained for accurate vegetation and water body analyses.

Next, data resampling was performed to achieve consistent spatial resolutions across different datasets. Sentinel-2 data, with a native resolution of 10 m, was resampled to align with the resolution of other ancillary datasets when necessary. Temporal aggregation was applied for rainfall data, generating monthly averages to account for short-term fluctuations and offer a more robust depiction of seasonal patterns in the region.

Additional preprocessing steps involved normalisation of indices, such as NDVI and NDWI values, to maintain comparability across different dates and data sources. Vegetation indices were computed after the removal of outliers and correction for atmospheric effects. For topographic data, elevation and slope values were extracted, ensuring they matched the spatial coverage of the study area.

Overall, these preprocessing techniques facilitated clean, standardised, and temporally aligned datasets, forming a robust foundation for subsequent analysis and the development of visualisation tools within the Google Earth Engine application.

3.3 Development of Google Earth Engine Application

The web application for monitoring and analysing changes in the Sundarbans forest was developed using Google Earth Engine (GEE), a cloud-based platform that enables efficient processing and analysis of large geospatial datasets. The application includes several key features that offer insights into the dynamic changes within the Sundarbans forest ecosystem. One of the core components is a split-panel visualisation tool that allows users to compare different indices, such as the Normalised Difference Vegetation Index (NDVI) and the Normalised Difference Water Index (NDWI), side by side. This functionality provides valuable insights into vegetation health, water presence, and changes over time within the forest. Additionally, the Rainfall Visualisation section allows users to explore rainfall patterns over the Sundarbans region, contributing to an understanding of hydrological influences on the ecosystem (Samanta et al., 2021; Tamiminia et al., 2020). Rainfall data is visualised using interactive maps and temporal charts, facilitating detailed analysis of seasonal trends and variations. The developed application can be accessed https://bit.ly/3NUSKUu. This interactive platform aims to support researchers, policymakers, and stakeholders in decision-making processes related to forest conservation and management by providing robust tools for data-driven insights.

3.4 Analytical Methods

Various algorithms and indices were employed to analyse and interpret the satellite imagery data. The primary methods used in this study are detailed below:

1. **Normalised Difference Vegetation Index (NDVI)**
   NDVI was calculated using the formula:

$$\text{NDVI} = \frac{(\text{NIR} - \text{Red})}{(\text{NIR} + \text{Red})}$$

The NDVI values were used to assess vegetation health and track changes in vegetation density over time.

2. **Normalised Difference Water Index (NDWI)**
   NDWI was calculated using the formula:

$$\text{NDWI} = \frac{(\text{Green} - \text{NIR})}{(\text{Green} + \text{Red})}$$

This index was employed to identify and monitor water bodies and changes in water coverage within the forest.

3. **Change Detection Techniques**
   Temporal change detection algorithms were used to assess differences in NDVI and NDWI values over specified time intervals, providing insights into vegetation and water dynamics (Ahmed & Akter, 2017; Ashok et al., 2021).
4. **Results**
   This section outlines the outcomes of the developed cloud-based web application for monitoring and visualising the Sundarbans forest dynamics using Google Earth Engine (GEE). The application provides comprehensive visualisation tools to assess vegetation health, water dynamics, and potential changes in forest cover. Key features of the application include NDVI and NDWI time-series visualisations, split-panel comparisons, rainfall visualisation, and probable forest change detection.

3.5 Web Interface Overview

The web application interface serves as a gateway for researchers and policymakers to access various tools for analysing forest changes. Users can interact with the platform to explore NDVI time-series trends, compare NDVI and NDWI indices side by side, and visualise rainfall patterns and probable forest cover changes. The gateway page (Fig. 1) provides a summary of the project's purpose and features, showcasing the ease of navigation and utility for stakeholders. Each application module, such as NDVI and NDWI split panels, is accessible directly through the interface, ensuring user-friendly interaction with the data and analytical outputs.

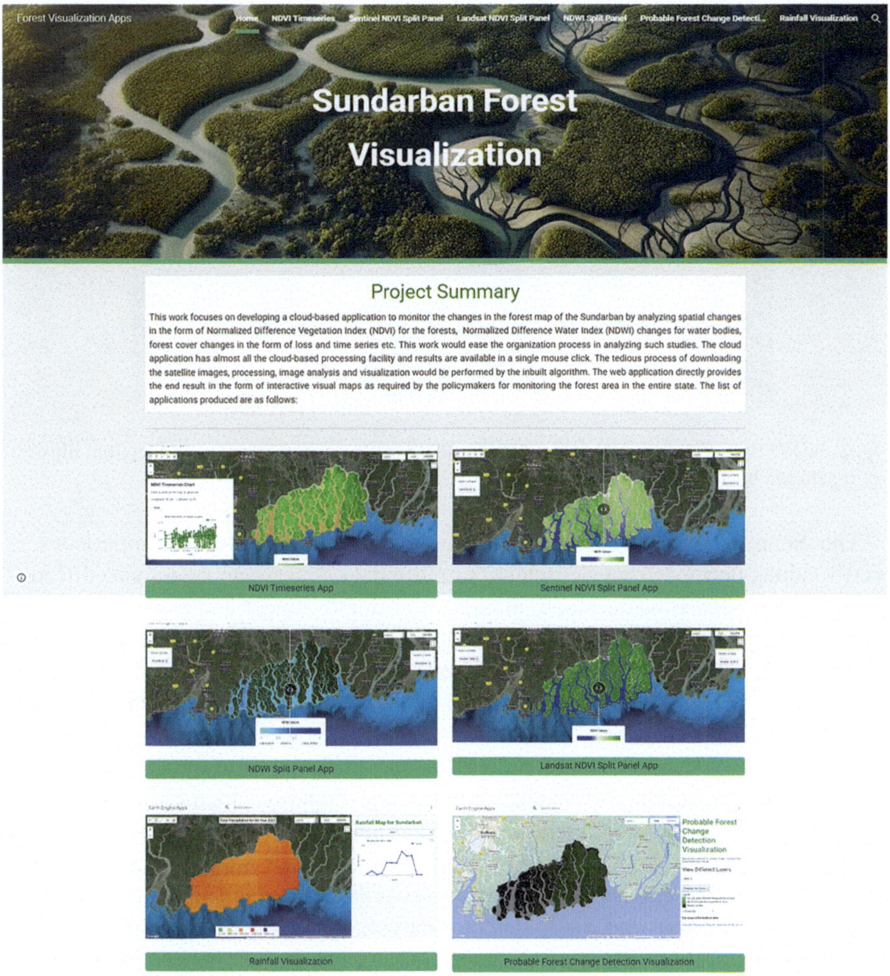

Fig. 1 Gateway of the Sundarbans forest visualisation application, highlighting project modules such as NDVI time-series, NDWI split panels, and probable forest cover change detection

3.6 NDVI Visualisation Panel

The Normalised Difference Vegetation Index (NDVI) visualisation panel within the web application offers a comprehensive platform for analysing vegetation health in the Sundarbans. This panel is structured into three distinct modules, each serving a specific purpose for vegetation monitoring. The NDVI Time Series Panel allows users to explore temporal trends in vegetation health by visualising NDVI values over an extended period. By enabling the analysis of seasonal variations and long--term changes, this panel provides critical insights into the dynamics of forest health.

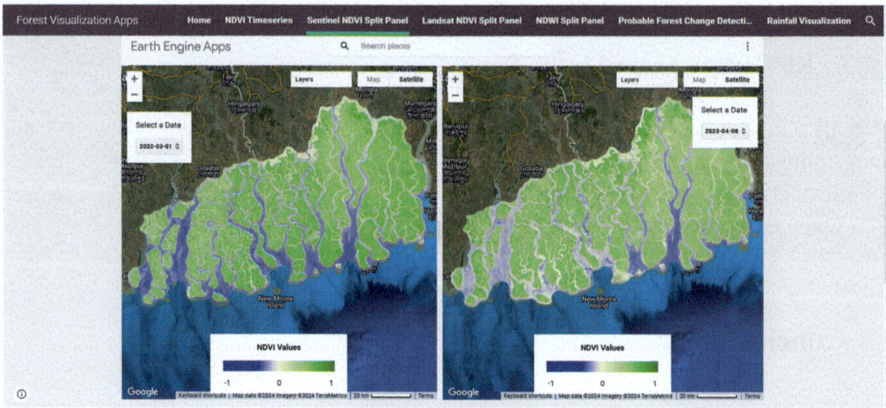

Fig. 2 NDVI Split Panel visualisation in the Sundarbans forest visualisation application, illustrating vegetation health comparisons between March 2022 and April 2023

The Sentinel NDVI Split Panel facilitates detailed temporal comparisons of NDVI values derived from Sentinel-2 satellite data. Users can select two different dates to view side-by-side maps, making it easier to detect and assess vegetation changes over time. Figure 2 demonstrates this feature, showcasing a comparison of NDVI values between March 2022 and April 2023. This tool is particularly valuable for identifying specific areas within the Sundarbans that have undergone significant changes in vegetation cover.

Similarly, the Landsat NDVI Split Panel enables temporal comparisons using Landsat data, leveraging its extensive historical coverage to monitor long-term changes in vegetation. The integration of Sentinel and Landsat datasets within the application ensures robust spatial and temporal analyses, enhancing the capabilities for forest management and conservation. Together, these three panels empower researchers and policymakers to detect, analyse, and address vegetation changes effectively, supporting informed decision-making for the sustainable management of the Sundarbans.

3.7 NDWI Split Panel

The NDWI (Normalised Difference Water Index) Split Panel in the web application provides a user-friendly interface for analysing water dynamics across the Sundarbans. This tool allows users to compare NDWI values between two selected dates, making it possible to monitor changes in water bodies and their extent over time. NDWI is particularly useful for identifying and classifying water-covered areas, marshy lands, and agricultural zones with shallow water presence.

The split-panel visualisation enables side-by-side comparison, allowing users to easily detect changes in water coverage and associated land use between different

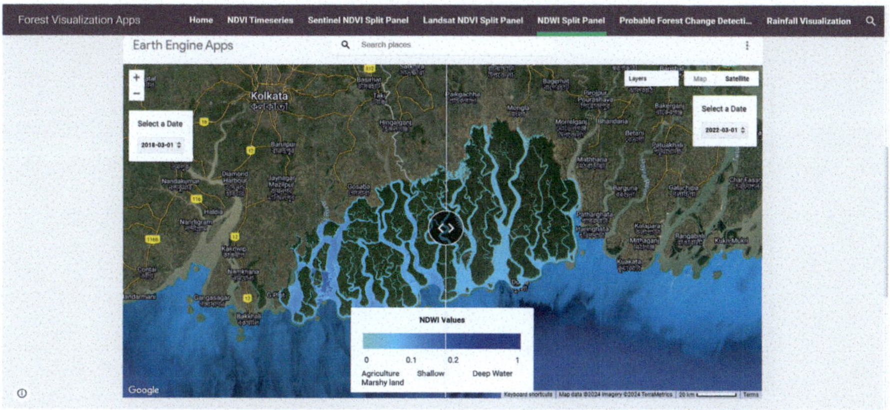

Fig. 3 NDWI split-panel visualisation in the Sundarbans forest visualisation application, comparing water body dynamics between March 2018 and March 2022

timeframes. Figure 3 showcases a comparison of NDWI values from March 2018 and March 2022. This analysis is essential for understanding hydrological shifts in the Sundarbans, which are influenced by seasonal changes, climate impacts, and anthropogenic activities. Such insights are critical for managing water resources, supporting conservation efforts, and mitigating the effects of flooding and sea-level rise.

3.8 Rainfall Visualisation Panel

The rainfall visualisation panel in the web application is designed to analyse and display precipitation patterns over the Sundarbans region. This panel provides an interactive platform where users can select specific years to view total precipitation data across the region, as well as monthly rainfall trends. The combination of spatial and temporal rainfall data facilitates a deeper understanding of hydrological dynamics and their influence on the Sundarbans ecosystem.

The spatial visualisation of total precipitation, shown as a heat map (Fig. 4), highlights varying rainfall intensities across the region, categorised into ranges from 0 to 1600 mm. This representation helps identify areas that experience significant rainfall disparities, which may affect vegetation health, water resources, and flood risks. Alongside this, the temporal graph plots monthly rainfall trends for the selected year, offering insights into seasonal variations and peaks in precipitation.

This dual approach enables policymakers, researchers, and conservationists to examine how rainfall patterns impact forest health and hydrology, providing valuable data for planning and management strategies aimed at sustaining the Sundarbans' fragile ecosystem.

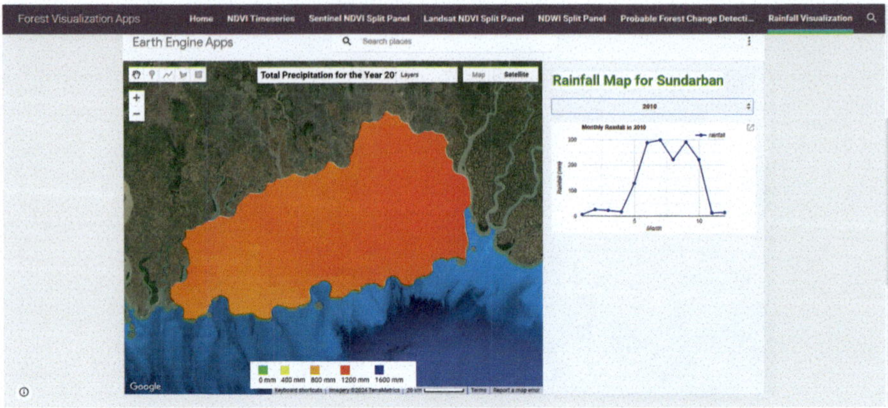

Fig. 4 Rainfall visualisation panel in the Sundarbans forest visualisation application, displaying total precipitation for the year 2010 and its corresponding monthly rainfall trends

4 Discussion

The development of a cloud-based web application using Google Earth Engine (GEE) for monitoring the Sundarbans has demonstrated the potential of integrating remote sensing technologies with cloud computing for effective environmental management. The study highlights the unique advantages of such an approach, particularly in addressing the challenges posed by dynamic ecosystems like the Sundarbans, which are highly sensitive to climatic and anthropogenic pressures.

The NDVI and NDWI visualisation tools within the application provide critical insights into vegetation health and water body dynamics. The ability to analyse temporal trends and compare spatial data using split panels ensures a nuanced understanding of changes in forest cover and water resources. For instance, the NDVI panels reveal areas of vegetation loss and regrowth, which are essential for planning conservation efforts. Similarly, the NDWI split panel effectively tracks changes in water bodies, shedding light on hydrological shifts that may result from climatic variations or human interventions.

The rainfall visualisation panel further strengthens the analytical capabilities of the application by linking precipitation patterns with changes in vegetation and water dynamics. The spatial distribution of rainfall intensity and the temporal trends in monthly precipitation provide a comprehensive view of hydrological influences on the Sundarbans ecosystem. This integration of hydrological data with vegetation and water indices offers a robust framework for understanding the interconnectedness of ecological variables.

Despite its strengths, the study also acknowledges some limitations. While the application effectively visualises and analyses historical and current data, its predictive capabilities could be enhanced by integrating machine learning algorithms to forecast future changes. Additionally, the resolution of some datasets, such as the rainfall data, may constrain the granularity of analyses for certain applications.

Overall, this study underscores the significance of cloud-based platforms like GEE in democratising access to advanced remote sensing tools, enabling researchers and policymakers to make data-driven decisions for sustainable forest management. The Sundarbans' ecological and socio-economic importance necessitates such innovative approaches to ensure its conservation and resilience against ongoing and future challenges. Future work can expand on this framework by incorporating additional datasets, higher-resolution imagery, and predictive modelling techniques to further enhance the application's utility.

5 Conclusion

This study presents a cloud-based web application developed using Google Earth Engine (GEE) for monitoring and visualising the Sundarbans, the world's largest contiguous mangrove forest. By integrating advanced remote sensing techniques with cloud computing capabilities, the application provides a comprehensive platform for analysing vegetation health, water dynamics, and rainfall patterns. The NDVI and NDWI visualisation tools, along with the rainfall analysis panel, enable researchers and policymakers to assess temporal and spatial changes in the Sundarbans, facilitating data-driven conservation and management efforts.

The application demonstrates the potential of remote sensing technologies in addressing complex environmental challenges (Singh et al., 2025a). It empowers stakeholders with interactive tools to explore vegetation trends, hydrological variations, and probable forest cover changes, which are critical for mitigating the impacts of climate change, sea-level rise, and anthropogenic pressures on this fragile ecosystem.

While the application offers significant advancements in forest monitoring, there is scope for further development, including the integration of predictive models and higher-resolution datasets to enhance its analytical capabilities. By continuing to build on such frameworks, future work can ensure more robust and effective solutions for the sustainable management of the Sundarbans and other ecologically significant regions.

In conclusion, the integration of remote sensing, cloud computing, and interactive visualisation in this study serves as a model for leveraging technology in environmental management. This approach not only advances our understanding of the Sundarbans but also sets a precedent for using innovative tools to protect and preserve vulnerable ecosystems worldwide.

Acknowledgements We gratefully acknowledge the University of the Sunshine Coast, Queensland, Australia, and the Asia-Pacific Network for Global Change Research (APN-GCR) for their valuable support. The financial assistance provided by these entities enabled us to pursue our research objectives and complete our work promptly.

References

Ahmad, Z., Shukla, A. K., Singh, V., Sharma, M., & Kumar, P. (2022). Thermodynamic analysis of solar powered trigeneration arrangement for cooling, power and drinking water generation. *Songklanakarin Journal of Science and Technology, 44*.

Ahmed, K. R., & Akter, S. (2017). Analysis of landcover change in southwest Bengal delta due to floods by NDVI, NDWI and K-means cluster with landsat multi-spectral surface reflectance satellite data. *Remote Sensing Applications, 8*, 168–181. https://doi.org/10.1016/j.rsase.2017.08.010

Ashok, A., Rani, H. P., & Jayakumar, K. V. (2021). Monitoring of dynamic wetland changes using NDVI and NDWI based landsat imagery. *Remote Sensing Applications, 23*, 100547. https://doi.org/10.1016/j.rsase.2021.100547

Das, G. K. (2022). *Climate change and coastal hazards in Sunderbans* (pp. 175–198). https://doi.org/10.1007/978-3-031-18846-6_9

Gopal, B., & Chauhan, M. (2006). Biodiversity and its conservation in the Sundarban Mangrove Ecosystem. *Aquatic Sciences, 68*, 338–354. https://doi.org/10.1007/s00027-006-0868-8

Harikesh, Singh, S., Shrivastava, V., & Sharma, V. (2020). *CityGML based 3D modeling of urban area using UAV dataset for estimation of solar potential* (pp. 355–367). https://doi.org/10.1007/978-3-030-37393-1_30

Islam, S. N. (2019). *Sundarbans a dynamic ecosystem: An overview of opportunities, threats and tasks* (pp. 29–58). https://doi.org/10.1007/978-3-030-00680-8_2

Kumar, P., & Singh, O. (2019a). Thermodynamic analysis of solid oxide fuel cell-gas turbine-organic Rankine cycle combined system. *International Journal of Material Science and Mechanical Engineering (JMSME), 6*, 98–102.

Kumar, P., & Singh, O. (2019b). Investigations on integrating SOFC with gas turbine for performance enhancement and sustainable energy. *Journal of Energy and Environmental Sustainability, 7*, 42–46. https://doi.org/10.47469/JEES.2019.v07.100076

Mandal, U. K., Maji, B., Mullick, S., Nayak, D. B., Mahanta, K. K., & Raut, S. (2019). *Global climate change and human interferences as risk factors, and their impacts on geomorphological features as well as on farming practices in sundarbans eco-region* (pp. 405–437). https://doi.org/10.1007/978-3-030-00680-8_14

Mitra, A. (2013). *Sensitivity of mangrove ecosystem to changing climate*. Springer India. https://doi.org/10.1007/978-81-322-1509-7

Miura, S., Amacher, M., Hofer, T., San-Miguel-Ayanz, J., Ernawati, & Thackway, R. (2015). Protective functions and ecosystem services of global forests in the past quarter-century. *Forest Ecology and Management, 352*, 35–46. https://doi.org/10.1016/j.foreco.2015.03.039

Mondal, M., Biswas, A., Haldar, S., Mandal, S., Mandal, P., Bhattacharya, S., & Paul, S. (2022). Climate change, multi-hazards and society: An empirical study on the coastal community of Indian Sundarban. *Natural Hazards Research, 2*, 84–96. https://doi.org/10.1016/j.nhres.2022.04.002

Neogi, S. B., Dey, M., Kabir, S. L., Masum, S. J. H., Kopprio, G., Yamasaki, S., & Lara, R. (2017). Sundarban mangroves: Diversity, ecosystem services and climate change impacts. *Asian Journal of Medical and Biological Research, 2*, 488–507. https://doi.org/10.3329/ajmbr.v2i4.30988

Samanta, S., Hazra, S., Mondal, P. P., Chanda, A., Giri, S., French, J. R., & Nicholls, R. J. (2021). Assessment and attribution of mangrove forest changes in the Indian Sundarbans from 2000 to 2020. *Remote Sensing (Basel), 13*, 4957. https://doi.org/10.3390/rs13244957

Sarkar, S. K. (2022). *Sundarban mangrove wetland (a UNESCO world heritage site): A comprehensive global treatise*. Elsevier.

Singh, H., & Pandey, A. C. (2021). Land deformation monitoring using optical remote sensing and PS-InSAR technique nearby Gangotri glacier in higher Himalayas. *Modeling Earth Systems and Environment, 7*, 221–233. https://doi.org/10.1007/s40808-020-00889-5

Singh, H., & Srivastava, S. K. (2024). From firestick to satellites: Technological advancement and indigenous cultural practice in managing forest fires in Australia. *The Historic Environment: Policy & Practice, 16*, 1–24. https://doi.org/10.1080/17567505.2024.2425246

Singh, H., & Srivastava, S. K. (2025). Identification of forest fire-prone region in Lamington National Park using GIS-based multicriteria technique: Validation using field and Sentinel-2-based observations. *Geocarto International, 40*. https://doi.org/10.1080/10106049.2025.2462484

Singh, H., Meraj, G., Singh, S., Shrivastava, V., Sharma, V., Farooq, M., Kanga, S., Singh, S. K., & Kumar, P. (2022a). Status of air pollution during COVID-19-induced lockdown in Delhi, India. *Atmosphere (Basel), 13*, 2090. https://doi.org/10.3390/atmos13122090

Singh, S., Singh, H., Sharma, V., Shrivastava, V., Kumar, P., Kanga, S., Sahu, N., Meraj, G., Farooq, M., & Singh, S. K. (2022b). Impact of forest fires on air quality in wolgan valley, new south Wales, Australia—A mapping and monitoring study using google earth engine. *Forests, 13*. https://doi.org/10.3390/f13010004

Singh, H., Ang, L.-M., Lewis, T., Paudyal, D., Acuna, M., Srivastava, P. K., & Srivastava, S. K. (2024). Trending and emerging prospects of physics-based and ML-based wildfire spread models: A comprehensive review. *Journal of Forestry Research, 35*, 135. https://doi.org/10.1007/s11676-024-01783-x

Singh, H., Ang, L.-M., Paudyal, D., Acuna, M., Srivastava, P. K., & Srivastava, S. K. (2025a). A comprehensive review of empirical and dynamic wildfire simulators and machine learning techniques used for the prediction of wildfire in Australia. *Technology, Knowledge and Learning*. https://doi.org/10.1007/s10758-025-09839-5

Singh, H., Ang, L.-M., & Srivastava, S. K. (2025b). Benchmarking artificial neural networks and U-net convolutional architectures for wildfire susceptibility prediction: Innovations in geospatial intelligence. *IEEE Transactions on Geoscience and Remote Sensing, 63*, 1–15. https://doi.org/10.1109/TGRS.2025.3529134

Singh, H., Ang, L.-M., & Srivastava, S. K. (2025c). Active wildfire detection via satellite imagery and machine learning: An empirical investigation of Australian wildfires. *Natural Hazards*. https://doi.org/10.1007/s11069-025-07163-w

Sos, J., Penglase, K., Lewis, T., Srivastava, P. K., Singh, H., & Srivastava, S. K. (2023). Mapping and monitoring of vegetation regeneration and fuel under major transmission power lines through image and photogrammetric analysis of drone-derived data. *Geocarto International, 38*. https://doi.org/10.1080/10106049.2023.2280597

Srivastava, S. K., Mukul, S. A., Banerjee, S., Barman, D., Kar, G., Hindersah, R., Candra, M., Arief, W., Arfin, M. A. S., Srivastava, C., Banerjee, S. A., Hindersah, G., Arief, R., & Khan, A. (2025). A hierarchical and inclusive zonation of Sundarbans for monitoring and management of the world heritage area. *APN Science Bulletin, 15*, 1–18. https://doi.org/10.30852/sb.2025.2896

Tamiminia, H., Salehi, B., Mahdianpari, M., Quackenbush, L., Adeli, S., & Brisco, B. (2020). Google Earth Engine for geo-big data applications: A meta-analysis and systematic review. *ISPRS Journal of Photogrammetry and Remote Sensing, 164*, 152–170. https://doi.org/10.1016/j.isprsjprs.2020.04.001

Watson, J. E. M., Evans, T., Venter, O., Williams, B., Tulloch, A., Stewart, C., Thompson, I., Ray, J. C., Murray, K., Salazar, A., McAlpine, C., Potapov, P., Walston, J., Robinson, J. G., Painter, M., Wilkie, D., Filardi, C., Laurance, W. F., Houghton, R. A., et al. (2018). The exceptional value of intact forest ecosystems. *Nature Ecology & Evolution, 2*, 599–610. https://doi.org/10.1038/s41559-018-0490-x

Yang, L., Driscol, J., Sarigai, S., Wu, Q., Chen, H., & Lippitt, C. D. (2022). Google earth engine and artificial intelligence (AI): A comprehensive review. *Remote Sensing (Basel), 14*, 3253. https://doi.org/10.3390/rs14143253

Zaman, M. S., & Chowdhury, T. H. (2024). The Sundarbans, the world's largest tidal halophytic mangrove Forest: Its economic and ecological significance. *Bangla Journal of Interdisciplinary Sciences, 2*(1), E1–E15.

Leaf Chlorophyll Concentration and Plant Diversity Assessment in a Mangrove Forest of Eastern India

Mukunda Dev Behera, Surbhi Barnwal, Somnath Paramanik, Sujit M. Ghosh, Soumit K. Behera, Bimal K. Bhattyacharya, Saroj K. Barik, and Prashant K. Srivastava

Abstract The leaf chlorophyll concentration is an important biophysical parameter of plant physiological state since it plays crucial roles in photosynthesis, physiology, and other biological functions in plants. It can be estimated directly using destructive sampling or indirectly using handheld devices. Leaves were collected from the Bhitarkanika mangrove ecosystem from the upper canopy of the trees for three species such as *Heritiera fomes*, *Excoecaria agallocha*, and *Avicennia officinalis* for laboratory analysis, while Soil Plant Analysis Development (SPAD)-502 was used to get an indirect estimation. Regression analysis was done to investigate the relationship between SPAD-based chlorophyll and lab-based Leaf Chlorophyll Content (LCC) using 15 samples for each three species *H. fomes, E. agallocha,* and *A. officinalis.* Two-pigment (chlorophyll a and b) indices such as PSNDa and PSNDb images and two soil line indices such as SAVI and (g) OSAVI images were derived using Airborne Visible/Infrared Imaging Spectrometer Next Generation (AVIRIS NG) Hyperspectral images. Classified image of three dominant mangrove species using six indices (red edge, chlorophyll, and soil line-based properties), and three red edge bands were derived from AVIRIS NG data using Random Forest (RF)

M. D. Behera (✉) · S. Barnwal · S. Paramanik · S. M. Ghosh
Centre for Ocean, River, Atmosphere and Land Sciences (CORAL), Indian Institute of Technology Kharagpur, Kharagpur, India
e-mail: mdbehera@coral.iitkgp.ac.in

S. K. Behera
CSIR–National Botanical Research Institute, Lucknow, India

B. K. Bhattyacharya
Space Applications Centre, ISRO, Ahmedabad, India

S. K. Barik
CSIR–National Botanical Research Institute, Lucknow, India

North-Eastern Hill University, Shillong, Meghalaya, India

P. K. Srivastava
Remote Sensing Laboratory, Institute of Environment and Sustainable Development, Banaras Hindu University, Varanasi, India

S. K. Srivastava et al. (eds.), *Earth Observation for Monitoring Mangrove Ecosystems*, https://doi.org/10.1007/978-3-032-16668-5_12

algorithm. Different spectra of *H. fomes* and *E. agallocha*, generated from AVIRIS, exhibit no significant variation in the pattern of reflectance around NIR region. Two mangrove species such as *H. fomes* and *E. agallocha* were classified as major or dominant species. In RF model, red edge-based indices exhibit highest contribution for generation of species distribution map.

Keywords Bhitarkanika WLS · Random Forest · Vegetation index · Hyperspectral imagery

1 Introduction

Mangrove forest represents a vegetation community which inhabits a diversified species of mangrove and unique herb, found in the intertidal zone of sea and land in the tropical and sub-tropical region of the world (Giri et al., 2011). It is one of the most carbon-rich ecosystems in the tropical region as it accumulates carbon at a faster rate than other tropical ecosystems due to the presence of carbon-absorbing dead roots. It stores about 3% of the world's tropical forest carbon, while it occupies <1% of the total area of tropical forest (Alongi, 2012). The Mangrove Ecosystem (ME)s have the ability to survive in a very critical environmental condition such as high temperature, high salinity, extreme tides, and strong wind conditions. The distribution and composition of mangrove species depend on the influencing environmental factors such as the distance from the sea or estuary bank, duration, and frequency of the tidal inundation and the composition of soil (Kuenzer et al., 2011). In addition to that, mangroves exhibit pigment variation between species and health conditions (Pastor-Guzman et al., 2015). The spatial variation in biophysical variables such as leaf area index and canopy chlorophyll content depends on the species composition and distribution (Zavoruev & Zavorueva, 2002; Gond et al., 1999).

Economically, they are highly important since livelihood of coastal villages are often very dependent on adjacent mangroves (Ghosh et al., 2020), either directly, because of the products like timber, wood, and medicines they derive from these habitats and are able to sell, or because of the coastal fisheries that are supported by mangroves, or the coastlines that are sheltered from storms. Apart from its ecological importance, there are some ecological values of the MEs as it protects the shoreline from the cyclones, floods, sea level rise, wave action and coastal erosion. Mangrove swamps act as traps for the sediments and sink for the nutrients. Additionally, Mangrove sediments can absorb and hold heavy metals thereby preventing the spread of metal pollution in coastal areas.

1.1 Importance of Leaf Chlorophyll Concentration

Leaf chlorophyll content is important biophysical parameter of plant physiological state since it plays crucial roles in photosynthesis, physiology, and other biological functions in plants (Verma et al., 2024). Consequently, the assessment of chlorophyll concentration could act as surrogates of mangrove vegetation health condition as well as for modeling ecosystem productivity at regional and global scale (Paramanik et al., 2025). Leaf chlorophyll concentration can be estimated directly using destructive sampling or indirectly using handheld devices (Paramanik et al., 2023). Destructive methods include laboratory analysis where the leaf chlorophyll concentration is extracted by chemical analysis (Priyanka et al., 2023). On the contrary, handheld devices such as Soil Plant Analysis Development (SPAD)-502 and Chlorophyll Content Meter (CCM)-200 are used in the field for nondestructive estimation of leaf chlorophyll concentration based on the transmittance of certain wavelengths (600–700 and 400–500 nm) (Minolta, 1989).

However, remote sensing offers an alternative to estimate leaf/canopy-level chlorophyll concentration (Singh et al., 2023, 2025). For vegetation analysis, narrow red edge spectral bands present between red and NIR range of electromagnetic spectrum, which indicates a sudden change in reflectance value and show strong correlation with the leaf chlorophyll content that depends on leaf type and canopy structure (Srivastava et al., 2021). Vegetation indices (VIs) use two or more spectral bands and enhance biophysical properties of different species based on their relative reflectance (Jagadish et al., 2024). Such vegetation indices are used as proxy, which are statistically linked with field estimated biophysical parameters (Paramanik et al., 2022). The VI approach is based on the statistical or empirical relationship between arithmetic combinations of two or more spectral bands and a particular leaf or canopy characteristic (i.e., chlorophyll concentration). The study objectives include (i) correlating the field-based chlorophyll concentration to that of laboratory-based chlorophyll concentration estimation and deriving a species-level classified map of mangroves using hyperspectral sensor-based high-resolution satellite data and machine learning algorithm. Such study may provide insights to automated index based classification and mapping of mangrove forests.

2 Study Area

This study was carried out in Dangamal and Bhitarkanika blocks of Bhitarkanika Wildlife Sanctuary (BWLS), situated on the eastern coastal Kendrapara district in Odisha state, India. Geographically it is located between 20°4′N and 20°5′N latitude and 86°49′E to 86°56′E longitude and occupies an area of 51.68 km^2 approximately. The study area covers the central region of the BWLS and is given maximum attention for conservation and management (Upadhyay & Mishra, 2014). Bhitarkanika mangrove ecosystem flourishes in the deltaic region, formed by the

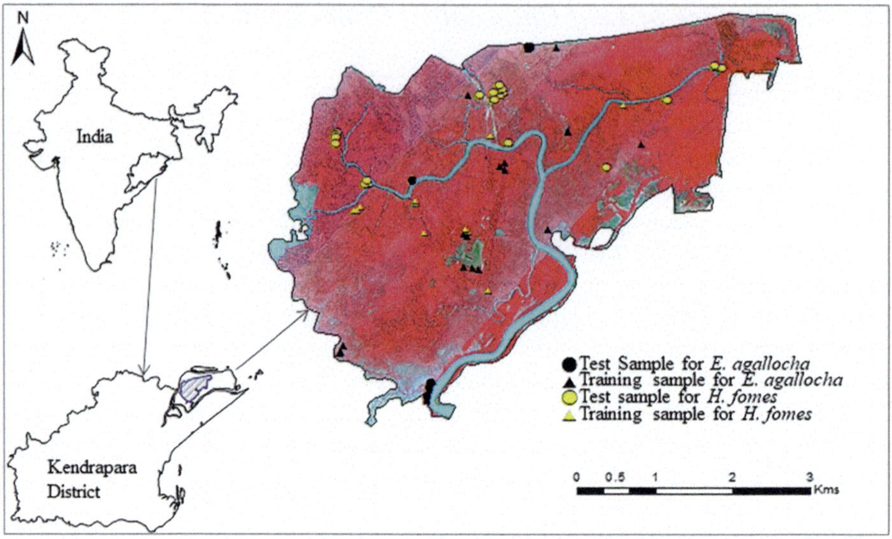

Fig. 1 FCC image showing Dangamala forest area of Bhitarkanika wildlife sanctuary. Field visit and reconnaissance survey were done and the location of some *E. agallocha* and *H. fomes* species-dominant plots are shown, along with their usage i.e. test and training site

rich alluvial deposits of Brahmani, Baitarani & the Dhamra River (Fig. 1; Dani et al., 1999).

The region is classified under the tropical warm and humid climate region and receives about 1670 mm average annual rainfall, where the maximum rainfall events occur between August and September. The average temperature in this region varies in between 10 °C in winter and 43 °C in summer and the relative humidity ranges from 70 to 84% throughout the year (Pattanaik et al., 2008). A total of 44 species of mangrove plants were recorded in the two blocks of BWLS (*Annexure*: Photo 1–6), among which 8 mangrove species are reported dominant such as *Heritiera fomes*, *Excoecaria agallocha*, *Cynometra ramiflora*, *Avicennia officinalis*, *Sonneratia apetala*, *Rhizophora mucronata*, *Aegiceras corniculatum* and *Ceriops decandra* (Upadhyay et al., 2008).

3 Material and Methods

3.1 Field Data Collection

SPAD-502 was used to get an indirect estimation of LCC. The value given by SPAD instrument is a unit less quantity. SPAD measurements were done for six species over Dangamal and Bhitarkanika forest blocks of BWLS for November and December. Plot-level sampling method was adopted with a grid size of 20 m × 20 m,

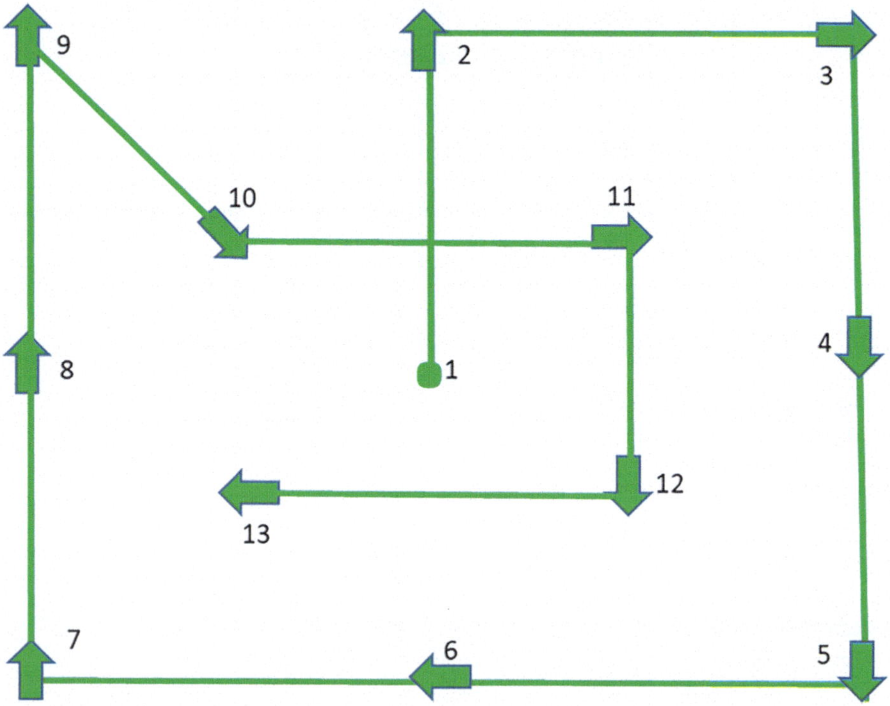

Fig. 2 Sampling protocol used for collecting chlorophyll data. Each dot represents a site (tree) where chlorophyll reading was taken; "S" indicates "starting point" and "E" indicates "ending point"

where 13 trees were selected in each plot to measure LCC using a sampling scheme (Fig. 2). LCC were taken for three leaves of upper, middle, and lower canopy, and from each leaf six readings were taken.

3.2 Leaf Chlorophyll Concentration Measurement

Leaf was collected from the upper canopy of the 15 mangrove trees for 3 species individually, i.e., *H. fomes*, *E. agallocha*, and *A. officinalis* in the field. Furthermore, it was immediately brought to the laboratory where 3 mL chlorophyll solution was created out of 80% acetone solution for all the samples. Moreover, absorbance value was recorded for all the samples at wavelength 665 and 649 nm using spectrophotometer. After that chlorophyll content (mg/g) was calculated using Arnon (1949) equation including absorption value at 665 and 659 nm. For the same leaf, six SPAD readings were taken from each leaf to cover the variation in the chlorophyll range.

3.3 Procedure for Preparation of Samples and Measurement of the Leaf Chlorophyll Concentration

The method of the chlorophyll extraction using acetone is described in two sections:

- Equipment list
- Preparation of leaf samples and measurement of the leaf chlorophyll concentration

3.3.1 Equipment List

- Konica Minolta SPAD-502
- Pipette
- Cuvettes
- Test tube rack
- Test tubes
- Bottle for 80% acetone solution
- Test tube rack
- Spectrophotometer (e.g., Thermo Fisher Genesys 50)
- Safety glasses
- Mortar and pestle
- Weighing machine

3.3.2 Preparation of Samples and Measurement of LCC

1. 30 leaves per species were selected from the upper canopy of the tree.
2. Six SPAD reading were taken from each leaf to cover the variation in the chlorophyll range.
3. Samples were taken into wet lab where it was kept in deep freeze immediately.
4. To get the best result, mortar and pestle were cooled down.
5. First of all 80% solution of acetone is prepared and kept it for cooling in the freeze.
6. Each sample weighed 0.1 mg and grinded using mortar and pestle (grinding was done with the help of liquid nitrogen) which was used to prepared 3 mL chlorophyll solution from 80% acetone.
7. 3 mL chlorophyll solution was prepared for each sample.
8. Solution was centrifuged in centrifuge machine.
9. 1 mL of the chlorophyll solution from each test tube was placed into a cuvette.
10. The spectrophotometer was used to measure the absorbance of the solution in the cuvette at 665, 649, and 680 nm for each sample.

3.3.3 Estimation of Chlorophyll Concentration

Chlorophyll contents were calculated using Arnon (1949). The following steps were used to calculate chlorophyll content in mg/g:

- Chl_a and Chl_b were calculated using the absorbance values at A_{665} and A_{649} using the following equations Arnon (1949):

$$\mathrm{Chla} = \left(\left(12.7^{*}663\right) - \left(2.63^{*}645\right)\right) * 3/100$$

$$\mathrm{Chlb} = \left(\left(22.9^{*}645\right) - \left(4.68^{*}663\right)\right) * 3/100$$

- Chl_a and Chl_b readings was added to obtain Chl_{total}:

$$\mathrm{Chl}_{\mathrm{total}} = \mathrm{Chla} + \mathrm{Chlb}$$

- For each sample, the mean of the SPAD readings was calculated.
- Finally to calculate the regression equation between the SPAD readings and the chlorophyll concentrations a linear equation was used.

3.4 Correlation Analysis

Regression analysis was done to investigate the relationship between SPAD-based chlorophyll and lab-based LCC Daughtry et al. (2000), using 15 samples for each three species *H. fomes*, *E. agallocha*, and *A. officinalis* individually. Regression curve and coefficient of determination is estimated for all three species.

3.5 Airborne Remote Sensing Imagery

Four continuous tiles of AVIRIS level 2 data were used to cover the study area for species identification, of which three tiles were taken on 25 December 2015, and one tile was taken on 28 December 2015. Atmospherically corrected AVIRIS level 2 data were provided by the airborne flights of hyperspectral sensor called AVIRIS NG over India in collaboration with Jet Propulsion Laboratory (JPL), NASA, which were used for further analysis. It has a very narrow contiguous bandwidth of 5 nm and spatial resolution of 4 m, and its spectral range covers 376–2500 nm with a total of 425 bands. Mean was calculated for upper, middle, and lower canopy LCC for each species to observe the variation in the chlorophyll level by averaging upper, middle, and lower canopy LCC for each species. Proportion of dominant trees was also calculated by counting the number of total trees and dominant trees in each plot.

3.6 Generation of Vegetation Indices (VIs)

Nine vegetation indices (VIs) were generated from AVIRIS bands, i.e., ReNDVI, ReCI, IRECI, PSNDa, PSNDb, SAVI, OSAVI, TCARI, and MCARI. Spectral Angle Mapper (SAM) classification was performed to discriminate mangrove forest and non-mangrove pixels (water body, river, grassland, and agriculture fallow land). Furthermore, this thematic map was applied in all VIs to remove non-mangrove pixels from the study area. All VIs values were extracted from each ground sampled plots to correlate with LCC. ENVI software was used to generate all VIs and SAM classified thematic map.

3.7 Spectra Generation

List of VIs layers derived from narrow bands of AVIRIS and VIs value extracted for 82 dominant plots, i.e., 50 plots of *H. fomes* and 32 plots of *E. agallocha* community, were used for generating RF-based model. Spectra for pure plot of *H. fomes* and *E. agallocha* plant species were generated using ENVI software to see if these two species can be separated from each other or not in visible and NIR wavelength region.

3.8 Random Forest (RF) Classification Algorithm

RF is a nonparametric ensemble classification algorithm that uses many classifiers or trees (Breiman, 2002). Hundreds of classifiers are built and their decisions combined usually by majority of the vote (Na et al., 2010). The RF classifier uses bagging to form an ensemble of classification and induction tree like tree classifiers (Kulkarni & Lowe, 2016). 2/3 of the total sample is used as training dataset, and remaining 1/3 of samples were used to estimate the out-of-bag (OOB) error (Fletcher, 2016). It can handle high-dimensional and intercorrelated data and is efficient to avoid the problem of overfitting (Belgiu & Drăguţ, 2016; Wang et al., 2018). It quantifies the level of importance of variables and indicates the percentage of contribution of each predictor variable used for generating model.

The field data points were divided into 70% (training points) and 30% (testing points) for model development and validation, respectively. Tenfold cross-validation approach was adopted to improve the performance of the model by tuning the "mtry" parameter, where the value of "mtry" was optimized based on the least root mean square error (RMSE). With satisfied model accuracy, RF algorithm-based model was applied on the whole image (four tiles) to classify the image data cube generated with the VIs. The "raster," "randomForest," and "caret" packages were used in this case using statistical software R.

4 Results and Discussion

4.1 Occurrence of Mangrove Species

Number of trees considered for chlorophyll concentration estimation was different for diverse mangrove species and depends on their occurrence and accessibility of the area. Majority of the obstructions occurred during field sampling was due to the presence of pneumatophores in the muddy floor. More number of *H. fomes* and *E. agallocha* trees were considered for leaf chlorophyll concentration estimation as they were found to be dominant in the field. However, very less number of trees was found for *A. officinalis*, *C. decandra*, *C. ramiflora* and *Xylocarpus mekongensis*, trees in the accessible site.

4.2 Analysis of Leaf Chlorophyll Concentration

Field visits were carried out on Bhitarkanika mangrove forest and total six species occurrence and their chlorophyll value (SPAD value) have been recorded. Average of upper-, middle-, and lower-layer canopy chlorophyll concentration for all species was calculated (Figs. 3, 4, and 5). Finally, average chlorophyll concentrations of all canopy layer of each species were calculated (Fig. 6) (Mishra et al. (2019)). Percentage of total number of trees of different species was also calculated. From the chlorophyll statistics of different canopy layer (upper, middle, and lower layer canopy) of each mangrove plants, it was observed that upper layer canopy LCC is

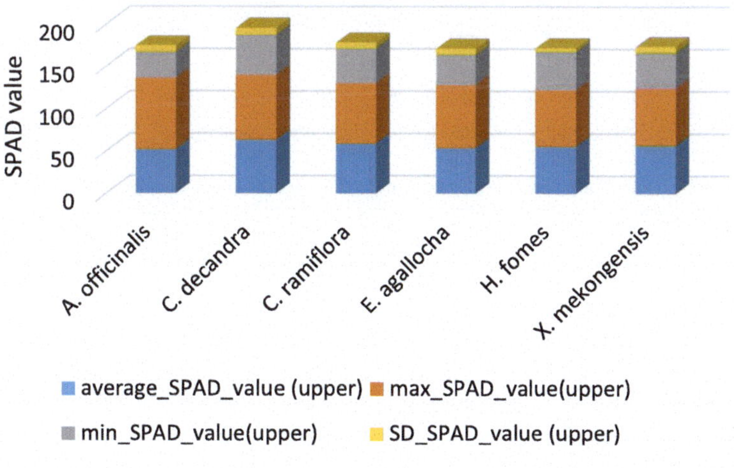

Fig. 3 Average, minimum, maximum, and standard deviation upper canopy chlorophyll concentration for six species

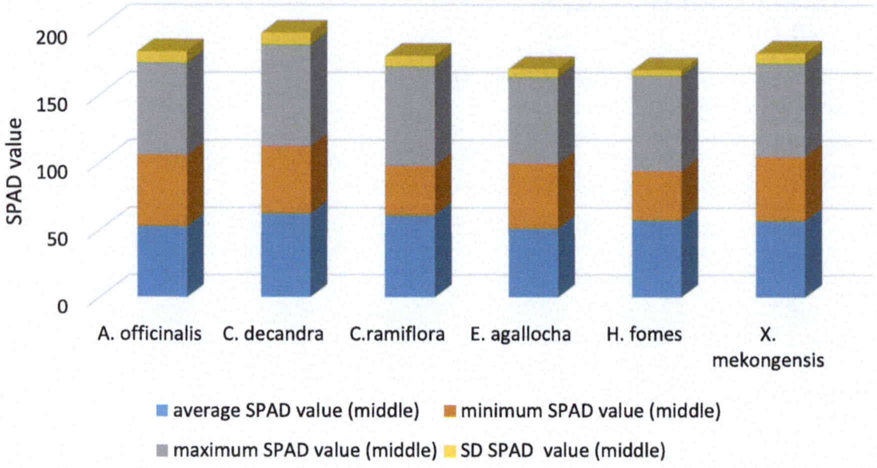

Fig. 4 Average, minimum, maximum, and standard deviation middle canopy chlorophyll concentration for six species

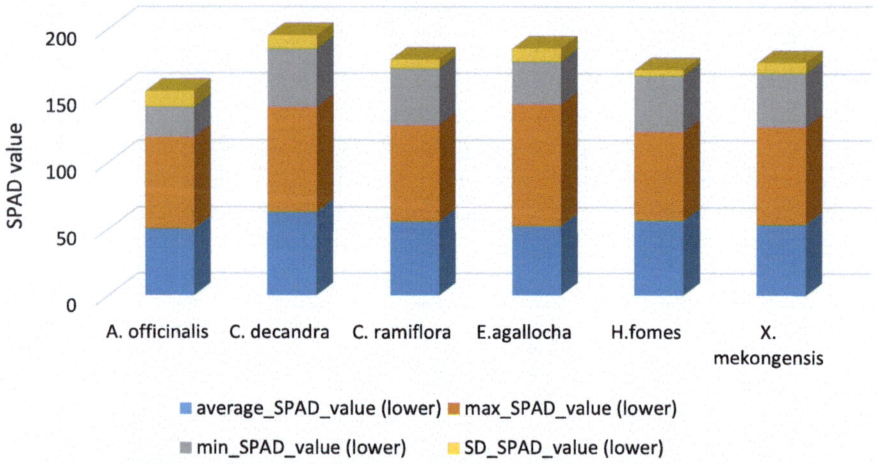

Fig. 5 Average, minimum, maximum, and standard deviation lower canopy chlorophyll concentration for six species

higher than lower and middle layer canopy because it receives more solar radiation compared to other layers.

Among six species, *X. mekongensis* was shown to have highest leaf chlorophyll concentration value, while *E. agallocha* have the lowest leaf chlorophyll concentration value, calculated by averaging leaf chlorophyll concentration value of the individual species of all the field plots. *A. officinalis*, *C. decandra*, and *C. ramiflora* showed the highest variation in the range of LCC, while *E. agallocha* LCC were confined to smaller range.

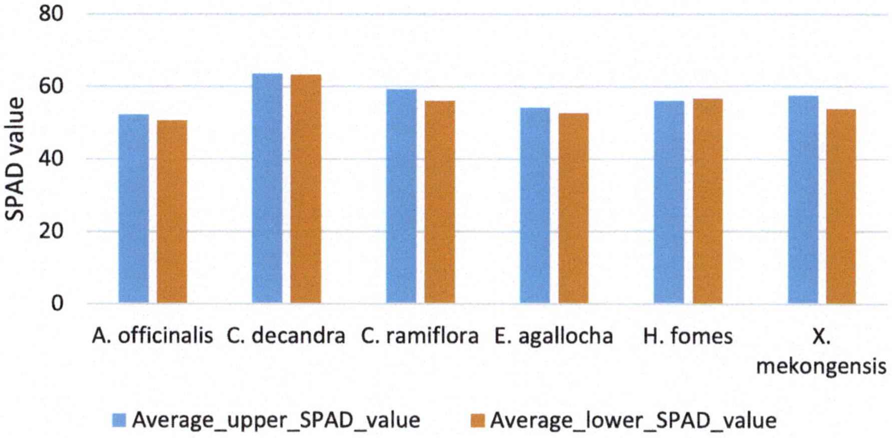

Fig. 6 Average of all layer canopy chlorophyll concentration for six species

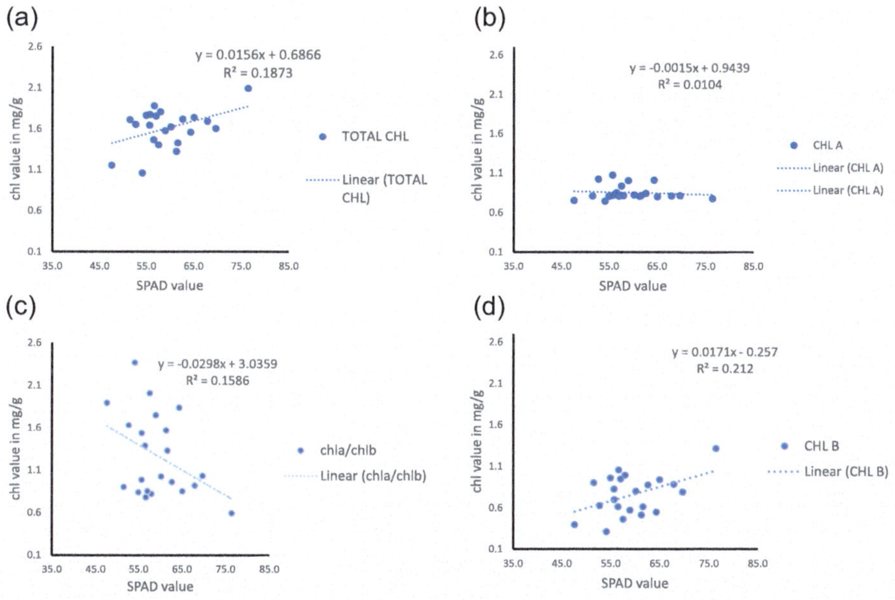

Fig. 7 Graph showing scatter plot between (**a**) SPAD readings and total chl (lab-based), (**b**) SPAD reading and chl a, (**c**) SPAD readings and chl a/chl b, (**d**) SPAD reading and chl b for *H. fomes*

4.3 Correlation Between SPAD and Chlorophyll Value

To investigate the relationship between SPAD and LCC, simple linear regression was applied, and the coefficient of correlation was computed for three mangrove species on a per-species basis (Nigam et al. (2019)) (Figs. 7, 8, and 9). Patterns of

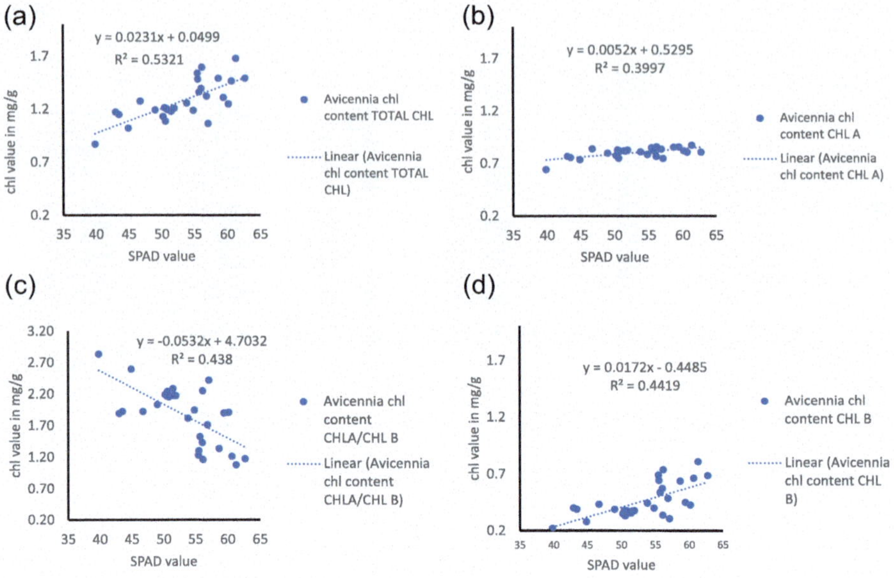

Fig. 8 Graph showing scatter plot between (**a**) SPAD readings and total chl (lab-based), (**b**) SPAD reading and chl a, (**c**) SPAD readings and chl a/chl b, (**d**) SPAD reading and chl b for *E. agallocha*

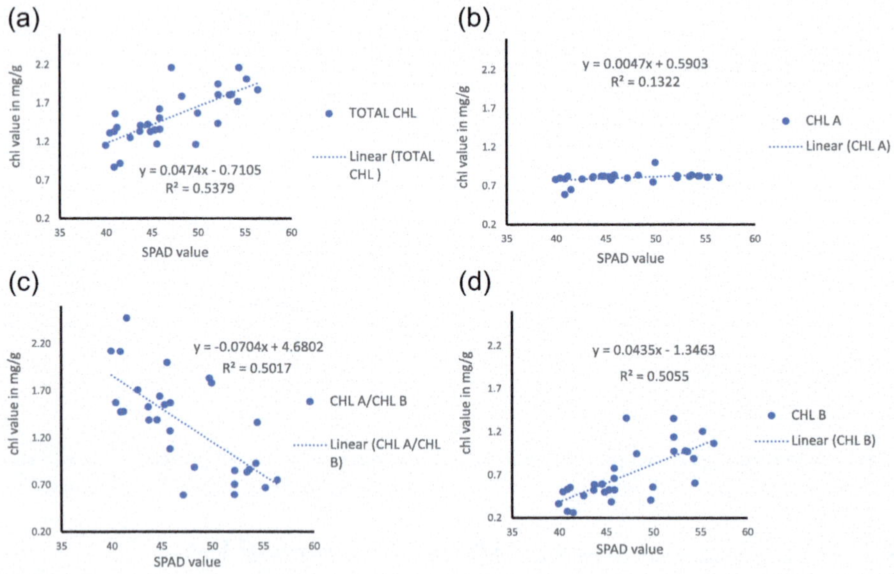

Fig. 9 Graph showing Scatter plot between (**a**) SPAD readings and total chl (lab-based), (**b**) SPAD reading and chl a, (**c**) SPAD readings and chl a/chl b, (**d**) SPAD reading and chl b for *A. officinalis*

SPAD data of 15 points for all species were observed and found that one value was extreme (outlier) for *E. agallocha* and *A. officinalis.* Coefficient of correlation was calculated after removing the extreme value. Highest value of coefficient of correlation was observed in *H. fomes* ($r = 0.72$), followed by *A. officinalis* (0.64) and *E. agallocha* (0.64).

Each curve depicts the correlation between SPAD observed chlorophyll concentration and the total chlorophyll concentration, chlorophyll a concentration, chlorophyll b concentration, and the ratio of chlorophyll a and chlorophyll b concentrations. While in each case a positive correlation was observed, the ratio of chlorophyll a and chlorophyll b concentrations demonstrated negative trend.

4.4 Vegetation Indices

The spatial pattern of ReCI and ReNDVI showed marginal variations (Fig. 10a, c). On the contrary, IRECI exhibited very less spatial variation, where higher values were mostly seen in the northern region, whereas, low to moderate values in the rest of the region (Fig. 10b). In comparison to IRECI, the spatial variation of RECI and

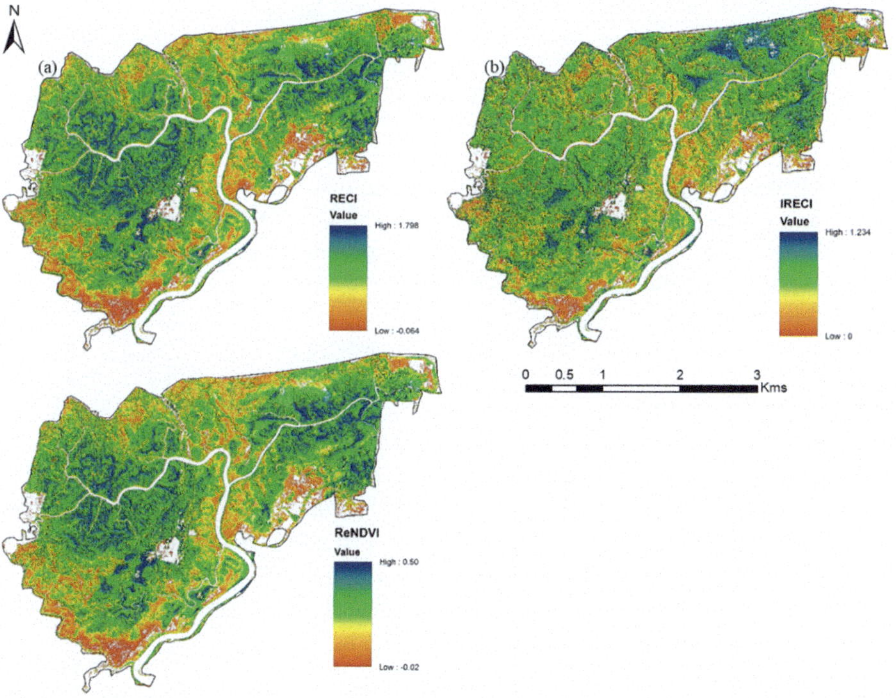

Fig. 10 Two chlorophyll indices (**a**) RECI and (**b**) IRECI images; and one vegetation index (**c**) ReNDVI images are shown, derived using AVIRIS Hyperspectral images

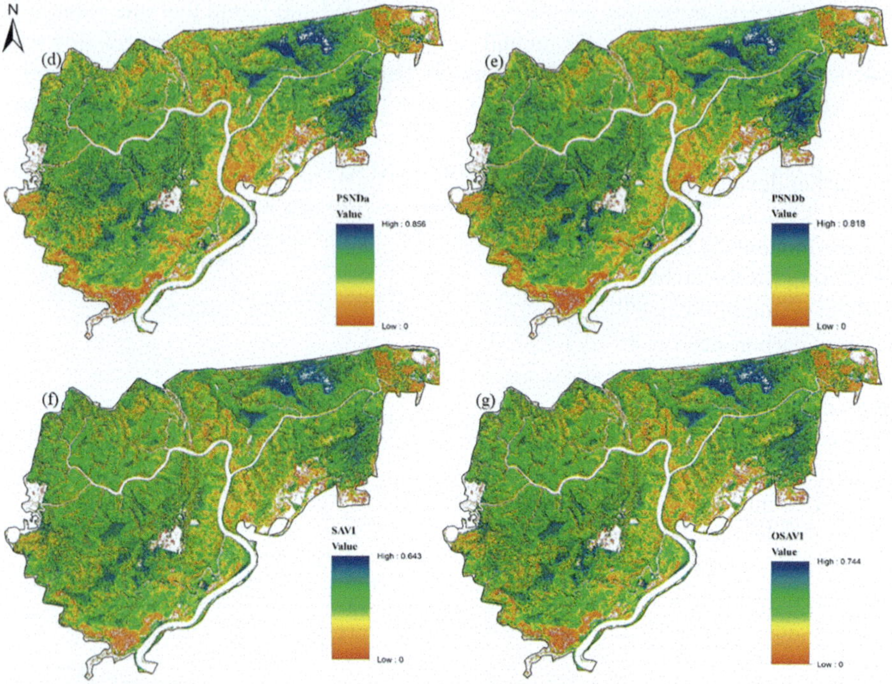

Fig. 11 Two pigment (chlorophyll a and b) indices (**d**) PSNDa and (**e**) PSNDb images; and two soil line indices (**f**) SAVI and (**g**) OSAVI images are shown, derived using AVIRIS-NG Hyperspectral images

ReNDVI were found more clustered. The spatial distribution of PSNDa, PSNDb, SAVI, and OSAVI (Fig. 11d–g) was observed similar as IRECI, where the maximum values were observed in the northern region and some parts of west-central regions and low to moderate values in rest of the regions.

4.5 Spectra Analysis

Spectra generated for two pure plots of each *H. fomes* and *E. agallocha* community using AVIRIS-NG imagery (Gupta et al., 2022). From a single plot, nine spectra were generated. In Fig. 12, spectra are shown for *A. officinalis*. From 18 spectra (9 spectra from a plot and 9 spectra of different plot), it was observed that there is consistency in the pattern of the reflectance value in NIR region, while in visible region due to similar chlorophyll range, it shows the similar variability pattern. Additionally, it was also observed that most of the spectra have a common peak at water absorption band around 900 nm. In NIR region, reflectance value for 18 spectra was shown to vary between 0.2 and 0.35.

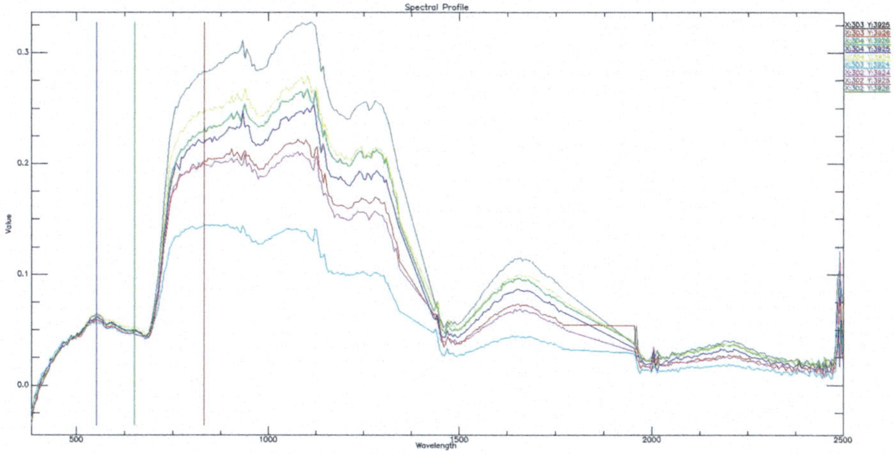

Fig. 12 Spectra generated from nine pixel of *A. officinalis* tree of same plot for Bhitarkanika forest, where *X*-axis represents the wavelength region and *Y*-axis represent the reflectance value

From the spectra analysis of *H. fomes* and *E. agallocha* pixel, generated from AVIRIS-NG imagery, we can conclude that same species shows random variation across certain part of EMR. It might be due to the incorporation of reflectance due to the underlying soil. So we need a spectroradiometer data to validate the spectra generated from AVIRIS-NG imagery.

4.6 Importance Ranking of All VIs

Importance ranking for all indices is as follows: ReNDVI, ReCI, PSNDb, PSNDa, IRECI, OSAVI, and SAVI in descending order, evaluated using RF-based model (Fig. 13). It revealed that two red edge-based indices such as ReNDVI and ReCI are given high to moderate importance in generating RF model, while soil line indices, i.e., SAVI and OSAVI have lowest contribution in making RF-based species map. IRECI was ranked between PSNDa and OSAVI. Pigment based indices such as PSNDa and PSNDb were found to moderately important for the classification. In addition to other indices, ReNDVI and ReCI were able to discriminate mangrove species to some extent as these indices incorporate a red edge band which is sensitive to chlorophyll variation. The overall accuracy was estimated as 95.45% with training pixel and kappa coefficient as 0.879.

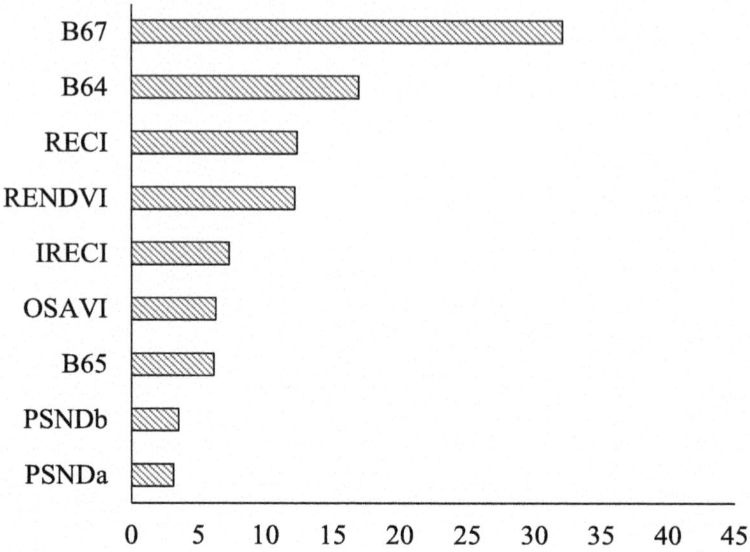

Fig. 13 Importance ranking of seven indices derived using random forest (RF) algorithm, with B67 depicting minimum and PSNDa showing maximum importance

4.7 *Maps Derived Using Random Forest*

Three species map was generated by implementing three set of data, i.e., (1) indices, (2) combination of two red edge band and indices, and (3) combination of six red edge bands and six indices in RF algorithm.

4.7.1 RF (Hyperspectral Indices Only)

RF algorithm was implemented to combination of six indices. Species map was classified into five classes among which three are dominant classes of mangrove, i.e., *H. fomes*, *E. agallocha*, and *A. officinalis*, and other two are mixed stands (*A. officinalis-E. agallocha* and *H. fomes-E. agallocha*). It was observed from the spatial distribution map of mangrove species derived from AVIRIS bands (Fig. 14a) that most of the region is occupied with *H. fomes*, in the form of homogeneous patch, wherever along Bhitarkanika boundary and southern west region of the study area, *E. agallocha* was observed dominantly (Behera et al., 2021). Overall accuracy of the map was observed to be 64%.

4.7.2 RF (Hyperspectral Indices and All Red Edge Bands)

OA increased from 64 to 67% in terms of accuracy and spatial variability when six red edge bands (B63, B64, B65, B65, B66, and B67) of AVIRIS-NG merged with indices, indicating that red edge bands and their indices are much more efficient to discriminate and classify mangrove species (Fig. 14b).

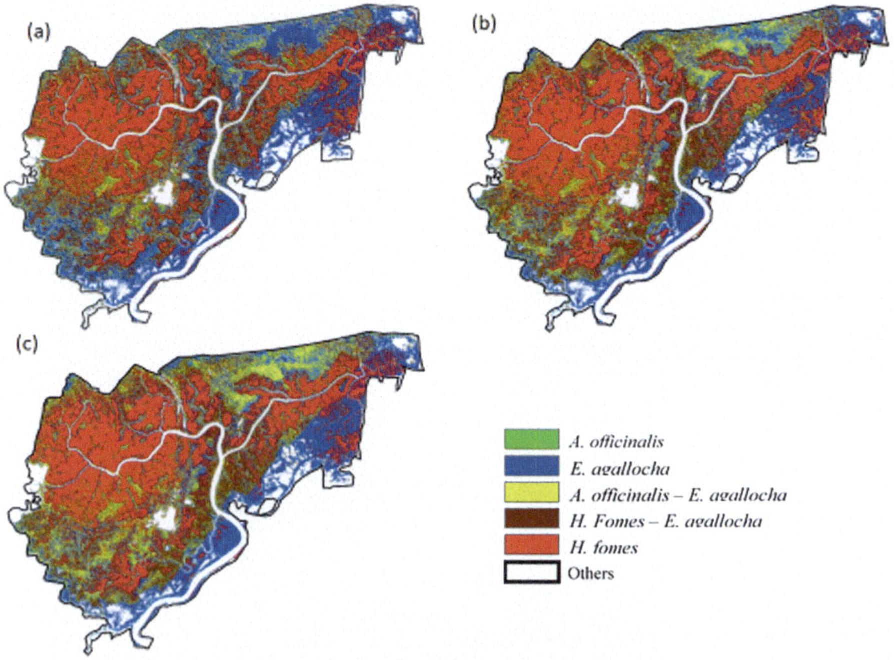

Fig. 14 (**a**) Classified image of three dominant mangrove species using (**a**) six indices (red edge, chlorophyll and soil line-based properties), (**b**) six indices and six red edge bands, (**c**) six indices, and three red edge bands derived from AVIRIS NG data, using random forest (RF) algorithm

4.7.3 RF (Hyperspectral Indices and Three Red Bands)

Overall accuracy of classified map (Fig. 14c) is further improved from 67 to 70% by integrating three red edge band (B63, B64, and b67) with the indices. Pattern of spatial distribution of mangrove remained similar relatively throughout the study area when compared with previous map (Fig. 14c).

5 Conclusions

From chlorophyll statistics, it was observed that mean LCC was the least for *E. agallocha*, and for all species, upper canopy LCC was greater than middle and lower LCC. Low to moderate correlation was observed between LCC and satellite-derived vegetation indices. The possible factors that contributed weaker correlation may include (1) leaf samples from top of the plants, (2) inclusion of SPAD reading from the lower branches, (3) contributions from background mud soils, (e) selection of plants and leaves, etc. Moreover, the SPAD reading was taken at the leaf level, while VIs were derived at the canopy level. Different spectra of *H. fomes* and *E.*

agallocha, generated from AVIRIS, exhibit no significant variation in the pattern of reflectance around NIR region. Two mangrove species such as *H. fomes* and *E. agallocha* were classified as major or dominant species. However, due to heterogeneous distribution of other species, it was not involved in the classification. In RF model, red edge-based indices exhibit highest contribution for generation of species distribution map.

Annexure

Field Photographs of H. fomes *Community*

Field Photo 1

Field Photo 2

Field Photo 3

Field Photo 4

Field Photographs of E. agallocha *Community*

Field Photo 5

Field Photo 6

References

Alongi, D. M. (2012). Carbon sequestration in mangrove forests. *Carbon Management, 3*(3), 313–322.

Behera, M. D., Barnwal, S., Paramanik, S., Das, P., Bhattacharya, B. K., Jagadish, B., Roy, P. S., Ghosh, S. M., & Behera, S. K. (2021). Species-level classification and mapping of a mangrove forest using Random Forest—Utilization of AVIRIS-NG and Sentinel data. *Remote Sensing, 13*(11), 2027. https://doi.org/10.3390/rs13112027

Belgiu, M., & Drăguţ, L. (2016). Random forest in remote sensing: A review of applications and future directions. *ISPRS Journal of Photogrammetry and Remote Sensing, 114*, 24–31.

Breiman, L. (2002). Manual on setting up, using, and understanding random forests v3. 1. *Statistics Department University of California Berkeley, CA, USA, 1*.

Dani, C. S., Kar, S. K., & Behura, B. K. (1999). Bhitarkanika—a unique mangrove ecosystem. Nature and Wildlife Conservation Society of Orissa, Bhubaneswar, India, 30–43.

Daughtry, C. S. T., Walthall, C. L., Kim, M. S., De Colstoun, E. B., & McMurtrey Iii, J. E. (2000). Estimating corn leaf chlorophyll concentration from leaf and canopy reflectance. *Remote Sensing of Environment, 74*(2), 229–239.

Fletcher, R. S. (2016). Using vegetation indices as input into random forest for soybean and weed classification. *American Journal of Plant Sciences, 7*(15).

Ghosh, S. M., Behera, M. D., & Paramanik, S. (2020). Canopy height estimation using Sentinel series images through machine learning models in a mangrove forest. *Remote Sensing, 12*(9), 1519. https://doi.org/10.3390/rs12091519

Giri, C., Ochieng, E., Tieszen, L. L., Zhu, Z., Singh, A., Loveland, T., et al. (2011). Status and distribution of mangrove forests of the world using earth observation satellite data. *Global Ecology and Biogeography, 20*(1), 154–159.

Gond, V., de Pury, D. G., Veroustraete, F., & Ceulemans, R. (1999). Seasonal variations in leaf area index, leaf chlorophyll, and water content; scaling-up to estimate fAPAR and carbon balance in a multilayer, multispecies temperate forest. *Tree Physiology, 19*(10), 673–679.

Gupta, A., Srivastava, P.K., Satish, K., Chauhan, A., Pandey, P.C.J.A.i.R.S.f.F.M., 2022. Challenges and Future Possibilities Toward Himalayan Forest Monitoring. 329-350.

Jagadish, B., Behera, M. D., Jaya Prakash, A., Paramanik, S., Ghosh, S. M., Patnaik, C., Das, A., Shekhar, S., Sharma, N., & Parida, B. R. (2024). Indicating saturation limits of multi-sensor satellite data in estimating aboveground biomass of a mangrove forest. *Journal of the Indian Society of Remote Sensing, 52*, 2483–2500. https://doi.org/10.1007/s12524-024-01968-1

Kuenzer, C., Bluemel, A., Gebhardt, S., Quoc, T. V., & Dech, S. (2011). Remote sensing of mangrove ecosystems: A review. *Remote Sensing, 3*(5), 878–928.

Kulkarni, A. D., & Lowe, B. (2016). Random forest algorithm for land cover classification.

Minolta. (1989). Chlorophyll meter SPAD-502. Instruction manual. l-22.

Mishra, M. K., Gupta, A., John, J., Shukla, B. P., Dennison, P., Srivastava, S. S., et al. (2019). Retrieval of atmospheric parameters and data-processing algorithms for AVIRIS-NG Indian campaign data. *Current Science, 116*(7), 00113891.

Na, X., Zhang, S., Li, X., Yu, H., & Liu, C. (2010). Improved land cover mapping using random forests combined with landsat thematic mapper imagery and ancillary geographic data. *Photogrammetric Engineering & Remote Sensing, 76*(7), 833–840.

Nigam, R., Tripathy, R., Dutta, S., Bhagia, N., Nagori, R., Chandrasekar, K., et al. (2019). Crop type discrimination and health assessment using hyperspectral imaging. *Current Science, 116*(7), 00113891.

Paramanik, S., Behera, M. D., & Dash, J. (2022). Symbolic regression based allometric model development of a mangrove forest using structural variables and digital spherical photography. *Applied Geography, 139*(102649), 1–11. https://doi.org/10.1016/j.apgeog.2022.102649

Paramanik, S., Rajdeep, N., Behera, M. D., Bhattacharya, B. K., & Dash, J. (2023). Species-level classification of mangrove forest using AVIRIS-NG hyperspectral imagery. *Remote Sensing Letters, 14*(5), 522–533. https://doi.org/10.1080/2150704X.2023.2215945

Paramanik, S., Behera, M. D., Rajdeep, N., Bhattacharya, B. K., & Dash, J. (2025). Leaf chlorophyll concentration estimation using absorption spectroscopy of AVIRIS-NG for a mangrove forest in India. *Journal of Photogrammetry, Remote Sensing and Geoinformation Science.* https://doi.org/10.1007/s41064-025-00332-3

Pastor-Guzman, J., Atkinson, P., Dash, J., & Rioja-Nieto, R. (2015). Spatiotemporal variation in mangrove chlorophyll concentration using Landsat 8. *Remote Sensing, 7*(11), 14530–14558.

Pattanaik, C., Reddy, C. S., Dhal, N. K., & Das, R. (2008). Utilisation of mangrove forests in Bhitarkanika wildlife sanctuary, Orissa. *Indian journal of traditional knowledge, 7*(4), 598–603.

Priyanka, Srivastava, P. K., & Rawat, R. (2023). Retrieval of leaf chlorophyll content using drone imagery and fusion with Sentinel-2 data. *Smart Agricultural Technology, 6*, 100353.

Singh, P., Srivastava, P. K., Verrelst, J., Mall, R. K., Rivera, J. P., Dugesar, V., & Prasad, R. (2023). High resolution retrieval of leaf chlorophyll content over Himalayan pine forest using Visible/IR sensors mounted on UAV and radiative transfer model. *Ecological Informatics, 75*, 102099.

Singh, P., Srivastava, P. K., Jha, P. K., Verrelst, J., Singh, P. N., & Prasad, R. (2025). Retrieval of crop traits using PROSAIL-based hybrid radiative transfer model and EnMAP hyperspectral data. *International Journal of Applied Earth Observation and Geoinformation, 143*, 104769.

Srivastava, P. K., Gupta, M., Singh, U., Prasad, R., Pandey, P. C., Raghubanshi, A. S., & Petropoulos, G. P. (2021). Sensitivity analysis of artificial neural network for chlorophyll prediction using hyperspectral data. *Environment, Development and Sustainability, 23*(4), 5504–5519.

Upadhyay, V. P., & Mishra, P. K. (2014), September). An ecological analysis of mangroves ecosystem of Odisha on the Eastern Coast of India. *Proceedings of the Indian National Science Academy, 80*(3), 647–661.

Upadhyay, V. P., Mishra, P. K., & Sahu, J. R. (2008). Distribution of mangrove species within Bhitarkanika national park in Orissa, India. *Trees for Life Journal, 3*, 4–1.

Verma, B., Singh, P., Prasad, R., Srivastava, P. K., & Dave, R. (2024). Leaf chlorophyll content retrieval for AVIRIS-NG imagery using different feature selection and wavelet analysis. *Advances in Space Research, 73*(2), 1304–1315.

Wang, X., Liu, T., Zheng, X., Peng, H., Xin, J., & Zhang, B. (2018). Short-term prediction of groundwater level using improved random forest regression with a combination of random features. *Applied Water Science, 8*(5), 125.

Zavoruev, V. V., & Zavorueva, E. N. (2002, November). Changes in the ratio between the peaks of red chlorophyll fluorescence in leaves of Populus balsamifera during vegetation. In *Doklady biochemistry and biophysics* (Vol. 387, No. 1, pp. 310–312). MAIK Nauka/Interperiodica.

Study of Carbon Dynamics in the Indian Mangroves Using Carbon Isotopes: A Comprehensive Review

Akancha Tripathi, Rina Kumari, and Parul Maurya

Abstract This review study offers an in-depth investigation of the carbon dynamics in the mangroves of India by using carbon isotopes. Mangrove ecosystems (MEs) are highly productive because of their remarkable ability to sequester carbon and hence significant in climate change mitigation efforts. This paper concludes some of the findings from various important studies carried out in the Indian MEs, exploring the distribution of mangroves in India, their ecological importance, and the different technologies employed in carbon isotopic analysis. Several investigations were carried out to check the potential of Indian mangroves to store carbon, considering various environmental as well as anthropogenic factors influencing the rate of carbon sequestration. This review article also explores the implications of carbon dynamics for the approaches to mitigate climate change and adaptation to the same, further emphasizing the need for mangrove conservation and restoration. This study concludes that Indian mangroves are significant sinks of carbon that have an enormous potential to reduce the impacts of climate change through carbon sequestration. To direct future research in this domain, the current issues facing carbon isotopes have been highlighted along with some potential paths for further study.

Keywords Carbon isotope · Carbon dynamics · Mangrove conservation · Indian mangroves

A. Tripathi · R. Kumari (✉) · P. Maurya
School of Environmental Sciences and Sustainable Development, Central University of Gujarat, Gandhinagar, India
e-mail: rina.sesd@cug.ac.in

S. K. Srivastava et al. (eds.), *Earth Observation for Monitoring Mangrove Ecosystems*, https://doi.org/10.1007/978-3-032-16668-5_13

1 Introduction

MEs provide a variety of social, ecological, and economic services and hence serve as extremely productive ecosystems found in the intertidal zones of tropical and subtropical regions (Pandey et al., 2024). Those ecosystems have the potential of storing greater quantities of carbon than most of the terrestrial forests, rendering them as one of the most carbon-rich regions of our planet. The global mean total forest stock of mangrove forests is 738.9 ± 27.9 Mg C ha^{-1}, out of which 8.7% is held in belowground roots, 14.8% in aboveground roots, and the rest, 76.5%, is retained by the soil. Thus, these forests comprise the greatest amount of organic carbon (OC) stores of any tropical marine or terrestrial ecosystem (Alongi, 2020). Mangrove forests have higher primary productivity than coral reefs and tropical humid evergreen forests due to exceptionally high carbon storage rates as well as faster rates of accretion of soil on the forest floor (Alongi, 2020).

Although these have huge ecological importance, these are severely impacted by global land conversion and deforestation, disturbing their role as carbon sinks significantly (Pandey et al., 2024). Spanning across 12 states and union territories, mangrove forests encompass an area of nearly 4740 km^2, which is approximately 3.3% of the world's total mangrove vegetation (Fig. 1) (Samanta, 2017; Ragavan & Mandal, 2018). Other than sequestering carbon, it has additional roles as well, such as sustaining biodiversity, securing coastlines, supporting local inhabitants, and providing them better standards of living (Rathor, 2024).

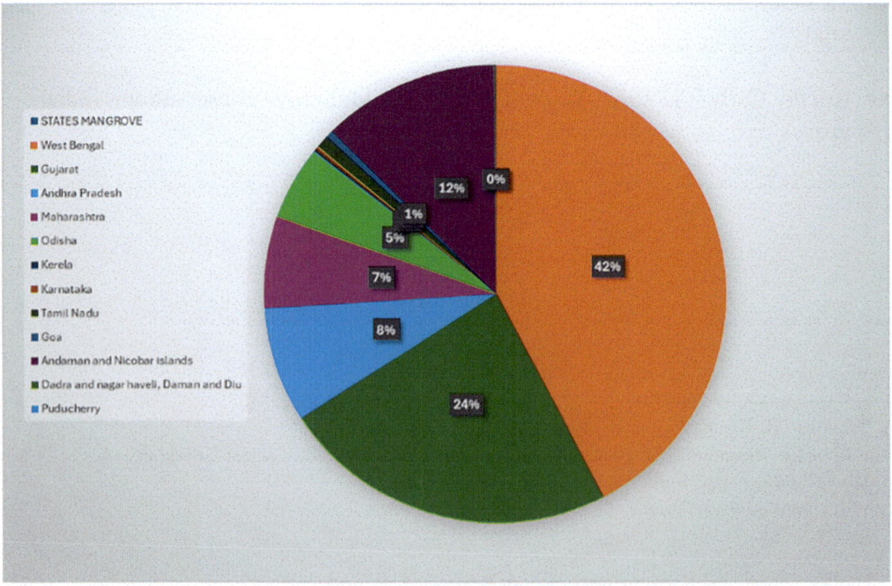

Fig. 1 Percentage distribution of mangroves in India (as per the Forest Survey of India, 2021)

1.1 Mangrove Forests and SDGs

As a holistic approach combining environmental, social, and economic goals in the context of global sustainability initiatives, restoration of mangroves is intimately correlated with the development of the Sustainable Development Goals (SDGs) of the United Nations. Restored or rehabilitated ecosystems of mangroves provide an array of essential ecosystem services, involving carbon sequestration, coastline protection, water purification, as well as habitat provision. These services are essential to tackle climate change (SDG 13), maintain marine life (SDG 14), and safeguard terrestrial biodiversity (SDG 15) (Chow, 2018). Mangrove's relevance in climate mitigation initiatives is illustrated by their ability to absorb enormous amounts of carbon dioxide, which reinforces worldwide objectives of decreasing greenhouse gas emissions (Sasmito et al., 2023). Also, sustainable aquaculture and fishing, which are crucial to sustaining income and promising food availability in a number of coastal areas, are facilitated by healthy mangrove forests. Therefore, they directly promote household income and food access, which helps accomplish SDGs 1 (No Poverty) and 2 (Zero Hunger) (Akram et al., 2023). By eliminating toxins while increasing water quality, rehabilitation efforts additionally promote clean water regimes and SDG 6 (Clean Water and Sanitation) (Bimrah et al., 2022).

Additionally, mangrove's potential to maintain stable coasts, minimize the possibility of catastrophes, and supply sturdy structures for coastal communities indicates their socioeconomic and cultural value and greatly contributes to SDG 11 (Sustainable Towns and Communities) (Su et al., 2021). The objectives of SDG 8 (Decent Work and Economic Growth) are in accordance with employment potential and sustainable revenue generation that result from ecotourism, local fisheries, and non-timber forest goods produced in rebuilt mangrove belts (Viswanathan, 2015). According to SDG 3 (Good Health and Well-Being) and growing equitable governance as stipulated in SDG 17 (Partnerships for the Goals), community engagement in managing mangroves and regeneration expands civic engagement, reinforces well-being and health by assisting mental and physical health, and minimizes vulnerability to natural hazards (Eyzaguirre et al., 2023). Mangrove ecosystem benefits are universal, so reestablishing these forests endorses a minimum of nine SDGs in their entirety: SDG 1, SDG 2, SDG 3, SDG 6, SDG 8, SDG 11, SDG 13, SDG 14, and SDG 15. It likewise promotes successful, multi-stakeholder collaboration, which is critical to accomplishing SDG 17 (Akram et al., 2023). Mangrove rehabilitation is an essential part of sustainable development initiatives throughout the globe, as it enables societies to address critical ecological problems while simultaneously creating complementary momentum across an array of SDG targets (Sasmito et al., 2023).

Table 1 Distribution and status of mangroves in India (as per Forest Survey of India, 2021)

S. no.	States/union territories	Open mangrove	Moderately dense mangrove	Very dense mangrove	Total
1	West Bengal	428	692	994	2114
2	Gujarat	1006	169	0	1175
3	Andhra Pradesh	192	213	0	405
4	Maharashtra	234	90	0	324
5	Odisha	84	94	81	259
6	Tamil Nadu	17	27	1	45
7	Goa	6	21	0	27
8	Karnataka	11	2	0	13
9	Kerala	4	5	0	9
10	Andaman and Nicobar Islands	49	168	399	616
11	Dadra and Nagar Haveli, Daman and Diu	3	0	0	3
12	Puducherry	2	0	0	2
	Total	2036	1481	1475	4992

2 Distribution and Status of Mangroves in India

The major proportion of Indian mangroves is spread across the eastern and western coastlines, with substantial quantities in Odisha and West Bengal (Sundarbans) (Table 1) (Ragavan & Mandal, 2018). The Indian Sundarbans serve as a habitat for various animal species that are threatened and constitute the world's largest mangrove site (Samanta, 2017). Although mangrove forests have experienced a worldwide pattern of loss, the mangrove cover in India has increased over the last decade. An overall rise of 875 km^2 during 1987–2017 with a mean yearly variation of 30.21 ± 81.72 km^2 was reported by the Forest Survey of India. According to recent figures, there has been an expansion of 181 km^2 between 2015 and 2017 and 112 km^2 between 2013 and 2015 (Ragavan & Mandal, 2018). India's second largest mangrove cover is in Gujarat, covering over 1058 km^2 (Pandey & Pandey, 2013). In India, mangroves are unequally distributed, having varying species diversity and densities. Odisha and West Bengal possess the greatest species diversity, while West Bengal and Daman Diu represent the highest and lowest densities of mangroves, respectively. The sustainable traits of MEs can be seen by the rising trend in mangrove forest cover, ensuring ecological well-being for future generations (Samanta, 2017).

3 Significance of Mangroves in India

The Indian MEs provide an array of essential ecological functions in addition to carbon sequestration. These forests perform the function of safeguarding coastal areas from storms and tidal waves due to their dense roots, effectively scattering wave energy. The roots of mangroves additionally assist in sediment accretion and carbon burial by trapping carbon-rich particles from water and depositing them in sediments (Cuenca-Ocay, 2019). Besides its ecological relevance, mangroves contribute to the production of marine fisheries by acting as nurseries for fish and other aquatic creatures. According to some studies, the contribution of mangroves to the commercial output of marine fisheries in India is estimated to be between 23 and 34% of commercial marine fisheries output in India, with values ranging from Rs. 1.46 to 2.15 lakhs per hectare in 2012–2013 prices.

The widespread impact of mangroves on fisheries is highlighted by comparing its contribution to estimates from other countries (Anneboina & Kumar, 2016). Mangroves also play an essential role in recycling nutrients, regulating hydrological cycles, and preserving coastal integrity (Rathor, 2024). Regardless of the advantages of Indian mangroves, they have suffered a loss of over 40% of their land within the past few decades as a consequence of environmental changes and increased human activities (Dhadse et al., 2021). The increased degradation of mangroves jeopardizes coastal resilience, enhancing the potential hazards of sea-level rise, storms, and erosion by liberating the accumulated carbon as CO emissions (Koo et al., 2024).

4 Carbon Dynamics in MEs

4.1 Factors Affecting Carbon Storage

A complicated interaction of physical, biological, ecological, and human influences governs the storage of carbon in MEs; all of these elements have a unique function in the storage, retention, and elimination of carbon in these coastal areas. In this case, different species of mangroves possess distinct abilities for storing carbon due to variations in yield of biomass, root morphology, and age structure. Particular species and older, robust stands may hold bigger carbon stocks by means of greater levels of primary efficiency and the development of intricate, flexible root structures that maintain and improve soil organic carbon buildup. These biological variables, like species composition, efficiency, and root structure, are vital (Sahu & Kathiresan, 2019). By impacting breakdown rates, material maintenance, and sediment deposition processes, physical as well as environmental variables, including soil type, hydrological conditions, tidal effects, and geomorphology, additionally have important parts. Soils in elevated clay content and lower oxygen availability are prone to slow breakdown and enhanced carbon preservation over time, while typical tidal flooding helps in the burial and stabilization of organic matter (Xiong et al., 2018; Xin et al., 2018). Both mangrove expansion and the reliability of carbon storage are additionally affected by climatic factors, such as local temperatures, rainfall patterns, and the intensity or frequency of disasters like storms and sea-level rise. In particular, higher salinity levels or flooding from sea-level rise could alter the diversity of species, shift breakdown rates, and undermine current carbon sinks if biophysical formation fails to keep up (Singh et al., 2022; Chambers et al., 2014). Based upon their severity and frequency, events like hurricanes and cyclones can right away lead to biomass destruction and erosion of soil that exposes carbon briefly yet often enables restoration of ecosystems and further sequestration over time (Griffiths & Mitsch, 2021). By eliminating carbon-rich foliage and modifying soil structure, driving up carbon release, and hindering the biogeochemical processes that enable carbon buildup, human-induced processes such as forest loss, conversion of land for farming or aquaculture, development, and environmental damage offer significant hazards. In some instances, these operations may transform mangrove areas from carbon sinks into net carbon sources (Wang et al., 2022). The future of organic carbon also depends on the biogeochemical reactions that constitute microbial activity, the cycling of nutrients, and the breakdown of organic matter. The pace of litter decay and the development of stable soil carbon pools are affected by salinity and availability of nutrients, as well as oxygen gradients (Ouyang et al., 2017). For efforts to sustain while enhancing the blue carbon function of the MEs, comprehensive management approaches that consider site-specific biological, environmental, and human-caused circumstances seriously are vital. The combination of all of these variables leads to substantial variations in time and space in mangrove carbon-storing capacity (Rovai et al., 2018; Koh et al., 2018).

4.2 Carbon Storage in the Mangroves

Relative to numerous other forest varieties, MEs possess unusually large capacities for storing carbon, which makes them among the most carbon-rich woods in the tropical regions. Based on studies, mangroves may hold a mean of 968 megagrams of carbon dioxide per hectare, and this is due to both substantial carbon buildup within their soils and aboveground biomass (Murdiyarso et al., 2009). The rich biomass of vegetation, deadwood, and especially thick, organic-rich soils that function as ongoing sinks for carbon by retaining organic matter and necro mass over a long period of time are the main reasons for such remarkable carbon storage (Inoue, 2018).

Using an array of operations, including the establishment of underground roots and allochthonous inputs that promote organic carbon stabilization—which often gets underestimated in carbon budgets—mangrove soils are exceptionally efficient at stabilizing carbon (Kida & Fujitake, 2020). Further, mangroves serve as vital ecosystems made up of blue carbon in the worldwide chain of carbon and climate change reduction efforts due to their disproportionately major contribution to carbon sequestration relative to their spatial space (Alongi, 2014). They serve not only as hubs for diversity but also as vital natural carbon repositories with major implications for environmental legislation and rehabilitation of ecosystems because of their ability to collect and store carbon over a longer period of time (Jennerjahn, 2020).

4.3 Pathways of Carbon in the Mangrove Ecosystem

Carbon flows along complicated pathways in MEs, involving fixation, change, storage, and exporting processes, each of which functions synergistically to regulate the carbon cycle in those environments. Mangrove vegetation's primary manufacturing procedure, the photosynthesis process, collects CO_2 from the environment and converts it into organic carbon, after which it enters the ecosystem's numerous abiotic and biotic components (Adame et al., 2024). When fixed, carbon circulates across belowground components like roots and sediments in addition to aboveground biomass like leaves and wood, where it can pile up as organic carbon stores (Yong et al., 2011). A substantial amount of the carbon through debris falls and drainage from roots finds its route into detrital food webs, which sustain a number of microbial and detritivore communities that break down and mineralize organic substances (Adame et al., 2024). Although wave movements allow the transportation of dissolved and particulate organic carbon to adjacent coastal waters, linking mangrove carbon dynamics to broader marine biogeochemical processes, respiration by both autotrophic and heterotrophic creatures leads to carbon losses primarily in the form of CO_2 emissions (Alongi, 2014). Also, the destiny of carbon stored under such anoxic situations is affected by particular pathways, such as limited methanogenesis, which is associated with sedimentary processes as anaerobic metabolism (Alongi, 2020). These interrelated routes emphasize mangroves as changing carbon systems that their role being vital blue carbon sinks in the worldwide carbon budget relying on the balance among carbon fixation, conversion, storage, and export (Adame et al., 2024; Alongi, 2020).

4.4 Carbon Dynamics

The extensive interaction of mechanisms such as biomass generation, breakdown, and carbon storage processes in MEs influences the carbon dynamics. The carbon that is stored within the sediment biomass of seagrass, mangrove, and peat swamp ecosystems is referred to as "blue carbon." The property of mangroves of storing carbon in coastal areas makes these habitats essential in minimizing the impacts of climate change (Salle et al., 2024). MEs have a diverse range of tidal dynamics, geomorphological settings, understory light intensity, and biological production, offering a broad variety of habitats such as forest gaps, tidal creeks and flats, and interior forests. The distinction between the carbon transfer and assimilation by fishes and invertebrates across several MEs varies and can be detected by stable isotopic compositions. The differences in small-scale carbon transfer and assimilation are suggested by some of the studies indicating the substantial drop of crabs' and snails' $\delta^{13}C$ values from tidal flats to inner woods (Tue et al., 2017). Organic carbon occurs in several forms in soil and water within the MEs, such as those found in zooplankton and phytoplankton, as well as soil organic carbon (SOC), dissolved organic carbon (DOC), and particulate organic carbon (POC). The primary productivity, dissolved oxygen, atmospheric temperature, litter biomass, water temperature, and tidal height are some of the environmental variables that regulate the dynamics of organic carbon. The movement of carbon across the mangrove is highly impacted by these interdependent components (Das et al., 2020).

5 Study of Mangroves Using Carbon Isotopes

5.1 Carbon Isotopes (CI) in the Mangrove Study

For the understanding of the movement of carbon within MEs, analyzing the carbon isotope is an essential requirement (Tue et al., 2011). The determination of carbon stocks, tracing of carbon sources, and comprehending the impact of climate change and human activities are made easier using this technology (Nguyen et al., 2016). For the identification of the origin of sedimentary organic carbon, the method involves evaluating stable nitrogen and carbon isotope ratios in mangrove leaves, particulate organic matter (POM), and surface sediments. A study from Vietnam indicates that the average $\delta^{13}C$ value of mangrove leaves was −28.32‰, which is representative of C3 terrestrial plants. The POM values at the upper estuary are found to steadily rise from −25.41‰ to −21‰ at the mouth of the estuary, showing the impact of terrestrial (phytoplankton) inputs in the upper river and estuary's mouth. A fall in the sources of phytoplankton followed by a rise in mangrove--derived organic carbon from the tidal creek to the mangrove forest is suggested by the mangrove-derived organic carbon from the tidal creek to the mangrove forest (Tue et al., 2011). Tracking carbon flow and distinguishing between primary producers in mangroves is done by this method (Tue et al., 2017).

5.2 Sample Collection and Processing Techniques

The first step in the carbon isotope analysis in MEs is strategic sample collection (Nguyen et al., 2016). Mangrove leaves, shrimp and crabs, sediment cores, squids and fish, and particulate organic matter are some typical samples (Hayase et al., 1999; Tue et al., 2018). Specific diameters and heights of polyvinyl chloride pipes are employed in the collection technique of sediment for taking out the cores, which are undisturbed. For example, in an Indonesian study, a PVC pipe with a height of

30 cm and a diameter of 47.46 mm was used to sample sediment. Total dissolved solids, pH, temperature, salinity, sediment texture, carbon stock, bulk density, and sequestration rates are some of the factors measured from these samples (Ariyanto & Pringgenies, 2024).

Applying the approach of the loss on-ignition (LOI) technique, scientists in Vietnam obtained samples of soil at multiple levels and evaluated them for organic matter content, bulk density, and soil organic carbon concentration (Pandey et al., 2024). Samples of various species are obtained for the analysis of carbon isotopes in marine animals for tracking of the food web design in mangrove brackish water habitats (Hayase et al., 1999). After drying and homogeneous grinding, the isotopic analysis of these samples is done by applying methods such as gas chromatography--combustion-isotope ratio mass spectrometry (Larsen et al., 2012).

5.3 *Advanced Analytical Techniques*

The degree of reliability and relevance of carbon isotopic research in mangroves has been substantially boosted by modern analytical techniques. The frequently used technique for the determination of amino acid carbon isotope ratios ($\delta^{13}C(AA)$) among different primary producers in mangrove habitats is gas chromatography-combustion-isotope ratio mass spectrometry (GC-C-IRMS). With the application of linear discriminant analysis for important amino acid $\delta^{13}C$ levels, this approach permits the experts to make distinctions between marine and terrestrial producers. The results from an experiment conducted in Spanish Lookout Caye, Belize, where the above approach was applied, found that in addition to differences in bulk $\delta^{13}C$ values, marine and terrestrial producers had different patterns of $\delta^{13}C(EAA)$ values as well. According to the results, $\delta^{13}C(EAA)$ values might give a further approach for the calculation of the proportion of terrestrial and marine sources in detrital food webs (Larsen et al., 2012). Tools such as ^{210}Pb chronology are utilized in the analysis of sediment and evaluation of the pace of organic carbon deposits at decadal periods. Investigations indicate the ^{210}Pb can be used for evaluating the rate of buildup of organic carbon in these habitats, even though the bio-/physical-turbation has potential effects when applied to ^{210}Pb chronology, specifically in Asian MEs where human-caused physical turbation by coastal fishery is rapid (Tateda et al., 2005).

5.4 *Carbon Isotope Studies in Indian Mangroves*

The mechanism of carbon sequestration and storage has come to light by investigations on carbon isotope studies in the Indian mangroves. The sedimentary organic carbon store in India's Coringa mangroves has been analysed as a component of blue carbon experiments. As per the results, the total organic carbon (TOC) of the area ranges from 0.63 to 2.76%, with an average of 1.30 ± 4.3%. Though the range of dry bulk density (DBD) spans from 0.63 to 2.58 gm cm^{-3} with an average of 1.33 ± 0.19 gm cm^{-3}, the value of total nitrogen (TN) varies from 0.04 to 0.19% with an average value of 0.08 ± 0.03%. The average sedimentary carbon stock in Coringa mangroves was 158 ± 32.60 Mg C ha^{-1}, equating to a below-ground sedimentary carbon stock of 1.86 Tg C, determining an equivalent CO_2 emission of 579.9 Mg CO_2e ha^{-1}. Important geographical differences in TOC, TN, DBD, and SCS within different areas: highly forested Coringa mangroves had the greatest carbon stock subsequent to modestly covered Gaderu mangroves, and the Matlapalem mangroves had the smallest carbon stock. The intricate carbon dynamics are affected by several drivers and have been proposed by the inconsistent connection between the amount of organic carbon and grain size fractions of sediment (Rao et al., 2022).

6 Carbon Stocks in Western Indian Sundarbans

Renowned for their wide range of luscious mangroves, the Indian Sundarbans are a vital means for the elimination of carbon dioxide from the atmosphere. Understanding the trend of carbon storage in the mangroves of the lower Gangetic belt demands proper measurement of biomass in these species (Mullick et al., 2023). Carbon stocks in the aboveground biomass (AGB) of prevalent species of mangroves have been studied in the western Indian Sundarbans. The greatest amount of aboveground storage of carbon of the species selected for the western portion, *Sonneratia apetala*, represents maximum aboveground carbon storage (26.53 tha^{-1}), followed by *Avicennia officinalis* (25.80 tha^{-1}), *Avicennia alba* (24.73 tha^{-1}), *Avicennia marina* (22.84 tha^{-1}), and *Excoecaria agallocha* (11.90 tha^{-1}) (Raha et al., 2013).

An average of 25 trees from 15 plots (10 m × 10 m) from intertidal mudflats on Chemaguri (southeast part of Sagar Island) were utilized to access the plant biomass of these species during low tide. A traditional technique applied to the stem, branch, and leaf biomass of each species was converted into carbon by multiplying with a factor of 0.45 as per standard procedure. The crucial role of diversity of species in accessing the total amount of carbon that is stored, and the ability of these ecosystems is highlighted by these investigations, which demonstrate the disparity in the ability of carbon storage among different mangrove species (Mullick et al., 2023).

7 Carbon Sequestration in Regional Mangrove Studies

Relevant knowledge on the ability to store carbon has been rendered by studies on carbon sequestration, especially in the mangrove regions of India. Experts examined the biomass and carbon sequestration power of two extensive mangrove patches in the Chettuva Estuary on India's southwest coast. The values for carbon stocks of aboveground biomass, root biomass, and sediment were figured out as 325.36 ± 54.33 t C ha^{-1}, 158.33 ± 25.41 t C ha^{-1}, and 85.6 ± 9.15 t C ha^{-1}, respectively. The overall mean carbon stock in these two mangrove patches was evaluated as 569.3 t C ha^{-1}, equivalent to 2089.33 t CO_2 ha^{-1}, and the whole amount of potential for the carbon sequestered in the study area was around 1119.68 t C, equivalent to 4109.24 t CO2. (41) (Varghese et al., 2021). In accordance with some of the investigations, a total of 8.116 million tons of carbon have been retained by the mangroves in Gujarat, which has the second largest mangrove cover in India (1058 km^2). This research concluded that the total amount of carbon trapped by mangrove soils is higher than that of mangrove plants (Pandey & Pandey, 2013). The mean concentration of the amount of carbon stored in leaf samples was 36.9467%, concluded by preliminary research in Mumbai for the Sewri, Wadala, and Mahul regions, characterized by high mangrove covering that has expanded since 2001. With an achievable yearly rise of approximately 50,000 tons of equivalent CO_2, the Godrej mangroves across the western bank of Thane Creek have been maintained since the 1940s and store about 600,000 tons of CO_2 annually (Patil et al., 2012).

8 Factors Influencing Carbon Dynamics in Indian Mangroves

8.1 Environmental Factors Influencing Carbon Storage in Indian Mangroves

The amount of carbon sequestered by Indian mangrove forests is greatly impacted by an assortment of environmental circumstances. An investigation compared the two types of mangroves and concluded that the natural mangroves had higher organic carbon values (2.2 ± 0.32%) compared to rehabilitated mangroves (0.9 ± 0.25%). Total dissolved solids, temperature, and sand had a negative relationship, whereas silt, salinity, and clay content had positive relationships for carbon stock, organic carbon, carbon dioxide equivalent, and carbon sequestration, according to principal component analysis. The distribution of sediment texture showed more silt in rehabilitated mangroves, while natural mangroves tended to have similar compositions of sand and silt (Ariyanto & Pringgenies, 2024).

The connection between organic carbon level and sediment particle size sections in the Coringa mangroves was unambiguous, showing that carbon dynamics are impacted by parameters other than sediment composition (Rao et al., 2022). The distributions of organic carbon (OC) and total phosphorus (P) in the sediments of three mangrove habitats in the Veli, Kochi, and Kannur regions of Kerala's southwest coast were examined in this research. Though the total P exhibited the largest percentages in Kochi (0.395%) in comparison to Veli (0.055%) and Kannur (0.323%), OC (4%) was somewhat elevated in Kannur's mangrove sediments compared to Veli (2.6%) and Kochi (3.7%). OC and P levels were evaluated statistically against several causal variables, which indicated how particle size influences their distributional arrangement in mangrove habitats (Badarudeen et al., 1998).

8.2 Anthropogenic Impacts on Carbon Dynamics

The interaction of carbon in Indian mangrove forests is substantially affected by human activities. Persistent spikes in development along the coast, including the transformation of wetlands for aquaculture and building for tourists, have ended up in the loss of mangrove habitats in India, despite their tremendous ecological and economic benefits. Around 40% of the territories of Indian mangroves have vanished over the last decade (Dhadse et al., 2021). Besides releasing retained carbon into the atmosphere, these changes decrease the potential ability of mangroves to hold onto carbon (Kooet al., 2024). The preservation and maintenance of mangroves is being seriously jeopardized by environmental changes and an increase in human activities (Dhadse et al., 2021). Based on an investigation, relationships associated with mangrove destruction are growing yearly in Indonesia's Tongke-tongke Mangrove Tourism Area. Owing to this interaction, mangrove forests are facing enormous invasion and changes that are conflicting with their capacity to manage the climate with respect to blue carbon. Climate adaptability, biodiversity, regional economies, and the well-being of coastal people are the main areas affected by the decline of the ecosystems of mangroves brought on by land conversion. In addition to professional initiatives, an extensive awareness of the complex dynamics of social interactions between humans and mangroves is essential for successful management of MEs for blue carbon (Salle et al., 2024).

8.3 Climate Change Implications and Carbon Dynamics

The movement of carbon across mangrove forests is greatly influenced by climate change. According to the studies, mangroves constitute the largest carbon-rich ecosystems worldwide, but there is uncertainty in the fact that the natural and rebuilt mangroves are going to be able to sustainably store carbon as a result of global climate change. The influences of worldwide disturbances (forest restoration, rising temperatures, and sea-level change) on the expansion of fine roots and the decomposition of soil organic matter (SOM) were examined in experimental mesocosms and field experiments. Based on research, SOM degradation jumped by 21% due to a spike in temperature from 27 °C to 31 °C; nevertheless, this effect was less noticeable in mangrove soils influenced by rising sea levels. SOM breakdown was drastically accelerated by 66% while in drought conditions, which appeared as a decrease in soil flooding. This effect was heightened by warmth. It's remarkable to mention that SOM degradation did not get impacted by elevated sea levels, either by itself or in connection with warming (simulated as a rise in flooding duration from 2 to 6 h per day). These outcomes show that instead of diminished SOM breakdown, formerly believed mangrove root development might be what promotes the continued existence of mangroves in the midst of rising sea levels by storing SOM. Further, fine root output reduces gradually with mangrove stand ages post-rehabilitation; this pattern has been observed in aboveground output but is unknown in belowground carbon (Arnaud, 2020).

8.4 Carbon Sequestration for Climate Change Mitigation

Due to their property of carbon storage, mangrove forests are vital toward minimizing the impacts of climate change. For the purpose of decreasing or preventing worldwide warming, the capture of carbon comprises the long-term retention of carbon dioxide in vegetation, soils, seas, and geologic structures. Emissions of greenhouse gases through a number of human actions have been effectively controlled by this method. Because the atmosphere currently contains 30% more carbon than it did 150 years ago, researchers have found an immediate link between developing worldwide temperatures and environmental carbon dioxide levels. Enhancing worldwide carbon storage via an assortment of methods is one proposed method of decreasing atmospheric carbon dioxide. In order to reduce the cumulative rate of increasing atmospheric CO_2, the capture of carbon includes storing CO_2 within persistent worldwide reservoirs that exist in forests, oceans, biomass, and sedimentary strata. For the worldwide cycle of carbon to be maintained, this process is vital. The entire world has to enhance net carbon storage while decreasing the release of carbon in order to successfully fight climate change. The pros and cons of each potential carbon capture method should be carefully assessed, though (Sethi & Syal, 2024). By boosting forest storage of carbon and promoting forest conservation, India's Reducing Emissions from Deforestation and Forest Degradation (REDD+) establishes attempts to avoid climate change. Besides addressing the root causes of destruction and forest loss, the approach provides an outline to strengthen forest sequestration and handle forests sustainably (Verma & Kumar Ghosh, 2022).

9 Remote Sensing and Advanced Technologies in Carbon Assessment

The examination of carbon fluxes in MEs has been significantly facilitated by breakthroughs in computerized and remote sensing technology. With the goal to figure out aboveground biomass and storage of carbon and anticipate mangrove distribution, lacking the need for tedious field research, a study that used both radar (Sentinel 1) and optical (Sentinel 2) remote sensing datasets compared multiple machines learning algorithms, including random forest, SVM, and CART. Employing a Kappa metric of 0.962, a random forest machine learning classifier achieved the highest level of precision of 0.984, according to the investigation. The aboveground biomass has been determined to be 125 Mg/ha employing this classifier, and the yearly carbon storage rate came out as 257 Mg/ha. These outcomes confirmed the accuracy of remote sensing techniques and were in accordance with past studies utilizing standard fieldwork. These innovations reinforce the vital role of MEs as blue-carbon storage systems and offer helpful devices for effectively tracking and preserving them worldwide. An additional investigation performed in Vietnam examined the success rate of restoration of mangroves in boosting carbon storage and recorded changes in mangrove cover utilizing remote sensing (Pandey et al., 2024). A greater awareness of the movement of carbon in MEs can be gained by integrating these innovative tools with standard field studies (Kotwal, 2024).

10 Challenges in Carbon Isotope Studies in Mangroves

The examination of carbon flows in MEs endures an array of barriers, given improvements made in carbon isotope studies. The shortage of a standardized approach for evaluating soil carbon pools is an important barrier causing problems with the estimation of carbon storage and turnover rates. Featuring different carbon pools—"active," "slow," and "passive"—soils are the biggest carbon deposit in terrestrial ecosystems. The quantity of soil carbon pools on various dates, input rates, and carbon turnover times are the primary subjects of the fluctuation of soil carbon cycles. Despite soil carbon cycles being able to be investigated by examining endogenous quantities of carbon isotopes, notably radiocarbon, these techniques have disadvantages. The size and turnover rates of unstable and stable soil organic matter pools can be identified with carbon-14 signatures, while the carbon-13 ($\delta^{13}C$) values of soil organic matter give an additional method to evaluate historical changes between C3 and C4 ecosystems over decadal to millennial periods. However, the absence of credible worldwide records that support techniques for assessing unstable and stable soil organic matter pools generates substantial errors in computing the amounts and turnover lengths of soil carbon pools. As an outcome, assessing shifts in soil carbon stores that could influence atmospheric CO_2 concentrations and the worldwide climate becomes extremely hard (Tao et al., 2004).

11 Conservation and Management Strategies for Indian Mangroves

By promoting the understanding of how crucial they are among everyone involved and preserving them with essential national laws and regulations, mangrove habitats can be saved. For the sake of mangrove preservation, the long-term sustainability of local community habitats along coastal areas, fishing communities, and the safeguarding of coastal stretches for marine habitats, particular laws such as Coastal Zone Regulation have been established (Dhadse et al., 2021). In India, mangrove collaborative projects involving local governments, NGOs, and the public showed potential. Severe flooding incidents, coastal loss, and increasing sea levels are some of the consequences of climate change hazards, which mangroves might be able to react to. The gradual growth of coastal areas is caused by the huge, raised root systems of mangroves that function as natural walls to catch runoff debris. Creative methods for management are needed since mangroves around the globe are deteriorating as a result of anthropogenic and other environmental activities (Noguchi et al., 2012). Mangrove regeneration is achievable, owing to study, and has been undertaken in several areas, but such attempts have generally failed as they are scattered. In India, recent projects try to outline and find mangrove restoration areas and assess potential profits on restoration. The future survival of these habitats relies on the focus on policy and management recommendations, as well as regulations and guidelines for mangrove renewal (Rathor, 2024).

12 Future Research Directions in Mangrove Carbon Dynamics

To further develop our knowledge of and oversight of these ecosystems, future studies on the motion of carbon in Indian mangroves should concentrate on a few essential components (Lokollo et al., 2024). The critical nature of mangroves for offering protection for the coastline, their role as carbon sinks, their possible future impacts, and the manner in which human activities and global warming influence their ability to store carbon need to be clarified (Koo et al., 2024). Long-term tracking studies will be required to monitor shifts in carbon dynamics caused by climate change and human activities (Arnaud, 2020). Engaging people as administrators or owners of MEs is vital for long-term sustainability when adopting blue carbon monetization at the district level (Lokollo et al., 2024).

Studies into the ability of multiple mangrove species to store carbon and their ability to adapt to shifting environmental conditions could offer significant data for restoration attempts (Mullick et al., 2023). There are plenty of possibilities for complete carbon estimates when cutting-edge technologies like machine learning and remote sensing merge with standard field measurements. It might be feasible to evaluate investigations and areas more effectively if carbon stock evaluation processes were standardized (Tao et al., 2004).

Also, exploring the link between carbon storage, the functioning of the ecosystem, and biodiversity in MEs might give a deeper understanding of these complicated settings (Varghese et al., 2021). At last, investigations into the monetary worth of the environmental benefits that mangroves offer—specifically, carbon sequestration—will assist in the development of sustainable management programs that find a balance among livelihood needs and conservation (Anneboina & Kumar, 2016).

13 Conclusions

In India, mangrove forests serve as vital carbon sinks with tremendous promise for reducing climate change through carbon sequestration (Varghese et al., 2021). Carbon sources, storage trends, and variables impacting carbon sequestration in these habitats have all been clarified because of the utilization of carbon isotopes for studies on carbon dynamics (Nguyen et al., 2016). The tremendous ability of Indian mangroves for storing carbon has been proven in multiple studies; estimates vary by location and range from 158 to 569.3 t C ha^{-1} (Rao et al., 2022). Conservation projects are essential because natural mangroves retain carbon at higher rates than restored ones (Ariyanto & Pringgenies, 2024). Salinity, hydrology, and sediment makeup are a few of the environmental variables that have considerable effects on mangrove carbon dynamics (Badarudeen et al., 1998).

The mangrove forests and their ability to sequester carbon are in danger from human-induced processes, especially land conversion and deforestation (Salle et al., 2024; Dhadse et al., 2021). The consequences of climate change, such as increasing sea levels and temperatures, considerably affect the dynamics of carbon in mangroves (Arnaud, 2020). Additionally, there are plenty of prospects for complete carbon evaluations and monitoring when modern technologies are paired with typical methods (Kotwal, 2024).

For MEs to be maintained ecologically, efficient strategies that involve residents and are reinforced by appropriate laws are essential (Rathor, 2024). The standardization of procedures, long-term surveillance, and grasping the complicated connections between environmental variables, human activity, and carbon dynamics in mangroves must be the key objectives of future research (Tao et al., 2004). India may profit from the variety of ecological and socioeconomic advantages that mangroves provide and substantially contribute to global efforts to reduce climate change by maintaining and enhancing these important ecosystems (Anneboina & Kumar, 2016).

Acknowledgments Akancha Tripathi is grateful to PRL, Ahmedabad for giving the learning platform on carbon isotopes.

Funding No funding is involved in this work.

Clinical Trial Not applicable.

Consent to Publish Not applicable.

Ethical Clearance Not required.

Consent to Participate Not applicable.

Competing Interest No competing interests.

References

Adame, M. F., Cormier, N., Taillardat, P., Iram, N., Rovai, A., Sloey, T. M., et al. (2024). Deconstructing the mangrove carbon cycle: Gains, transformation, and losses. *Ecosphere, 15*(3), e4806.

Akram, H., Hussain, S., Mazumdar, P., Chua, K. O., Butt, T. E., & Harikrishna, J. A. (2023). Mangrove health: A review of functions, threats, and challenges associated with mangrove management practices. *Forests, 14*(9), 1698.

Alongi, D. M. (2014). Carbon cycling and storage in mangrove forests. *Annual Review of Marine Science, 6*(1), 195–219.

Alongi, D. M. (2020). Global significance of mangrove blue carbon in climate change mitigation. *The Sciences, 2*(3), 67.

Anneboina, L. R., & Kumar, K. K. (2016). *Contribution of mangroves to marine fisheries in India* (No. 2016-145).

Ariyanto, D., & Pringgenies, D. (2024). Carbon sequestration rate in sediment mangroves from natural and rehabilitated mangroves. *Global Journal of Environmental Science & Management (GJESM), 10*(4).

Arnaud, M. A. S. (2020). *Belowground carbon and hydrological dynamics of mangrove forests* (Doctoral dissertation, University of Leeds).

Badarudeen, A., Padmalal, D., & Sajan, K. (1998). Organic carbon and total phosphorus in the sediments of some selected MEs of Kerala, southwest coast of India. *Journal Geological Society of India, 51*(5), 679–684.

Bimrah, K., Dasgupta, R., Hashimoto, S., Saizen, I., & Dhyani, S. (2022). Ecosystem services of mangroves: A systematic review and synthesis of contemporary scientific literature. *Sustainability, 14*(19), 12051.

Chambers, L. G., Davis, S. E., Troxler, T., Boyer, J. N., Downey-Wall, A., & Scinto, L. J. (2014). Biogeochemical effects of simulated sea level rise on carbon loss in an Everglades mangrove peat soil. *Hydrobiologia, 726*, 195–211.

Chow, J. (2018). Mangrove management for climate change adaptation and sustainable development in coastal zones. *Journal of Sustainable Forestry, 37*(2), 139–156.

Cuenca-Ocay, G. (2019). MEs' role in climate change mitigation. *Davao Research Journal, 12*(2), 72–75.

Das, N., Mondal, A., Gangopadhay, S., Banerjee, R., Batabyal, S., Ghosh, P. B., & Mandal, S. (2020). Environmental variables regulating organic carbon dynamics of Sundarban mangrove ecosystem, India. *International Journal of Chemical and Environmental Sciences, 2*(1), 7–17.

Dhadse, S., Majumdar, P., Valluri, A., & Sakhare, P. (2021). Status, threats and conservation of the wetland ecosystem dominated with mangroves in India. *Plantae Scientia, 4*(6), 277–286.

Eyzaguirre, I. A., Iwama, A. Y., & Fernandes, M. E. (2023). Integrating a conceptual framework for the sustainable development goals in the mangrove ecosystem: A systematic review. *Environmental Development, 47*, 100895.
Griffiths, L. N., & Mitsch, W. J. (2021). Estimating the effects of a hurricane on carbon storage in mangrove wetlands in Southwest Florida. *Plants, 10*(8), 1749.
Hayase, S., Ichikawa, T., & Tanaka, K. (1999). Preliminary report on stable isotope ratio analysis for samples from Matang mangrove brackish water ecosystem. *Japan Agricultural Research Quarterly, 33*, 215–221.
Inoue, T. (2018). Carbon sequestration in mangroves. In *Blue carbon in shallow coastal ecosystems: Carbon dynamics, policy, and implementation* (pp. 73–99). Springer Singapore.
Jennerjahn, T. C. (2020). Relevance and magnitude of 'Blue Carbon' storage in mangrove sediments: Carbon accumulation rates vs. stocks, sources vs. sinks. *Estuarine, Coastal and Shelf Science, 247*, 107027.
Kida, M., & Fujitake, N. (2020). Organic carbon stabilization mechanisms in mangrove soils: A review. *Forests, 11*(9), 981.
Koh, H. L., Teh, S. Y., Kh'Ng, X. Y., & Raja Barizan, R. S. (2018). Mangrove forests: Protection against and resilience to coastal disturbances. *Journal of Tropical Forest Science, 30*(5), 446–460.
Koo, H., Kleemann, J., Cuenca, P., Noh, J. K., & Fürst, C. (2024). Implications of landscape changes for ecosystem services and biodiversity: A national assessment in Ecuador, Ecosystem Services, *Elsevier*, vol. 69(C).
Kotwal, S. (2024). Applying remote sensing, Google Earth engine, and machine learning to predict the carbon sequestration potential of Sundarbans mangroves. *Asian Journal of Environment & Ecology, 23*(8), 140–150.
Larsen, T., Wooller, M. J., Fogel, M. L., & O'Brien, D. M. (2012). Can amino acid carbon isotope ratios distinguish primary producers in a mangrove ecosystem? *Rapid Communications in Mass Spectrometry, 26*(13), 1541–1548.
Lokollo, F. F., Wawo, M., Ceanturi, A., Lewerissa, Y. A., Hulopi, M., Handayani, L. D., et al. (2024). Blue carbon potential of MEs and its management to promote climate change mitigation in Indonesia. *Jurnal Ilmu Kehutanan, 18*(2), 208–218.
Mullick, P., Sengupta, G., Biswas, S., Pal, N., Pramanick, P., Ahmed, S., et al. (2023). Carbon storage potential of dominant mangroves in Western Indian Sundarbans.
Murdiyarso, D., Donato, D., Kauffman, J. B., Kurnianto, S., Stidham, M., & Kanninen, M. (2009). Carbon storage in mangrove and peatland ecosystems: A preliminary account from plots in Indonesia.
Nguyen, P. T., Tue, N. T., Quy, T. D., & Thai, N. D. (2016). Quantifying organic carbon storage and sources in sediments of Dong Rui mangrove forests, Tien Yen district, Quang Ninh province using carbon stable isotope. *Vietnam Journal of Earth Sciences, 38*(4), 317–326.
Noguchi, Y., DasGupta, R., & Shaw, R. (2012). Cooperative management of MEs in India. In *Ecosystem-based adaptation* (world; Vol. 12, pp. 63–84). Emerald Group Publishing Limited. https://doi.org/10.1108/S2040-7262(2012)0000012010
Ouyang, X., Lee, S. Y., & Connolly, R. M. (2017). The role of root decomposition in global mangrove and saltmarsh carbon budgets. *Earth-Science Reviews, 166*, 53–63.
Pandey, C. N., & Pandey, R. (2013). Carbon sequestration in mangroves of Gujarat, India. *International Journal of Botany and Research, 3*(2), 57–70.
Pandey, B., Koirala, S., Aung, H., Li, R., Timilsina, A., Htun, S. T. T., et al. (2024). Understanding carbon storage dynamics in Ayeyarwady delta's mangrove ecosystem in Myanmar: Insights for restoration efforts. *Environmental Research Communications, 6*(2), 025006.
Patil, V., Singh, A., Naik, N., Seema, U., & Sawant, B. (2012). Carbon sequestration in mangroves ecosystems.
Ragavan, P., & Mandal, R. N. (2018). Status of Indian mangroves. *Oceanography & Fisheries Open Access Journal, 7*(4), 95–99.

Raha, A. K., Bhattacharyya, S. B., Zaman, S., Banerjee, K., Sengupta, K., Sinha, S., et al. (2013). Carbon census in dominant mangroves of Indian Sundarbans. *The Journal of Energy and Environmental Science (Photon), 127*, 345–354.

Rao, K., Ramanathan, A. L., & Raju, N. J. (2022). Assessment of blue carbon stock of Coringa mangroves: Climate change perspective. *Journal of Climate Change, 8*(2), 41–58.

Rathor, S. (2024). Sustaining India's mangroves: Evaluating law and policies for ecological restoration. *Vidya—A Journal of Gujarat University, 3*(1), 105–112.

Rovai, A. S., Twilley, R. R., Castañeda-Moya, E., Riul, P., Cifuentes-Jara, M., Manrow-Villalobos, M., et al. (2018). Global controls on carbon storage in mangrove soils. *Nature Climate Change, 8*(6), 534–538.

Sahu, S. K., & Kathiresan, K. (2019). The age and species composition of mangrove forest directly influence the net primary productivity and carbon sequestration potential. *Biocatalysis and Agricultural Biotechnology, 20*, 101235.

Salle, S., Soma, A. S., Paembonan, S. A., & Rahmat, S. (2024, December). Social interaction dynamics and mangrove ecosystem management for blue carbon, case study: Tongke-tongke mangrove tourism area Indonesia. In *IOP conference series: Earth and environmental science* (Vol. 1430, No. 1, p. 012025). IOP Publishing.

Samanta, B. (2017). Mangroves in India: A geographical perspective.

Sasmito, S. D., Basyuni, M., Kridalaksana, A., Saragi-Sasmito, M. F., Lovelock, C. E., & Murdiyarso, D. (2023). Challenges and opportunities for achieving Sustainable Development Goals through restoration of Indonesia's mangroves. *Nature Ecology & Evolution, 7*(1), 62–70.

Sethi, P., & Syal, J. (2024). Carbon sequestration and climate change. *Contemporary Advances in Science and Technology, 7*(1), 15–28.

Singh, M., Schwendenmann, L., Wang, G., Adame, M. F., & Mandlate, L. J. C. (2022). Changes in mangrove carbon stocks and exposure to sea level rise (SLR) under future climate scenarios. *Sustainability, 14*(7), 3873.

Su, J., Friess, D. A., & Gasparatos, A. (2021). A meta-analysis of the ecological and economic outcomes of mangrove restoration. *Nature Communications, 12*(1), 5050.

Tao, Z., Shen, C. D., Yi, W. X., & Gao, Q. Z. (2004). Progresses in the study of carbon isotopes tracing of the soil carbon dynamics. *Advances in Earth Science, 19*(5), 793.

Tateda, Y., Nhan, D. D., Wattayakorn, G., & Toriumi, H. (2005). Preliminary evaluation of organic carbon sedimentation rates in Asian mangrove coastal ecosystems estimated by {210} Pb chronology. *Radioprotection, 40*(1), S527.

Tue, N. T., Hamaoka, H., Sogabe, A., Quy, T. D., Nhuan, M. T., & Omori, K. (2011). The application of δ13C and C/N ratios as indicators of organic carbon sources and paleoenvironmental change of the mangrove ecosystem from Ba Lat Estuary, Red River, Vietnam. *Environmental Earth Sciences, 64*(5), 1475–1486.

Tue, N. T., Hamaoka, H., Sogabe, A., Quy, T. D., Nhuan, M. T., & Omori, K. (2011). Sources of sedimentary organic carbon in MEs from Ba Lat estuary, Red River, Vietnam. In *Interdisciplinary studies on environmental chemistry-marine environmental modeling & analysis* (pp. 151–157). Terrapub.

Tue, N. T., Quy, T. D., Nhuan, M. T., Dung, L. V., & Thai, N. D. (2017). Tracing carbon transfer and assimilation by invertebrates and fish across a tropical mangrove ecosystem using stable isotopes. *Marine Ecology, 38*(5), e12460.

Tue, N. T., Nguyen, P. T., Quan, D. M., Dung, L. V., Quy, T. D., Nhuan, M. T., & Thai, N. D. (2018). Sedimentary composition and organic carbon sources in mangrove forests along the coast of northeast Vietnam. *Regional Studies in Marine Science, 17*, 87–94.

Varghese, M., Vinod, K., Raju, A. K., Sreekumar, K. M., Thobias, P. A., Ansar, C. P., et al. (2021). Biomass and carbon stocks in two mangrove patches of Chettuva Estuary, south-west coast of India. *Journal of the Marine Biological Association of India, 63*(1), 122–127.

Verma, P., & Kumar Ghosh, P. (2022). REDD+ strategy for forest carbon sequestration in India. *The Holistic Approach to Environment, 12*(3), 117–130.

Viswanathan, P. K. (2015). Multifunctional benefits of community-based mangrove restoration in Gujarat: An analysis. In *Nature, economy and society: Understanding the linkages* (pp. 243–265). Springer India.

Wang, Q., Kang, Q., Zhao, B., Li, H., Lu, H., Liu, J., & Yan, C. (2022). Effect of land-use and land-cover change on mangrove soil carbon fraction and metal pollution risk in Zhangjiang Estuary, China. *Science of the Total Environment, 807*, 150973.

Xin, K., Yan, K., Gao, C., & Li, Z. (2018). Carbon storage and its influencing factors in Hainan Dongzhangang mangrove wetlands. *Marine and Freshwater Research, 69*(5), 771–779.

Xiong, Y., Liao, B., Proffitt, E., Guan, W., Sun, Y., Wang, F., & Liu, X. (2018). Soil carbon storage in mangroves is primarily controlled by soil properties: A study at Dongzhai Bay, China. *Science of the Total Environment, 619*, 1226–1235.

Yong, Y., Baipeng, P., Guangcheng, C., & Yan, C. (2011). Processes of organic carbon in MEs. *Acta Ecologica Sinica, 31*(3), 169–173.

Quantitative Assessment and Prediction of Changes in Mangrove Cover of Sundarbans Through Markov Chain Analysis

Shariq Nawaz Khan, Pashupati Nath Singh, Upasana Yadav, and Prashant K. Srivastava

Abstract The Sundarbans Mangrove Ecosystem (ME), the world's largest contiguous mangrove forest, is undergoing significant changes due to increasing human activities and climate-related stressors. This study provides a detailed spatiotemporal analysis of land cover changes across the Indian Sundarbans using MODIS decadal land cover data (2000–2020) combined with Markov Chain modeling to forecast future scenarios up to 2050. The methodological framework incorporates remote sensing classification, spatial statistics, transition matrix analysis, and predictive simulation within a Geographic Information System (GIS) environment. Results show a notable decline in dense mangrove areas and a growth in sparse vegetation and non-vegetative land cover types, highlighting ongoing degradation. Transition probability analysis suggests continued deforestation trends under a business-as-usual approach. Future land cover maps for 2030, 2040, and 2050 project a considerable decline in ecological resilience, with high-risk zones identified for biodiversity loss and coastal vulnerability. The findings stress the urgent need for proactive conservation efforts, adaptive coastal zone management, and strengthened community-based strategies. This integrated approach demonstrates the effectiveness of medium-resolution satellite data and Markov modeling for long-term mangrove monitoring and decision-making, especially in data-limited, ecologically sensitive areas like the Sundarbans.

Keywords MODIS · Deforestation · GIS · Markov modeling · Sundarbans

---

S. N. Khan · U. Yadav (✉)
Amity School of Applied Sciences, Amity University, Lucknow, India
e-mail: uyadav@lko.amity.edu

P. N. Singh · P. K. Srivastava
Remote Sensing Laboratory, Institute of Environment and Sustainable Development, Banaras Hindu University, Varanasi, India

S. K. Srivastava et al. (eds.), *Earth Observation for Monitoring Mangrove Ecosystems*, https://doi.org/10.1007/978-3-032-16668-5_14

1 Introduction

The Sundarbans represent the largest contiguous mangrove ecosystem on the planet, covering approximately 10,000 km^2 within the deltaic confluence of the Ganges, Brahmaputra, and Meghna rivers (Samanta et al., 2021). Geographically unique and ecologically vibrant, this expansive region traverses the borders of Bangladesh (approximately 62%) and India (approximately 38%), offering a mosaic landscape shaped by tidal influences, saline estuaries, intertidal mudflats, and continually shifting island formations (Mukhopadhyay et al., 2015). Acknowledged by UNESCO as a World Heritage Site, the Sundarbans are globally recognized for their extraordinary biodiversity, complex ecological interactions, and substantial provisioning of critical ecosystem services such as coastal protection, biodiversity conservation, carbon sequestration, and livelihood support for local populations (Chanda et al., 2016). Focusing on India, the Sundarbans mangrove forest occupies around 4263 km^2 within the southern extent of West Bengal. This region provides indispensable habitats for numerous globally threatened and iconic species, notably *Panthera tigris tigris* (Royal Bengal Tiger), *Crocodylus porosus* (estuarine crocodile), and *Platanista gangetica* (Ganges River dolphin) (Ghosh et al., 2015). The Indian Sundarbans ecosystem further supports an extensive and diverse faunal assemblage, encompassing more than 350 bird species, an array of reptiles and amphibians, and a rich diversity of aquatic organisms. Such biodiversity underpins regional fisheries and traditional practices such as honey collection, timber harvesting, and medicinal plant utilization, sustaining intricate socio-economic webs vital for local communities (Barik et al., 2018).

Ecologically, the Sundarbans exhibit pronounced vegetation zonation patterns driven primarily by gradients in salinity, sedimentation rates, tidal dynamics, inundation frequency, and soil physicochemical properties (Barik et al., 2018). Prominent mangrove species, including *Heritiera fomes* (Sundari), *Avicennia marina* (Baen), *Excoecaria agallocha* (Gewa), *Ceriops decandra* (Goran), and *Sonneratia apetala* (Keora), showcase distinct ecological tolerances and adaptations to varying saline conditions and inundation regimes, thus serving as bioindicators for monitoring environmental change and stress (Mukhopadhyay et al., 2015). This zonation not only fosters biodiversity through niche differentiation but also enhances ecosystem resilience by providing structural complexity and functional redundancy, thereby mitigating the impacts of environmental disturbances and supporting robust ecological functionality (Chanda et al., 2016). However, despite their ecological robustness and global recognition, the Sundarbans face increasing threats from anthropogenic pressures and climate-related stressors. Rising sea levels, intensified cyclonic activity, altered sediment flux from upstream dam constructions, increased salinity intrusion, habitat fragmentation, and unsustainable resource extraction collectively challenge the ecological integrity and sustainability of this invaluable ecosystem (Samanta et al., 2021). These pressures necessitate precise and continuous monitoring, informed management interventions, and comprehensive predictive modeling to support adaptive conservation strategies.

Despite their substantial ecological significance and the vital socio-economic roles, the Sundarbans mangroves face escalating threats from both anthropogenic activities and climate-induced stressors. Rising sea levels and increased frequency and severity of tropical cyclones, notably Cyclones Aila (2009), Amphan (2020), and Yaas (2021), critically threaten the structural integrity and ecological balance of this sensitive mangrove system (Samanta et al., 2021). Moreover, ongoing salinity intrusion, exacerbated by reduced freshwater inflows resulting from upstream river management practices, including dam construction and water diversion, directly compromises the viability of mangrove species and associated biodiversity (Mukhopadhyay et al., 2015). Anthropogenic pressures compound these natural stressors, with illegal logging, overfishing, unsustainable resource extraction, unplanned aquaculture development, and agricultural encroachment causing extensive habitat degradation, fragmentation, and loss of biodiversity. The cumulative impact of these activities accelerates ecological deterioration, undermining ecosystem services and posing significant challenges to the resilience and sustainability of the Sundarbans ecosystem (Samanta et al., 2021). Addressing these complex challenges necessitates rigorous, spatially explicit monitoring methods capable of capturing dynamic changes at relevant ecological scales.

Remote sensing technologies, particularly satellite datasets from the Moderate Resolution Imaging Spectroradiometer (MODIS), offer powerful tools to systematically monitor and quantify spatiotemporal variations in mangrove cover and condition across large geographic extents (Singh et al., 2025). Advanced remote sensing technologies, such as MODIS, combined with predictive spatial modeling approaches like Markov Chain Analysis (MCA) (Singh et al., 2015a, 2015b), offer robust methodologies for understanding historical land cover dynamics, forecasting future ecological conditions, and guiding policy and management strategies tailored to safeguarding the Sundarbans ecosystem (Halder et al., 2021). These remote sensing data enable robust forecasting of future land cover scenarios and provide critical insights for decision-makers. Such predictive capabilities are indispensable for identifying vulnerable areas, anticipating ecological outcomes, and implementing informed management actions to ensure the long-term ecological integrity and resilience of the Sundarbans transboundary mangrove ecosystem (Halder et al., 2021).

MEs such as the Sundarbans are indispensable for biodiversity conservation, coastal protection, and the maintenance of marine and estuarine food webs. Numerous studies underscore the value of mangroves in buffering against natural hazards, stabilizing shorelines, and sequestering carbon (Mukhopadhyay et al., 2015; Samanta et al., 2021). In both India and Bangladesh, these forests are essential for the sustenance of local communities, offering resources such as fish, honey, fuelwood, and medicinal plants. The Sundarbans also exhibit salinity gradients that influence species composition and distribution. Studies reveal that species like *Heritiera* (Sundari) dominate in lower salinity zones, while salt-tolerant species like *Avicennia* (Baen) and *Excoecaria* (Gewa) are prevalent in high-salinity zones (Barik et al., 2018). The dynamic nature of this ecosystem necessitates integrated management approaches, considering the transboundary ecological and socio-economic significance (Thomas et al., 2017). In view of the above, the objectives of this chapter are:

- To assess decadal-scale land cover changes in the Sundarbans mangrove forest using MODIS datasets
- To apply Markov Chain Analysis for predictive modeling of future land cover scenarios across the India-Bangladesh border
- To evaluate the ecological implications of observed and projected land cover changes for the entire Sundarbans ecosystem

2 Literature Review

2.1 MODIS Datasets in Land Cover Assessment

MODIS has emerged as a crucial instrument for evaluating the temporal evolution of MEs. Land cover dynamics at broad spatial scales can be consistently monitored thanks to their high temporal frequency. Classifying the main land cover types in the Indian and Bangladeshi Sundarbans has been made easier by earlier research that successfully used MODIS land cover phenology data to examine transitions in mangrove areas by capturing seasonal variations in vegetation characteristics (Kanan & Pirotti, 2022; Samanta et al., 2021).

Several studies have leveraged MODIS datasets to assess long-term vegetation trends and detect disturbances (Singh et al., 2022). For example, Pastor-Guzman et al. (2018) highlighted the value of MODIS-derived phenological data in tracking mangrove dynamics. However, limitations in spatial resolution necessitate the integration of MODIS with higher-resolution data and field validation for accurate land cover classification and change detection (Halder et al., 2021). The details are presented in (Table 1).

2.2 Markov Chain Analysis for Land Cover Prediction

MCA is widely used for modeling land cover transitions and projecting future scenarios based on observed historical trends (Singh et al., 2018; Singh & Srivastava, 2025). It operates on probabilistic principles, generating transition matrices that inform changes over time (Mukhopadhyay et al., 2015). In the context of the Sundarbans, Markov Chain modeling has been successfully implemented in both India and Bangladesh. Halder et al. (2021) used the CA-Markov model to project mangrove cover changes in the Indian Sundarbans, while Mukhopadhyay et al. (2015) applied the model to predict species zonation changes in the Bangladeshi part. These studies demonstrate the model's ability to simulate spatial and temporal dynamics, albeit with some limitations in capturing complex ecological processes.

Table 1 Detailed overview of the Indian Sundarbans ecosystem

Aspect	Description	Key examples or indicators	Importance/role	Threats and challenges	References
Geographical area	Approximately 4263 km^2 in West Bengal, India	Coastal delta, tidal estuaries, intertidal mudflats	Unique estuarine ecosystem, biodiversity hotspot	Sea-level rise, erosion, cyclones	Samanta et al. (2021)
Biodiversity and habitats	Rich biodiversity including globally threatened species	Royal Bengal Tiger, estuarine crocodile, Ganges River dolphin	Habitat protection, conservation of species	Habitat fragmentation, illegal hunting, habitat loss	Ghosh et al. (2015); Barik et al. (2018)
Vegetation structure	Zonation influenced by salinity, tides, sediments	Sundari, Baen, Gewa, Goran, Keora	Supports ecosystem resilience, stability	Salinity intrusion, changes in sediment dynamics	Mukhopadhyay et al. (2015); Barik et al. (2018)
Ecological services	Coastal protection, carbon sequestration, fisheries support	Mangrove carbon stocks, fish and shellfish nurseries	Mitigation of climate impacts, livelihood support	Climate change, unsustainable extraction practices	Chanda et al. (2016)
Socio--economic significance	Supports the livelihoods of local communities	Fishing, honey collection, traditional medicines	economic resilience, cultural value	Overexploitation, economic pressures	Ghosh et al. (2015)
Management and monitoring	Remote sensing (MODIS), Markov Chain modeling	Land cover dynamics, predictive land use scenarios	Enhanced conservation and management planning	Need for high-resolution, continuous data	Halder et al. (2021)

Integrating Markov Chain models with cellular automata (CA) enhances spatial realism, enabling the incorporation of neighborhood influence and landscape patterns into the predictions (Mustak et al., 2022; Maurya et al., 2021). This hybrid approach is particularly suited to dynamic ecosystems like the Sundarbans, where spatially explicit changes occur due to environmental and anthropogenic drivers (Kundu et al., 2021).

3 Materials and Methods

3.1 Study Area

The present study focuses on the Indian Sundarbans, located in the southernmost part of West Bengal, India, within the North and South 24 Parganas districts (Fig. 1). This region constitutes the western segment of the larger Sundarbans delta and spans approximately 4263 km^2 of protected mangrove forest interspersed with inhabited islands, estuarine networks, tidal creeks, and mudflats (Roy et al., 2018; Ramsar Convention Secretariat, 2019). Geographically, the area lies between latitudes 21°32′N to 22°40′N and longitudes 88°05′E to 89°00′E, forming a dynamic interface between land and sea along the northern Bay of Bengal. The Indian Sundarbans comprises a complex coupled human–natural system under the joint jurisdiction of the Forest Department and local administrative bodies (Chaudhuri & Choudhury, 2021). It includes designated zones such as the Sundarban Biosphere Reserve, the Sundarban Tiger Reserve, and multiple protected areas like Sajnekhali, Lothian Island, and Haliday Island Wildlife Sanctuaries. The region is characterized by a tropical monsoon climate with high humidity, seasonal rainfall (mostly from June to September), and frequent cyclonic disturbances that significantly influence mangrove structure and function. This area is ecologically unique due to its pronounced salinity gradient, high tidal amplitude, and dynamic sedimentary processes that continuously reshape its geomorphology. The dominant land cover types

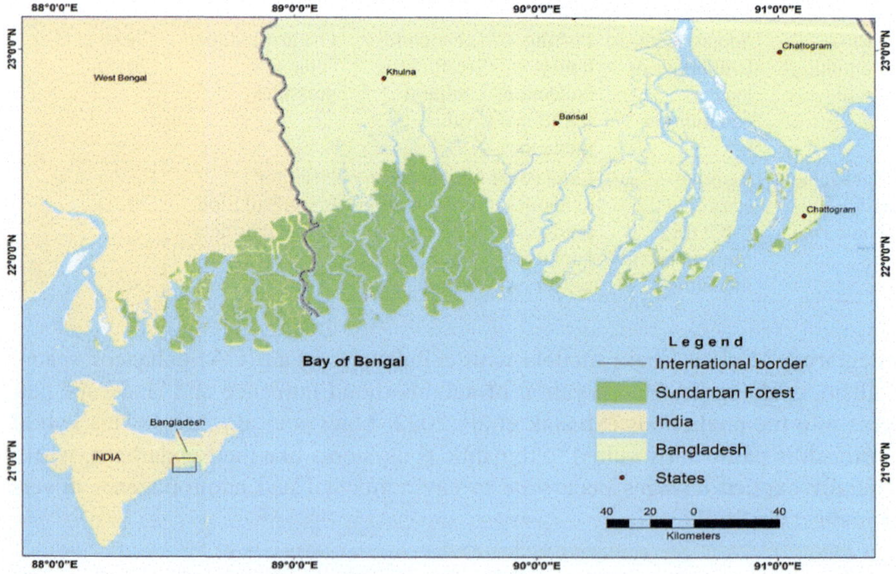

Fig. 1 Map illustrating the transboundary extent of the Sundarbans mangrove forest along the Bay of Bengal, covering southern West Bengal (India) and southwestern Bangladesh

include dense mangrove forests, sparse mangrove patches, mudflats, aquaculture zones, and human settlements (Mitra et al., 2011).

Given its ecological importance and vulnerability, the Indian Sundarbans serve as an ideal natural laboratory for assessing long-term land cover dynamics using remote sensing. The integration of MODIS datasets and Markov Chain predictive modeling in this study allows for the temporal assessment of landscape transitions and spatial forecasting of future land use scenarios.

3.2 Data Sources and Acquisition

To assess land cover dynamics in the Indian Sundarbans, the MODIS MCD12Q1 Land Cover Type product (Collection 6) was employed, obtained from NASA Earthdata (2023) via the LP DAAC archive (NASA LP DAAC, 2019). This dataset offers annual global land cover classifications at a 500-m resolution and is widely validated for long-term environmental monitoring (Friedl et al., 2010). The phenological consistency and land surface dynamics captured by the MODIS products have been further evaluated in previous studies (Ganguly et al., 2010; Tan et al., 2011; Gray et al., 2019). Data were acquired for the years 2001, 2011, and 2021—chosen deliberately to represent decadal intervals that help capture temporal land use transformations and evaluate long-term ecological changes. Five biologically relevant land cover classes—mangrove forest, open water, built-up regions, farming, and sparse/barren terrain—were taken from the International Geosphere-Biosphere Programme (IGBP) methodology, which served as the basis for the classification system that was used. The significance of these classes to the distinct ecological and socio-economic structure of the Sundarbans region led to their selection (Rahman et al., 2020; Ghosh et al., 2018b). A reclassification procedure was carried out in QGIS utilizing raster calculator logic and a pixel-based mapping technique in order to expedite the classification and eliminate redundancy (QGIS Development Team, 2022). To remove noise and guarantee uniformity across time periods, spatial overlays were employed. The MCD12Q1 product's quality assurance layers were consulted in order to reduce cloud cover and misclassification mistakes. Additional cloud and shadow masking approaches in satellite-based land cover products have been widely discussed in the literature (Zhu et al., 2019).

The classification reasoning was based on a comprehensive assessment of peer-reviewed literature and environmental studies relevant to the Indian Sundarbans, ensuring ecological relevance and geographic accuracy. Critical information about land cover typologies, historical land use changes, and regional phenology was obtained from these sources. Finally, in order to do MCA and gain a better understanding of the changes in land cover between the chosen years, transition probability matrices were developed. The focus on 2001, 2011, and 2021 offers a methodical framework for examining decadal changes and serves as the analytical basis for the findings and debate that follow.

3.3 Preprocessing of MODIS Data

The MCD12Q1 MODIS land cover data utilized in this study underwent necessary preprocessing before classification to guarantee temporal and spatial dependability. The first step was reprojecting the data to the WGS84 coordinate system (EPSG:4326) and using a pre-established vector border to extract the extent of the Indian Sundarbans. Under the direction of the quality assurance layers integrated into the MCD12Q1 product, annual composites that reduced seasonal and cloud-induced variability were chosen in order to preserve temporal consistency. No additional atmospheric correction was necessary because the dataset offers outputs that have been corrected for surface reflectance. QGIS and Python-based libraries like GDAL and Rasterio, Scikit-learn were used for all raster processing and modelling operations, including reprojection, clipping, and quality filtering (GDAL Developers, 2022; Python Software Foundation, 2022; Pedregosa et al., 2011). The foundation for consistent classification and change detection over the 3 chosen years—2001, 2011, and 2021—was established by these preprocessing processes.

3.4 Methodological Overview Framework

This study employed a comprehensive methodological approach that included prediction modeling using MCA, quantitative analysis of spatial and temporal changes, and land cover classification based on MODIS data. The objective was to predict future land cover dynamics and to understand the decadal evolution of mangrove landscapes in the Indian Sundarbans.

3.4.1 Land Cover Classification and Change Detection

Land cover classification was performed using the MODIS MCD12Q1 (Collection 6) dataset, which provides annual global land cover data at 500 m resolution. For ecological applicability to the Sundarbans, the original IGBP land cover classes were reclassified into five ecologically significant categories: mangrove forest, open water, built-up areas, farming, and sparse/barren land. Classification was conducted using pixel-based logic in QGIS, which reduced complexity while enhancing thematic accuracy. To ensure reliability, thematic validation was performed using published land use datasets and field-reported mangrove zones (Bera & Chatterjee, 2019). Standard accuracy assessment techniques were applied, including confusion matrix generation and the calculation of overall accuracy and the Kappa coefficient (Congalton & Green, 2019). These ensured the validity of the classification outputs across time periods.

To quantify land cover transitions over two decades, a post-classification comparison technique was applied across the years 2001, 2011, and 2021—a proven method for land use land cover (LULC) change detection (Lu et al., 2004). Transition matrices were generated using Python and R programming environments (Core Teams, 2023), and key spatial metrics—net gain/loss, change intensity, and percentage change—were computed. Spatial raster processing and matrix computation in R were supported using the terra package (Hijmans, 2023). These metrics enabled the identification of hotspot zones of mangrove degradation or regeneration. Spatial dynamics were visualized using QGIS temporal raster stacks and overlay tools.

To assess stability and transformation, persistence and change intensity were calculated using:

$$\text{Changed Pixels} = \text{Total Pixels} - \text{Unchanged Pixels} \tag{1}$$

$$\text{Persistence}\,(\%) = (\text{Unchanged Pixels} / \text{Total Pixels}) \times 100 \tag{2}$$

$$\text{Change Intensity}\,(\%) = (\text{Changed Pixels} / \text{Total Pixels}) \times 100 \tag{3}$$

These calculations provided an empirical basis for evaluating the long-term spatial resilience of land cover categories in the Sundarbans landscape.

3.4.2 Predictive Modeling Using Markov Chain Analysis

To forecast future land cover dynamics, a first-order Markov Chain model was applied, if future land cover states depend solely on the current distribution and its associated transition probabilities (Malhi et al., 2020). The model is governed by:

$$P_t + 1 = P_t \times T \tag{4}$$

where:

P_t = state vector of land cover at time t.

T = transition probability matrix derived from classified maps (2001–2011).

P_{t+1} = predicted land cover at time t + 1.

The transition matrix M satisfies the constraints:

$$\sum P_{ij} = 1 \text{and} P_{ij} \geq 0 \tag{5}$$

To improve spatial realism, the Markov model was conceptually integrated with cellular automata (CA), allowing neighborhood interactions and spatial suitability to influence future projections. The CA-Markov modelling framework was implemented using the TerrSet geospatial monitoring system (Eastman, 2022). This hybrid model is represented as:

$$J_{t+1} = f\left(i_t, N, P_{\text{uit}}, P_t\right) \tag{6}$$

where:

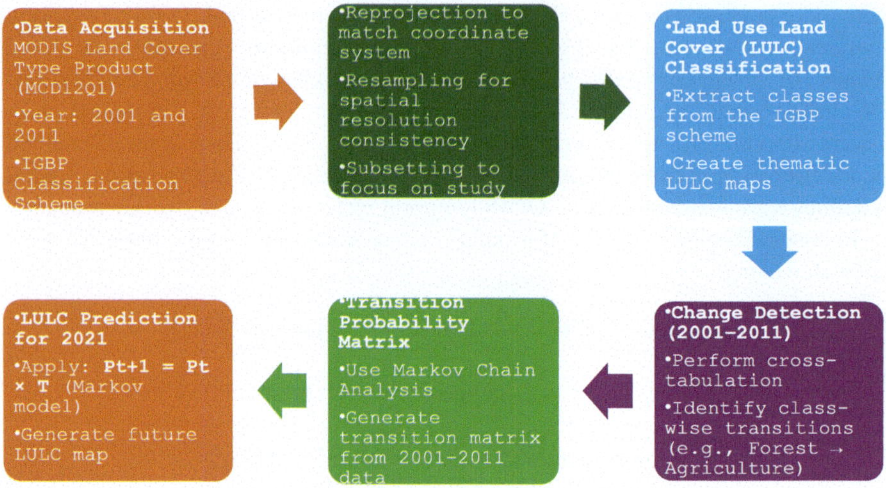

Fig. 2 Workflow illustrating the stepwise methodology for land cover classification and future LULC prediction using MODIS data and Markov Chain modeling

i_t = current land cover state.

N = neighborhood configuration.

S_{uit} = spatial suitability map.

P_t = land cover probability at time t.

Using this framework (Fig. 2), the land cover map for 2021 was simulated based on 2001–2011 transitions and compared to the actual 2021 MODIS-derived classification. Validation showed a high degree of accuracy, with a Kappa coefficient >0.86 and overall classification accuracy >90%, indicating excellent agreement between predicted and observed results.

Despite its utility, the model assumes stationarity and spatial homogeneity, which can oversimplify real-world complexities such as policy shifts, anthropogenic interventions, and climate anomalies (Mustak et al., 2022; Singh et al., 2015a, 2015b). Nonetheless, this approach provides a robust, data-driven foundation for ecological forecasting, spatial planning, and conservation policy formulation in highly sensitive landscapes such as the Sundarbans.

4 Results and Discussion

There were significant changes in all land cover categories between the classified LULC maps for 2001, 2011, and 2021 (Figs. 3, 4, and 5). Table 2 displays the different class cover codes and the related description of the MODIS land cover type provided. Significantly, cropland decreased from 2146.75 km^2 in 2001 to 1796.75 km^2 in 2021. On the other hand, the built-up area grew from 563.50 to 930.75 km^2, a sign of increased urbanization activity. A slight increase of 7.50 km^2

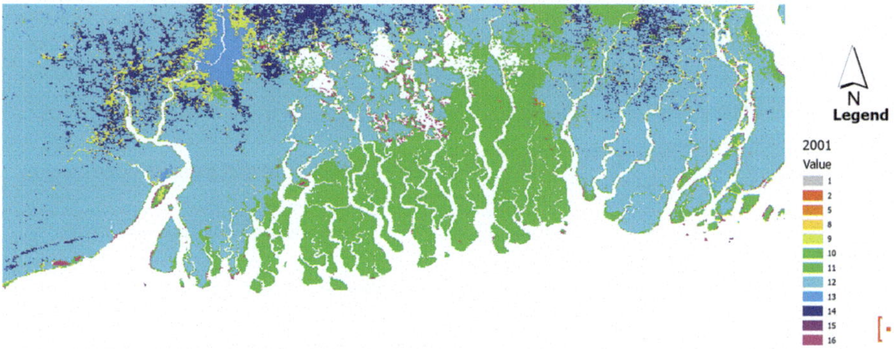

Fig. 3 MODIS land cover map of 2001

Fig. 4 MODIS land cover map of 2011

Fig. 5 Markov chain predicted map of 2021

in water bodies suggests either reservoir extension or better surface water retention. The amount of forest area decreased marginally from 228.25 to 227.00 km^2 but stayed largely constant. There were conflicting patterns in grasslands and bare areas. Grasslands grew by about 70.50 km^2, while barren land decreased from 168.50 to

Table 2 MCD12Q1 International Geosphere-Biosphere Programme (IGBP) legend and class descriptions (NASA LP DAAC, MCD12Q1 User Guide Version 6)

Name	Value	Description
Evergreen needleleaf forests	1	Dominated by evergreen conifer trees (canopy >2 m). Tree cover >60%
Evergreen broadleaf forests	2	Dominated by evergreen broadleaf and palmate trees (canopy >2 m). Tree cover >60%
Deciduous needleleaf forests	3	Dominated by deciduous needleleaf (larch) trees (canopy >2 m). Tree cover >60%
Deciduous broadleaf forests	4	Dominated by deciduous broadleaf trees (canopy >2 m). Tree cover >60%
Mixed forests	5	Dominated by neither deciduous nor evergreen (40–60% of each) tree type (canopy >2 m). Tree cover >60%
Closed shrublands	6	Dominated by woody perennials (1–2 m height) >60% cover
Open shrublands	7	Dominated by woody perennials (1–2 m height) 10–60% cover
Woody savannas	8	Tree cover 30–60% (canopy >2 m)
Savannas	9	Tree cover 10–30% (canopy >2 m)
Grasslands	10	Dominated by herbaceous annuals (<2 m)
Permanent wetlands	11	Permanently inundated lands with 30–60% water cover and >10% vegetated cover
Croplands	12	At least 60% of area is cultivated cropland
Urban and built-up lands	13	At least 30% impervious surface area including building materials, asphalt, and vehicles
Cropland/natural vegetation mosaics	14	Mosaics of small-scale cultivation 40–60% with natural tree, shrub, or herbaceous vegetation
Permanent snow and ice	15	At least 60% of area is covered by snow and ice for at least 10 months of the year
Barren	16	At least 60% of area is non-vegetated barren (sand, rock, soil) areas with less than 10% vegetation
Water bodies	17	At least 60% of area is covered by permanent water bodies
Unclassified	255	Has not received a map label because of missing inputs

121.25 km^2, suggesting that some areas of the study region had better vegetation. The area of mining and industrial land grew from 48.00 to 90.00 km^2, presumably as a result of development operations. By 2021, plantations had grown from 36.75 km^2 in 2001 to 50.50 km^2. Scrublands and wetlands saw very little change. The entire region being analysed stays at 4536 km^2. The pressure of urban growth and shifting land dynamics in the area is highlighted by this extensive change in land use, particularly the reduction of agricultural land and the increase of built-up and industrial sectors. The 2021 prediction map and the years 2001–2011 both showed a steady rise in built-up areas. This pattern highlights the wider pressure of urbanization that characterizes India's peri-urban areas (Kafy et al., 2022). The expansion of built-up land has mostly come at the expense of vegetation and agricultural classes, which have significantly decreased. Infrastructure development, growing housing demand, and shifting livelihoods—particularly in the vicinity of urban peripheries—are probably the main causes of this transition (Wang et al., 2021). The accuracy assessment for the 2021 classification was conducted using the Kappa confusion matrix, based on 500 m × 500 m pixel resolution.

These maps show how the study region's agricultural zones have decreased in tandem with the growth of urban and industrial sectors. Table 3 displays the quantitative results derived from the change detection analysis based on satellite image classification.

Table 4 shows the number of pixels categorized by type, converted to area (km^2). For example, 35 pixels in Class 2 (cropland) were correctly classified, resulting in 8.75 km^2 (35 × 0.25 km^2). Throughout the confusion matrix, this pixel-to-area conversion has been used consistently. With a Kappa coefficient of 0.8615 and an overall classification accuracy of 90.86%, the reference and forecasted land cover classes were compared. A high level of confidence in the land cover prediction model is indicated by a Kappa value above 0.80, which generally shows significant agreement between the reference data and the categorized map. Rapid urbanization and shifting agricultural patterns are the main drivers of the study area's dynamic land use change, according to an examination of LULC trends from 2001 to 2011 and the projected 2021 scenario. The high overall accuracy (above 85%) and Kappa coefficients (above 0.8) of the random forest algorithm's classification outputs demonstrated strong results, demonstrating the algorithm's efficacy in distinguishing between different forms of land cover (Belgiu & Drăguț, 2016).

Relative stability over time was shown by the enduring classes, such as woods and water bodies, indicating little human interference or environmental and legal protection. Agricultural regions, on the other hand, showed less persistence, suggesting seasonal fluctuations, land conversion, or potential abandonment (Giri et al., 2016). These changes were further highlighted by the 2021 prediction analysis, which predicted a sustained increase in the urban footprint and a decrease in vegetation and agricultural area.

These temporal and spatial changes are consistent with South Asian urbanization trends, where economic shifts and population pressure are causing land to be diverted more and more to non-agricultural uses (Zhang et al., 2019; Jain et al., 2022). These trends also suggest possible threats to ecological services, such as diminished soil health, changed hydrological cycles, and biodiversity loss. Temporal expansion to incorporate multi-seasonal data is recommended in order to improve comprehension of LULC dynamics. Particularly in areas where agriculture predominates, this would aid in capturing intra-annual fluctuations. Furthermore, incorporating socio-economic factors like land tenure systems, economic drivers, and population density may offer a more thorough understanding of the factors influencing changes in land cover. Assessing how LULC change affects microclimatic factors like temperature and precipitation—which are essential for creating climate-resilient urban infrastructure—may be one of the future directions. To help with proactive policymaking and sustainable land resource planning, predictive models such as CA-Markov chains or artificial neural networks could also be used to estimate future LULC patterns. Furthermore, to increase categorization accuracy and promote greater community involvement in land management, field validation

Table 3 Area-wise LULC (2001–2021)

Class	Pixels (Y1)	Pixels (Y2)	Pixels (Pred.)	Actual change (px)	Predicted change (px)	Area (Y1, km^2)	Area (Y2, km^2)	Area (Pred., km^2)	Actual change (km^2)	Predicted change (km^2)
2	3	0	0	−3	0	0.75	0	0	−0.75	0
3	39	144	161	+105	+17	9.75	36	40.25	+26.25	+4.25
4	0	0	0	0	0	0	0	0	0	0
5	0	0	0	0	0	0	0	0	0	0
6	2	1	3	−1	+2	0.5	0.25	0.75	−0.25	+0.5
7	0	0	0	0	0	0	0	0	0	0
8	0	0	0	0	0	0	0	0	0	0
9	3	87	73	+84	−14	0.75	21.75	18.25	+21	−3.5
10	5357	14,312	19,636	+8955	+5324	1339.25	3578	4909	+2238.75	+1331
11	1822	4896	6310	+3074	+1414	455.5	1224	1577.5	+768.5	+353.5
12	50,304	48,496	48,420	−1808	−76	12,576	12,124	12,105	−452	−19
13	110,737	101,946	95,598	−8791	−6348	27,684.25	25,486.5	23,899.5	−2197.75	−1587
14	3375	3402	3457	+27	+55	843.75	850.5	864.25	+6.75	+13.75
15	11,354	8855	8025	−2499	−830	2838.5	2213.75	2006.25	−624.75	−207.5
16	11	7	7	−4	0	2.75	1.75	1.75	−1	0

Table 4 Kappa analysis confusion matrix (prediction vs reference), all values shown in square kilometers

Ref/Pred	2	5	8	9	10	11	12	13	14	15	16	17	Total
2	8.75	0	0	4.0	0	0	0	0	0	0	0	0	12.75
5	0	0	0	0	0	0	0.75	0	0	0	0	0	0.75
8	0	0	0	2.75	0	3.25	12.0	0.25	0	0	0	0	18.25
9	2.5	0	15.75	2208.25	247.75	405.75	1494.0	0	529.75	0	5.25	0	4908.0
10	0	0	6.0	162.0	484.0	389.25	516.5	0	5.0	0	14.75	0	1577.5
11	24.75	0	0	243.5	178.75	10,835.0	579.0	0	19.5	0	199.5	0	12,079.0
12	0	0.25	0	733.5	203.25	238.25	22,023.5	0	698.25	0	2.5	0	24,200.5
13	0	0	0	10.5	1.5	0	1.75	850.5	0	0	0	0	864.25
14	0	0	0	208.75	23.0	3.25	810.75	0	960.5	0	0	0	2006.25
15	0	0	0	0	0	0	0	0	0	1.75	0	0	1.75
16	0	0	0	4.75	85.75	196.75	48.25	0	0.5	0	304.0	0	639.0
17	0	0	0	0	0	0	0	0	0	0	0	45,367.5	45,367.5

and participatory GIS mapping are advised. These bottom-up methods guarantee that local perspectives influence spatial planning choices and improve the relevance of policies. Significant LULC changes over two decades are mapped in the study, along with the consequences for ecosystem services, urban growth, and future land management techniques. The results support the increasing demand for data-driven planning instruments to support regional resilience and sustainable development.

5 Conclusion

This chapter highlights the alarming trend of land cover change in the Sundarbans mangrove forest, caused by climate change, rising sea levels, human activities, and natural disasters. The use of MODIS decadal land cover data and Markov Chain Analysis provides a solid, scientific basis for tracking historical land cover shifts and predicting future landscape changes. Results show a steady decline in dense mangrove areas and an increase in transitional and degraded zones, especially in ecologically sensitive buffer regions. Transition probability matrices suggest that if current trends continue, the region will lose vital mangrove cover, decreasing ecosystem services like coastal protection, carbon storage, and biodiversity protection. Spatial predictions for 2030, 2040, and 2050 show major changes in forest structure, affecting ecological health and local livelihoods. The Markov-based forecasts indicate that freshwater and moderately saline mangrove species will be increasingly replaced by salt-tolerant types, driven by rising salinity and hydrological disruption. Ultimately, this study emphasizes the value of combining remote sensing and probabilistic modeling to predict environmental changes. It recommends integrating these tools into national mangrove management plans, focusing on restoring degraded areas, regulating human activities, and encouraging community participation. Strengthening cross-border cooperation between India and Bangladesh could also promote joint conservation efforts across the Sundarbans landscape.

References

Barik, J., Mukhopadhyay, A., Ghosh, T., Mukhopadhyay, S. K., Chowdhury, S. M., & Hazra, S. (2018). Mangrove species distribution and water salinity: An indicator species approach to Sundarbans. *Journal of Coastal Conservation, 22*(2), 361–368. https://doi.org/10.1007/s11852-017-0574-y

Belgiu, M., & Drăguţ, L. (2016). Random forest in remote sensing: A review of applications and future directions. *ISPRS Journal of Photogrammetry and Remote Sensing, 114*, 24–31. https://doi.org/10.1016/j.isprsjprs.2016.01.011

Bera, K., & Chatterjee, S. (2019). Land use/land cover change and its impact on mangrove forests in Indian Sundarban using remote sensing and GIS. *Spatial Information Research, 27*, 577–589. https://doi.org/10.1007/s41324-019-00254-2

Chanda, A., Mukhopadhyay, A., Ghosh, T., Akhand, A., Mondal, P., Ghosh, S., & Hazra, S. (2016). Blue Carbon Stock of the Bangladesh Sundarban Mangroves: What could be the scenario after a century? *Wetlands, 36*, 1033–1045. https://doi.org/10.1007/s13157-016-0819-7

Chaudhuri, P., & Choudhury, A. (2021). Human–mangrove interactions and management in the Indian Sundarbans: An analysis from a socio-ecological perspective. *Environmental Challenges, 5*, 100257. https://doi.org/10.1016/j.envc.2021.100257

Congalton, R. G., & Green, K. (2019). *Assessing the accuracy of remotely sensed data: Principles and practices* (3rd ed.). CRC Press.

Core Team. (2023). *R: A language and environment for statistical computing*. https://www.r--project.org/

Eastman, J. R. (2022). *TerrSet geospatial monitoring and modeling system*. Clark Labs, Clark University.

Friedl, M. A., et al. (2010). MODIS Collection 5 global land cover: Algorithm refinements and characterization of new datasets. *Remote Sensing of Environment, 114*(1), 168–182.

Ganguly, S., Friedl, M. A., Tan, B., Zhang, X., & Verma, M. (2010). Land surface phenology from MODIS: Characterization of the Collection 5 global land cover dynamics product. *Remote Sensing of Environment, 114*(8), 1805–1816.

GDAL Developers. (2022). *GDAL: Geospatial data abstraction library*. https://gdal.org

Ghosh, A., Schmidt, S., Fickert, T., & Nüsser, M. (2015). The Indian Sundarban mangrove forests: History, utilization, conservation strategies and local perception. *Diversity, 7*(2), 149–169. https://doi.org/10.3390/d7020149

Ghosh, T., Mukhopadhyay, A., & Hazra, S. (2018b). Changes in the extent of the Sundarbans mangrove forest in India (1985–2015). *Journal of Coastal Conservation, 22*(4), 763–775. https://doi.org/10.1007/s11852-018-0604-7

Giri, C., Long, J., Tieszen, L., & Graesser, J. (2016). Global mangrove forest distribution, density, and change from 2000 to 2012. *Science Reports, 6*, 34395. https://doi.org/10.1038/srep34395

Gray, J. M., Song, B., & Zhang, X. (2019). Monitoring land surface phenology from MODIS. *International Journal of Remote Sensing, 40*(5–6), 1764–1786.

Halder, S., Samanta, K., & Das, S. (2021). Monitoring and prediction of dynamics in Sundarban forest using CA–Markov Chain model. In P. K. Shit et al. (Eds.), *Spatial modeling in forest resources management* (pp. 425–437). Springer. https://doi.org/10.1007/978-3-030-56542-8_18

Hijmans, R. J. (2023). *terra: Spatial data analysis*. R package version 1.7-65.

Jain, M., Naeem, S., Singh, P. K., & Gupta, R. K. (2022). Satellite-based assessment of vegetation dynamics and its relation with climatic variables in the Sundarbans. *Environmental Monitoring and Assessment, 194*, 1–16. https://doi.org/10.1007/s10661-022-09876-2

Kafy, A. A., Faisal, A. A., & Jahan, N. (2022). Monitoring and modeling of urban sprawl using remote sensing and machine learning techniques. *Journal of Urban Management, 11*(1), 35–49.

Kanan, A. H., & Pirotti, F. (2022). A comparative assessment of land use-land cover dynamics between Bangladesh and Indian Sundarbans from 1975–2020: A geospatial and statistical-based approach. *ISPRS Archives, XLIII-B3-2022*, 625–634. https://doi.org/10.5194/isprs-archives-XLIII-B3-2022-625-2022

Kundu, K., Halder, P., & Mandal, J. K. (2021). Detection and prediction of Sundarban reserve forest using the CA-Markov chain model and remote sensing data. *Earth Science Informatics, 14*(3), 1503–1520. https://doi.org/10.1007/s12145-021-00648-9

Lu, D., Mausel, P., Brondízio, E., & Moran, E. (2004). Change detection techniques. *International Journal of Remote Sensing, 25*(12), 2365–2407. https://doi.org/10.1080/0143116031000139863

Malhi, R. K. M., Anand, A., Srivastava, P. K., Kiran, G. S., Petropoulos, G. P., & Chalkias, C. (2020). An integrated spatiotemporal pattern analysis model to assess and predict the degradation of protected forest areas. *ISPRS International Journal of Geo-Information, 9*(9), 530.

Maurya, S., Srivastava, P. K., Yaduvanshi, A., Anand, A., Petropoulos, G. P., Zhuo, L., & Mall, R. K. (2021). Soil erosion in future scenario using CMIP5 models and earth observation datasets. *Journal of Hydrology, 594*, 125851.

Mitra, A., Chowdhury, R., & Banerjee, K. (2011). *Dynamics of Sundarbans estuarine ecosystem: Ecomorphological and ecotoxicological perspectives*. Springer. https://doi.org/10.1007/978-3-642-15777-6

Mukhopadhyay, A., Hazra, S., & Ghosh, T. (2015). Changes in mangrove species assemblages and future prediction of the Bangladesh Sundarbans using Markov chain model and cellular automata. *Environmental Science: Processes & Impacts, 17*(6), 1110–1121. https://doi.org/10.1039/C4EM00611A

Mustak, S., Baghmar, N. K., Singh, S. K., & Srivastava, P. K. (2022). Multi-scenario based urban growth modeling and prediction using earth observation datasets towards urban policy improvement. *Geocarto International, 37*(27), 18275–18303.

NASA Earthdata. (2023). *MCD12Q2 MODIS/Terra+Aqua land cover dynamics yearly L3 global 500 m SIN grid V006*. NASA LP DAAC. https://www.earthdata.nasa.gov/data/catalog/lpcloud-mcd12q2-006

NASA LP DAAC. (2019). *MCD12Q1 user guide Version 6*, NASA Land Processes Distributed Active Archive Center (LP DAAC). [Online]. https://lpdaac.usgs.gov/documents/101/MCD12_User_Guide_V6.pdf

Pastor-Guzman, J., Dash, J., & Atkinson, P. M. (2018). Remote sensing of mangrove forest phenology and its environmental drivers. *Remote Sensing of Environment, 205*, 71–84. https://doi.org/10.1016/j.rse.2017.11.009

Pedregosa, F., et al. (2011). Scikit-learn: Machine learning in python. *Journal of Machine Learning Research, 12*, 2825–2830. https://jmlr.org/papers/v12/pedregosa11a.html

Python Software Foundation. (2022). *Python programming language*. https://www.python.org

QGIS Development Team. (2022). *QGIS geographic information system*. Open Source Geospatial Foundation. https://qgis.org

Rahman, A., Aggarwal, S. P., Netzband, M., Fazal, S., & Rana, A. (2020). Monitoring urban sprawl using remote sensing and GIS techniques: A case study of Delhi, India. *Journal of the Indian Society of Remote Sensing, 48*(2), 289–304. https://doi.org/10.1007/s12524-019-01067-0R

Ramsar Convention Secretariat. (2019). *The Sundarbans (Ramsar Site No. 2370)*. Ramsar Sites Information Service. https://rsis.ramsar.org/ris/2370

Roy, C., Dasgupta, R., Mukhopadhyay, A., Hazra, S., & Ghosh, T. (2018). Changing environment in the Indian Sundarbans: Challenges and opportunities for sustainable development. *Frontiers in Marine Science, 5*, 224. https://doi.org/10.3389/fmars.2018.00224

Samanta, S., Hazra, S., Mondal, P. P., Chanda, A., Giri, S., French, J. R., & Nicholls, R. J. (2021). Assessment and attribution of mangrove forest changes in the Indian Sundarbans from 2000 to 2020. *Remote Sensing, 13*(24), 4957. https://doi.org/10.3390/rs13244957

Singh, H., & Srivastava, S. K. (2025). Identification of forest fire-prone region in Lamington National Park using GIS-based multicriteria technique: Validation using field and Sentinel-2-based observations. *Geocarto International, 40*. https://doi.org/10.1080/10106049.2025.2462484

Singh, P., Sinha, V. S. P., & Nayak, P. C. (2015a). Modeling land use/land cover change using cellular automata and Markov chain models. *Egyptian Journal of Remote Sensing and Space Science, 18*(1), 81–89. https://doi.org/10.1016/j.ejrs.2015.02.002

Singh, S. K., Mustak, S., Srivastava, P. K., Szabó, S., & Islam, T. (2015b). Predicting spatial and decadal LULC changes through cellular automata Markov chain models using earth observation datasets and geo-information. *Environmental Processes, 2*(1), 61–78.

Singh, S. K., Laari, P. B., Mustak, S. K., Srivastava, P. K., & Szabó, S. (2018). Modelling of land use land cover change using earth observation data-sets of Tons River Basin, Madhya Pradesh, India. *Geocarto International, 33*(11), 1202–1222.

Singh, S., Singh, H., Sharma, V., Shrivastava, V., Kumar, P., Kanga, S., Sahu, N., Meraj, G., Farooq, M., & Singh, S. K. (2022). Impact of forest fires on air quality in wolgan valley, new south Wales, Australia—A mapping and monitoring study using google earth engine. *Forests, 13*. https://doi.org/10.3390/f13010004

Singh, H., Srivastava, P. K., Prasad, R., & Srivastava, S. K. (2025). Tracking post-fire vegetation regrowth and burned areas using bitemporal Sentinel-1 SAR data: A Google Earth engine approach in heath vegetation of Mooloolah River National Park, Queensland, Australia. *Remote Sensing (Basel), 17*, 2031. https://doi.org/10.3390/rs17122031

Tan, B., Morisette, J. T., Wolfe, R. E., et al. (2011). An enhanced TIMESAT algorithm for estimating vegetation phenology metrics from MODIS data. *IEEE Transactions on Geoscience and Remote Sensing, 49*(8), 2923–2933.

Thomas, N., Lucas, R., Bunting, P., Hardy, A., Rosenqvist, A., & Simard, M. (2017). Distribution and drivers of global mangrove forest change, 1996–2010. *PLoS One, 12*(6), e0179302. https://doi.org/10.1371/journal.pone.0179302

Wang, J., Xiao, X., & Li, X. (2021). Mapping land cover changes using MODIS time series and deep learning in coastal areas. *Remote Sensing of Environment, 259*, 112417. https://doi.org/10.1016/j.rse.2021.112417

Zhang, X., Friedl, M. A., Schaaf, C. B., & Strahler, A. H. (2019). *MODIS MCD12Q2 C6 vegetation phenology product user guide*. NASA LP DAAC. https://lpdaac.usgs.gov/products/mcd12q2v006/

Zhu, Z., Wang, S., & Woodcock, C. E. (2019). Improvement and expansion of the Fmask algorithm: Cloud, cloud shadow, and snow detection for Landsats 4–7, 8, and Sentinel 2 images. *Remote Sensing of Environment, 159*, 269–277.

Key Challenges and Future Research Need

Remote Sensing Platforms and Sensors for Mangrove Monitoring

Pashupati Nath Singh, Prashant K. Srivastava, Amol Kumar Garg, and Sk. Mustak

Abstract Mangroves represent highly productive coastal ecosystems that play a pivotal role in maintaining biodiversity, shoreline stabilization, carbon sequestration, and providing ecosystem services critical to climate change adaptation. Despite their importance, mangroves worldwide are experiencing rapid degradation due to anthropogenic activities, climate-related disturbances, and environmental stressors. Therefore, accurate and timely monitoring of MEs is essential for conservation and sustainable management. Remote sensing (RS) technologies have emerged as indispensable tools, offering extensive coverage and repeatability across inaccessible coastal landscapes. This chapter presents a comprehensive overview of various RS sensors—such as optical, synthetic aperture radar (SAR), light detection and ranging (LiDAR), hyperspectral, thermal infrared, and UAV-based platforms—and discusses their unique strengths, limitations, and suitability for monitoring MEs. Specific attention is given to sensor-derived parameters like vegetation indices, canopy structure, biomass estimation, hydrological assessment, species discrimination, and stress detection. To contextualize the practical application of these technologies, this chapter features the Sundarbans mangrove ecosystem as a representative study area. Sundarbans, the world's largest contiguous mangrove forest, faces constant threats from sea-level rise, cyclones, and anthropogenic pressures, making it an ideal example to illustrate remote sensing's operational effectiveness. The integration of multi-sensor and multi-platform RS approaches, supported by advancements in artificial intelligence and machine learning, is highlighted as the future direction for enhancing monitoring precision and ecosystem resilience. Ultimately, this synthesis not only provides critical insights into current

P. N. Singh · P. K. Srivastava (✉)
Remote Sensing Laboratory, Institute of Environment and Sustainable Development, Banaras Hindu University, Varanasi, Uttar Pradesh, India
e-mail: prashant.iesd@bhu.ac.in

A. K. Garg · S. Mustak
Department of Geography, School of Environment and Earth Sciences, Central University of Punjab, Bathinda, Punjab, India

S. K. Srivastava et al. (eds.), *Earth Observation for Monitoring Mangrove Ecosystems*, https://doi.org/10.1007/978-3-032-16668-5_15

RS capabilities but also emphasizes the importance of adopting integrative technologies for improved decision-making in mangrove conservation globally.

Keywords Mangrove monitoring · Remote sensing sensors · Sundarbans · Ecosystem management

1 Introduction

Mangrove Ecosystems (ME)s represent highly productive and biologically diverse ecosystems predominantly located along tropical and subtropical coastlines. Covering approximately 137,760 km^2 globally, mangrove forests provide vital ecosystem services such as shoreline stabilization, coastal protection from storm surges, flooding, and erosion control (Alongi, 2008; Giri et al., 2011). These evergreen woody plants exhibit distinct morphological, anatomical, and physiological adaptations, such as aerial roots (pneumatophores) for oxygen uptake in anaerobic substrates, salt-excreting glands on leaves, and viviparous seedlings capable of floating and dispersing over long distances (Tomlinson, 2016; Alongi, 2008). Globally, the mangrove flora comprises approximately 70 recognized species, distributed across multiple genera and families, predominantly Avicenniaceae, Rhizophoraceae, Combretaceae, Acanthaceae, and Sonneratiaceae (Table 1) (Duke et al., 2007). The Most Common mangrove genera are *Rhizophora* (red mangroves), *Avicennia* (black mangroves), and *Excoecaria* (Tomlinson, 2016). MEs also play a crucial role in supporting marine biodiversity by serving as nursery grounds for numerous marine organisms, thus underpinning fisheries and local livelihoods (Barbier et al., 2011). Furthermore, mangroves exhibit exceptional carbon sequestration capacities, significantly contributing to climate change mitigation strategies through their substantial carbon storage potential, particularly within their sediment layers (Donato et al., 2011). Despite their ecological and economic importance, MEs have experienced significant global decline over recent decades, driven by multiple factors, including deforestation for agriculture, aquaculture, and urban development, coupled with climate-related threats such as sea-level rise, increased storm intensity, and altered hydrological regimes (Friess et al., 2019). Consequently, there is an urgent necessity

Table 1 Key mangrove genera and their adaptive traits (adapted from Tomlinson, 2016; Duke et al., 2007)

Genus	Family	Common name	Key characteristics and adaptations
Rhizophora	Rhizophoraceae	Red mangrove	Prominent prop roots, viviparous seedlings
Avicennia	Acanthaceae	Black mangrove	Pneumatophores (aerial roots), salt excretion glands on leaves
Sonneratia	Lythraceae	Mangrove apple	Large pneumatophores, succulent leaves
Bruguiera	Rhizophoraceae	Large-leaf mangrove	Buttress roots, vivipary, knee-shaped roots
Excoecaria	Euphorbiaceae	Milky mangrove	Toxic latex, salt-tolerant foliage

for regular and precise monitoring of mangrove areas to enable effective conservation and sustainable management practices.

Remote sensing (RS) technologies have emerged as invaluable tools for monitoring mangroves, offering extensive spatial coverage, systematic temporal revisits, and objective, quantifiable datasets (Heumann, 2011; Kuenzer et al., 2011; Singh et al., 2025a, 2025b, 2025c). Various RS platforms and sensors, including optical imagery, synthetic aperture radar (SAR), light detection and ranging (LiDAR), hyperspectral imaging, thermal infrared sensors (TIRS), and unmanned aerial vehicle (UAV)-based technologies, provide essential capabilities to monitor mangrove structural parameters, extent, ecological health, and response to environmental changes (Lucas et al., 2007; Singh et al., 2025b). To illustrate the practical applicability of these RS technologies, this chapter specifically references the Sundarbans mangrove ecosystem as a representative study area (Fig. 1). Located across the coastal regions of India and Bangladesh, the Sundarbans is the largest contiguous mangrove forest globally and is a recognized UNESCO World Heritage Site (also a critical Ramsar wetland) (Giri et al., 2007). Due to its geographical positioning, the Sundarbans is particularly vulnerable to climate-induced threats, including frequent tropical cyclones, coastal erosion, rising sea levels, increased saline intrusion, and anthropogenic pressures such as exploitation of forest resources (Rahman et al., 2021). This vulnerability underscores the significance and complexity involved in using RS technologies for comprehensive ecosystem monitoring, making the Sundarbans an ideal reference ecosystem for evaluating different sensors and monitoring approaches.

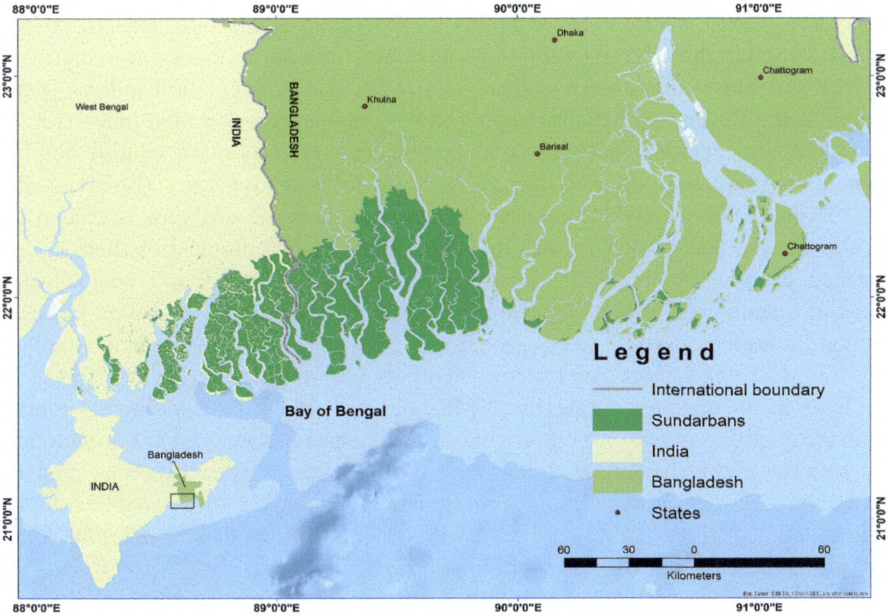

Fig. 1 A map illustrating the location of the study area

2 Fundamentals of Remote Sensing for Mangrove Monitoring

RS involves acquiring information about Earth's surface by measuring electromagnetic (EM) radiation reflected or emitted by targets from a distance, typically utilizing airborne or spaceborne platforms. At its core, RS leverages the EM spectrum, which ranges from short wavelengths (gamma rays, X-rays, ultraviolet) through visible and infrared wavelengths to long wavelengths (microwave and radio waves) (Jensen, 2015). Different parts of the EM spectrum offer distinct capabilities for environmental monitoring; visible (0.4–0.7 μm), near-infrared (NIR, 0.7–1.3 μm), and short-wave infrared (SWIR, 1.3–3 μm) regions are extensively used for vegetation analysis, enabling quantification of parameters such as canopy reflectance, chlorophyll concentration, and biomass (Campbell & Wynne, 2011).

RS datasets are characterized by four primary resolution types: spatial, spectral, temporal, and radiometric resolutions. Spatial resolution pertains to the pixel size or the smallest distinguishable ground feature on imagery; higher spatial resolution enables detailed analysis at finer scales. Spectral resolution refers to the ability of a sensor to differentiate between wavelengths; hyperspectral sensors, for instance, capture data in hundreds of narrow spectral bands, facilitating detailed ecological characterization, including species discrimination (Adam et al., 2010). Temporal resolution, representing revisit frequency, is critical for monitoring dynamic coastal environments like mangroves, especially in capturing phenological changes and disturbances such as cyclones or anthropogenic degradation. Radiometric resolution signifies the sensor's sensitivity in distinguishing differences in energy levels, enhancing detection of subtle variations in vegetation conditions (Jensen, 2015). The specific biophysical and ecological attributes that RS monitors in mangroves typically include canopy density, leaf area index (LAI), vegetation indices (e.g., normalized difference vegetation index (NDVI), enhanced vegetation index (EVI), soil-adjusted vegetation index (SAVI)), species diversity, biomass, canopy height, inundation frequency, salinity impacts, and disturbance dynamics (Lucas et al., 2007; Kuenzer et al., 2011). Accurate assessment of these attributes is crucial as they directly reflect mangrove health, productivity, and ecological resilience. For instance, canopy structure and biomass data derived from LiDAR and radar sensors assist in quantifying carbon sequestration potential and structural health, whereas vegetation indices derived from optical and hyperspectral imagery help monitor physiological status, stress, and species diversity within MEs (Giri et al., 2011).

Given the increasing complexity of MEs and the variability in ecological conditions across different regions, employing a multi-sensor and integrated RS approach has become highly beneficial. Integrating multiple sensor types, such as optical, SAR, and LiDAR leverages their complementary strengths, thus enabling more robust, detailed, and accurate ecosystem assessments than single-sensor methodologies alone (Heumann, 2011; Lucas et al., 2007).

3 Study Area: The Sundarbans Mangrove Ecosystem

The Sundarbans, spanning the deltaic regions of the Ganges, Brahmaputra, and Meghna rivers, represent the largest contiguous mangrove forest in the world. Covering approximately 10,000 km^2, this transboundary ecosystem is shared between southwestern Bangladesh (~60%) and southeastern India (~40%) (Giri et al., 2007). Ecologically, they are characterized by an intricate network of tidal waterways, mudflats, and small islands interspersed with dense mangrove vegetation (Dasgupta et al., 2019). Major environmental stressors include sea-level rise, saline water intrusion, land subsidence, frequent tropical cyclones (e.g., Aila, Amphan), and land-use changes that exacerbate mangrove degradation (Rahman et al., 2021; Hazra et al., 2020). Given its ecological richness and vulnerability to both natural and human-induced changes, the Sundarbans serves as an ideal case study for evaluating the performance of various RS sensors and monitoring platforms. Its dynamic landscape, high biodiversity, and exposure to extreme climatic events offer a complex yet critical setting to explore how integrated sensor-based monitoring can aid in conservation, disaster response, and sustainable ecosystem management (Fig. 1).

4 Classification of Sensors Used in Mangrove Monitoring

RS technologies employed in mangrove monitoring can be broadly classified into passive and active sensors based on their operational characteristics and interaction with EM radiation (Jensen, 2015). Passive sensors capture naturally emitted or reflected solar radiation from Earth's surface without generating their own energy source. Optical sensors, including multispectral and hyperspectral systems, and thermal infrared sensors, are typical examples, extensively utilized for monitoring vegetation health, phenology, and structural attributes (Kuenzer et al., 2011). Active sensors, in contrast, emit their EM radiation and measure the reflected energy returned from the target surfaces. SAR sensors, operating in microwave frequencies, and LiDAR systems, employing laser pulses, fall within this category, allowing for structural characterization and biomass estimation regardless of weather conditions (Fig. 2) (Lucas et al., 2007).

Further classification can be based on platforms supporting these sensors, including satellite-based, airborne, UAV-based, and ground-based platforms. Satellite platforms offer systematic, large-area coverage with consistent revisit intervals, suitable for long-term ecological and environmental monitoring at regional to global scales. Airborne and UAV-based platforms provide enhanced flexibility, capturing high-resolution imagery suited to detailed site-specific assessments, such as post-disaster analysis or precise species-level mapping (Fig. 2). Ground-based or in situ platforms encompass instruments such as spectroradiometers and environmental

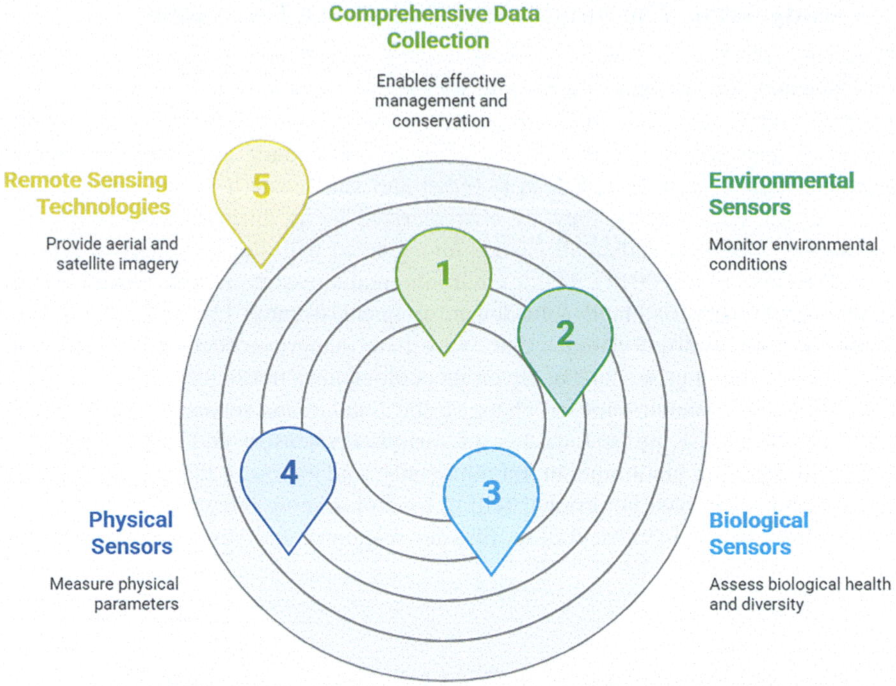

Fig. 2 A conceptual framework that illustrates integrated sensor-based methods for comprehensively monitoring MEs

sensors, serving primarily for calibration and validation of remotely sensed data, thereby improving data accuracy and ecological interpretation (Heumann, 2011).

5 Optical Remote Sensing Sensors

Optical RS systems use sunlight as energy to detect reflected EM radiation in the visible, near-infrared, and short-wave infrared regions. These sensors operate passively, capturing surface reflectance patterns that vary depending on vegetation type, health, moisture content, and canopy structure (Jensen, 2015). Mangrove vegetation, characterized by dense canopy, waxy leaves, and unique structural traits, reflects strongly in the NIR region while absorbing significantly in the red and blue wavelengths—features that are exploited by vegetation indices such as normalized difference vegetation index (NDVI), enhanced vegetation index (EVI), and SAVI (soil-adjusted vegetation index).

Several key optical satellite sensors have been widely used in mangrove monitoring:

- **Landsat series (e.g., Landsat 5 TM, Landsat 7 ETM+, Landsat 8 OLI/TIRS, Landsat 9 OLI-2):** Offering medium spatial resolution (30 m) and multispectral coverage since the 1970s, the Landsat program provides one of the longest continuous Earth observation datasets.
- **Sentinel-2 (A and B):** With a spatial resolution ranging from 10 to 60 m across 13 spectral bands, Sentinel-2 provides enhanced vegetation monitoring capability due to its high revisit frequency (5 days) and red-edge bands critical for stress and species discrimination.
- **Moderate Resolution Imaging Spectroradiometer (MODIS):** Though coarser in spatial resolution (250–1000 m), MODIS offers high temporal resolution (daily observations) suitable for long-term phenological and regional mangrove dynamics analysis.

These optical sensors are especially effective in mapping mangrove spatial distribution, quantifying degradation, detecting land cover changes, and monitoring ecosystem health through time-series vegetation index analysis (Kuenzer et al., 2011; Giri et al., 2011). However, their limitations include susceptibility to cloud cover, particularly in tropical coastal zones, which often necessitates multi-temporal compositing or fusion with radar data.

The Sundarbans mangrove forest, known for its ecological complexity, dynamic hydrological regime, and frequent exposure to tropical cyclones, has been extensively studied using optical RS technologies. Researchers have widely employed satellite datasets such as Landsat and Sentinel-2 to assess temporal changes in mangrove extent, canopy health, and post-disturbance recovery. A landmark study by Giri et al. (2007) used multi-temporal Landsat data (1973–2000) to map mangrove forest dynamics across the Indian and Bangladeshi Sundarbans, revealing significant forest loss driven by both natural and anthropogenic factors. Similarly, Rahman et al. (2021) utilized Landsat imagery (1990–2020) to conduct a detailed spatiotemporal analysis of the Bangladeshi Sundarbans, identifying degradation linked to salinity intrusion, reduced freshwater inflow, and human-induced pressures. Optical vegetation indices such as NDVI, EVI, and red-edge bands from Sentinel-2 have proven valuable in evaluating mangrove health by detecting subtle stress indicators, chlorophyll variability, and seasonal vegetation dynamics. These indices have been used to quantify the impacts of tidal flooding, storm surges—such as those from Cyclone Amphan—and changing rainfall patterns on canopy greenness and density (Hazra et al., 2020; Dasgupta et al., 2019). In addition, MODIS-derived time-series analyses have supported regional-scale phenological monitoring, enabling the detection of long-term trends in productivity and the mangrove ecosystem's response to climatic variability, despite limitations in spatial resolution.

6 Synthetic Aperture Radar (SAR Sensors)

SAR is an active RS technology that operates in the microwave region of the EM spectrum, emitting its own energy and measuring the backscattered signal from Earth's surface. SAR systems can penetrate clouds, operate both day and night, and provide consistent observations in regions with frequent cloud cover and dense vegetation—making them ideal for MEs typically located in humid, tropical coastal zones (Fig. 4) (Jensen, 2015). SAR backscatter intensity is influenced by surface roughness, dielectric properties, vegetation structure, and moisture content, enabling assessment of biophysical properties such as canopy density, above-ground biomass (AGB), and inundation patterns.

Key SAR sensors used in mangrove monitoring include:

- Sentinel-1 (C-band, 5.405 GHz): Provides dual-polarization (VV, VH) data with a spatial resolution of 10 m and a revisit time of 6–12 days, widely used for land cover classification, flood mapping, and vegetation structure estimation.
- ALOS PALSAR (L-band, 1.27 GHz): Offers deeper vegetation penetration than C-band SAR, enabling improved retrieval of biomass and structural parameters due to its longer wavelength and sensitivity to larger-scale canopy features.

SAR has demonstrated high utility in mangrove monitoring, including biomass estimation, degradation detection, and inundation extent mapping (Lucas et al., 2007). Multi-polarization SAR data and time-series analysis have been particularly effective in quantifying structural differences among mangrove species, detecting illegal logging, and monitoring responses to hydrological changes such as tidal flooding and storm surges. SAR-based RS can operate independently of weather and light conditions. Numerous studies have employed SAR data to investigate the region's vegetation structure, biomass distribution, and hydrodynamic processes. For instance, Lucas et al. (2007) demonstrated the effectiveness of multi-frequency and multi-polarization SAR in differentiating mangrove species and estimating above-ground biomass in tropical forests, including the Sundarbans, highlighting the sensitivity of L-band SAR to variations in canopy height and density. In a more localized study, Islam et al. (2018) utilized ALOS PALSAR imagery to produce biomass distribution maps of the Sundarbans, revealing pronounced spatial heterogeneity in forest condition and confirming the utility of L-band SAR in structurally complex MEs. Additionally, Sentinel-1 C-band SAR data has proven valuable for monitoring flood dynamics and tidal inundation across the Sundarbans delta. Time-series analyses during extreme weather events have enabled the mapping of inundation extents, identification of erosion-prone zones, and assessment of hydrological shifts associated with sea-level rise and land subsidence. Collectively, these studies affirm the indispensable role of SAR in monitoring MEs, especially in atmospherically challenging environments where optical RS is constrained.

7 LiDAR (Light Detection and Ranging)

LiDAR is also an active remote sensing technique that measures distance to a target by emitting laser pulses and calculating the time delay of their return after reflecting off surfaces. This enables the generation of precise three-dimensional point clouds representing terrain elevation, vegetation height, and canopy structure (Lefsky et al., 2002). LiDAR systems can be airborne (ALS), spaceborne, or terrestrial (TLS), depending on the platform, with airborne systems being most common for large-scale forest mapping.

In MEs, LiDAR's ability to capture vertical forest structure makes it a powerful tool for estimating canopy height, crown geometry, gap fraction, basal area, and, most critically, AGB and carbon stocks. Unlike passive sensors, LiDAR can penetrate through gaps in the canopy, enabling the mapping of understory features and surface topography, which is particularly useful for flood modeling in low-lying mangrove regions (Simard et al., 2006). LiDAR-derived metrics are often used in conjunction with field data and allometric equations to improve the accuracy of biomass and carbon estimations. Integration of LiDAR data with multispectral or SAR imagery further enhances classification accuracy and ecological monitoring by fusing horizontal and vertical information dimensions.

Although LiDAR applications in the Sundarbans remain comparatively limited relative to optical and SAR studies, emerging research has demonstrated its exceptional value for high-resolution structural mapping in this complex deltaic landscape. Airborne LiDAR surveys across parts of the Indian Sundarbans have effectively captured detailed canopy height variations, vegetation gaps, and microtopographic elevation, providing critical insights into forest health, biomass distribution, and exposure to tidal inundation and cyclonic impacts (Fig. 3). Complementing these efforts, NASA's Global Ecosystem Dynamics Investigation (GEDI) mission has enabled the use of spaceborne LiDAR to characterize vertical forest structure, enhancing carbon stock estimation and forest degradation assessments across the Sundarbans. GEDI-derived elevation data have further been utilized to delineate inundation gradients and salinity transitions, which are vital for understanding mangrove zonation and hydrological stress. Given the Sundarbans' low-lying geomorphology and vulnerability to sea-level rise, such structural and elevation information is indispensable for projecting future ecological resilience. As LiDAR technologies continue to evolve—through spaceborne missions and UAV-mounted platforms—their integration into mangrove monitoring frameworks in the Sundarbans is anticipated to expand, offering unmatched precision for conservation planning and adaptive ecosystem management.

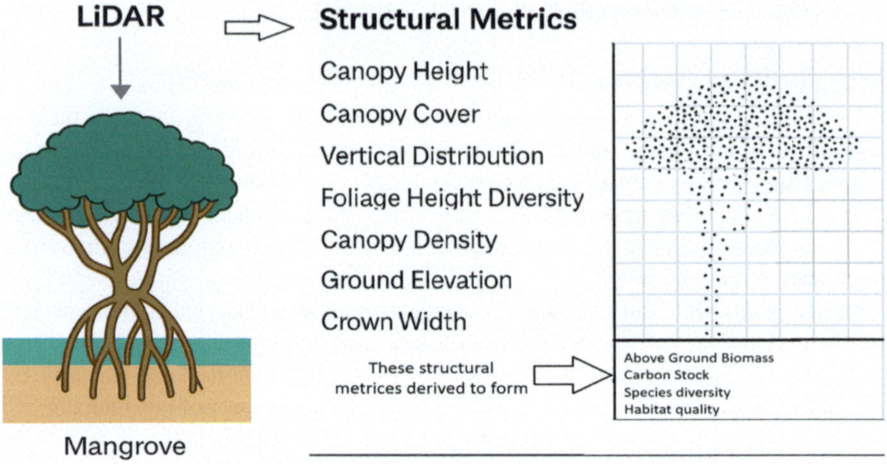

Fig. 3 Ecological assessment of MEs, including biomass estimation, carbon stock quantification, and habitat biodiversity analysis using LiDAR

8 Hyperspectral Sensors

Hyperspectral remote sensing involves the collection of reflectance data across hundreds of narrow and contiguous spectral bands, typically ranging from the visible to shortwave infrared region of the electromagnetic spectrum. Unlike multispectral sensors, which capture broad-band spectral information, hyperspectral sensors enable detection of subtle spectral signatures associated with vegetation biochemistry, physiology, and structural traits (Govender et al., 2007). This spectral richness facilitates highly detailed analyses of vegetation types, species-level classification, and early stress detection.

Major hyperspectral sensors used in ecological studies include:

- Hyperion (aboard EO-1): With 220 bands and a 30 m spatial resolution, Hyperion has been widely used in vegetation classification and environmental stress mapping.
- UAV-based hyperspectral cameras: These offer ultra-high spatial and spectral resolution at flexible, low-altitude deployments, allowing for detailed, site-specific monitoring of plant health and diversity, even in fragmented mangrove patches.

Hyperspectral data have proven especially valuable in differentiating between mangrove species with similar canopy structures but distinct physiological properties, as well as in detecting early signs of nutrient deficiency, salinity stress, or disease that may not be visible in conventional RGB or multispectral imagery (Dahdouh-Guebas et al., 2004). In the Sundarbans, hyperspectral imaging has been increasingly explored for mapping species composition and identifying stress gradients across mangrove zones. Field-calibrated spectral libraries have been developed to distinguish dominant species such as *Avicennia*, *Rhizophora*, and *Sonneratia*, based on leaf pigment content and moisture-related reflectance characteristics. UAV-mounted hyperspectral sensors have been deployed over restoration sites and cyclone-affected zones to detect physiological stress, salinity exposure, and pigment degradation—indicators that precede visual symptoms of decline.

9 Thermal Infrared Sensors

Thermal infrared (TIR) remote sensing measures the long-wave radiation emitted by Earth's surface and is primarily used to estimate land surface temperature (LST), thermal inertia, and evapotranspiration. In vegetated landscapes like mangroves, thermal data can help assess plant stress, soil moisture conditions, and hydrological dynamics, as temperature anomalies often indicate physiological stress, water deficits, or altered tidal flow (Fig. 4) (Schmugge et al., 2002).

Key thermal sensors include:

- Moderate Resolution Imaging Spectroradiometer (MODIS): Provides daily global LST data at 1 km resolution, useful for analyzing seasonal and interannual temperature trends.

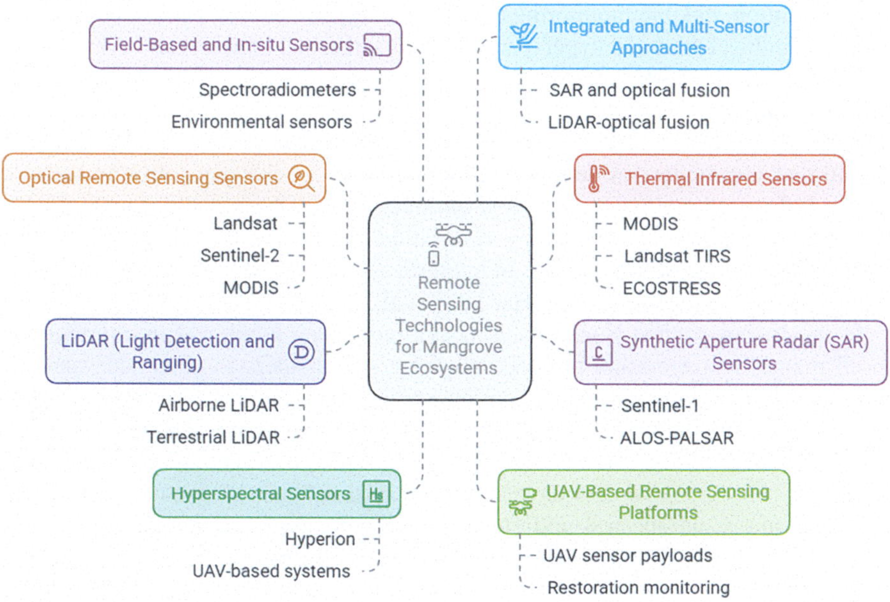

Fig. 4 Remote sensing technologies classified by their ecological applications in mangrove systems, ranging from canopy health and biomass estimation to hydrological mapping and restoration tracking

- Landsat Thermal Infrared Sensor (TIRS): Onboard Landsat 8 and 9, offers moderate spatial resolution (100 m) LST data suitable for finer-scale analysis.
- ECOsystem Spaceborne Thermal Radiometer Experiment on Space Station (ECOSTRESS): Captures high-resolution (~70 m) temperature data to assess plant water stress and evapotranspiration.

Thermal datasets are particularly effective when integrated with vegetation indices or SAR data to diagnose ecosystem stress conditions and model mangrove productivity under climate variability. In the Sundarbans, thermal infrared sensors have been utilized to monitor surface temperature variations, particularly during dry seasons and post-cyclonic periods. Studies using MODIS and Landsat TIRS have detected hotspots in degraded mangrove patches, correlating with reduced canopy cover and impaired evapotranspiration rates. ECOSTRESS data have enabled mapping of plant water stress at diurnal scales, helping identify zones affected by saline intrusion or restricted tidal flushing. These observations are crucial for assessing the hydrological health and vulnerability of mangroves to climatic extremes and altered freshwater inflows, especially in regions already experiencing sea-level rise.

10 UAV-Based Remote Sensing Platforms

Unmanned aerial vehicles (UAVs), or drones, have become indispensable tools for high-resolution, flexible monitoring of MEs. Equipped with payloads such as RGB cameras, multispectral and hyperspectral sensors, thermal imagers, and even miniature LiDAR units, UAVs offer centimeter-level spatial resolution ideal for micro-habitat mapping, phenological tracking, and vegetation structure assessment (Paneque-Gálvez et al., 2014). Compared to satellite platforms, UAVs offer several operational advantages: on-demand deployment, rapid revisit capability, and access to cloud-obscured regions through low-altitude flights. These qualities make UAVs particularly suitable for monitoring restoration plots, detecting illegal encroachment, and assessing storm damage in mangroves, where conditions change rapidly and spatial heterogeneity is high (Harikesh et al., 2020).

In the Indian context Sundarbans, UAVs have been employed in community-based mangrove restoration projects, allowing stakeholders to document survival rates, canopy development, and planting success in near real time. Post-disaster surveys following Cyclone Amphan utilized drone-mounted cameras to assess fallen trees, defoliation, and damage to embankments and habitats at unprecedented resolution (Fig. 4). UAV data have also supported habitat classification efforts by capturing fine-scale spectral and structural information in mixed-species zones. Given the fragmented and highly vulnerable nature of the Sundarbans landscape, UAVs offer a powerful and scalable tool for localized monitoring, rapid response planning, and participatory ecosystem management.

11 Field-Based and In Situ Sensors

Field-based and in situ sensors are critical for calibrating and validating RS outputs, ensuring that satellite- or UAV-derived parameters accurately reflect on-ground ecological conditions. These tools play an essential role in bridging the scale gap between point-based field measurements and spatially extensive remote observations, particularly in heterogeneous ecosystems like mangroves (Kuenzer et al., 2011). Among the most commonly used field instruments are:

- **Spectroradiometers**, which measure reflectance across specific wavelengths and are used to develop spectral libraries for identifying mangrove species and validating satellite-derived reflectance indices (Dahdouh-Guebas et al., 2004).
- **Chlorophyll meters** (e.g., SPAD-502), which estimate leaf chlorophyll content, a proxy for plant health and nitrogen status (Padilla et al., 2014).
- **Leaf Area Index (LAI) meters,** such as the LAI-2000 or ceptometers, which quantify foliage density, supporting models of canopy structure, transpiration, and carbon assimilation (Ponton et al., 2014).
- **Environmental sensors** measuring parameters like soil/water salinity, pH, dissolved oxygen, and water level, which directly influence mangrove physiology, zonation, and productivity (Rahman et al., 2021).

These instruments are especially valuable for validating remotely sensed vegetation indices such as NDVI and EVI, as well as canopy height and biomass models derived from LiDAR or SAR data (Lucas et al., 2007). When integrated into field campaigns, such sensors enhance the interpretability and precision of RS datasets used in mangrove monitoring.

In the Sundarbans, field-based observations have been crucial for enhancing the ecological accuracy of RS-derived products. Spectroradiometric measurements of dominant mangrove species such as *Avicennia officinalis*, *Excoecaria agallocha*, and *Sonneratia apetala* have supported the development of spectral libraries tailored to the region, significantly improving species-level classification using hyperspectral imagery (Dahdouh-Guebas et al., 2004). In parallel, chlorophyll content and LAI measurements have been employed to validate vegetation indices from Sentinel-2 and UAV datasets, allowing researchers to monitor phenological shifts linked to seasonal salinity changes, drought stress, and nutrient limitations—factors that affect photosynthetic performance and canopy vigor. For instance, studies have recorded reduced chlorophyll values during pre-monsoon dry periods, indicating early physiological stress before visible canopy decline. Moreover, salinity sensors and tidal gauges have been instrumental in linking prolonged saline water intrusion, resulting from upstream freshwater scarcity and tidal asymmetry—with alterations in species composition and seedling mortality patterns. These in situ tools support the validation of satellite-based flood and vegetation stress mapping and provide valuable ecological context for understanding long-term ecosystem responses to environmental gradients.

12 Integrated and Multi-sensor Approaches

Given MEs' structural complexity and dynamic hydrology, no single sensor type can capture all necessary ecological parameters. Multi-sensor integration leverages the unique strengths of different RS systems—optical, radar, LiDAR, hyperspectral, thermal, and field-based—to generate more comprehensive, accurate, and scalable ecosystem assessments (Kuenzer et al., 2011; Lucas et al., 2007). Some widely applied integration approaches include:

- **SAR-optical fusion**, which combines backscatter sensitivity to structure and moisture (SAR) with spectral information on canopy condition (optical), improving classification and degradation mapping (Rahman et al., 2021).
- **LiDAR-optical integration**, in which 3D structural data from LiDAR (e.g., canopy height, gap fraction) is coupled with reflectance-based vegetation indices to enhance **biomass and carbon stock estimation** (Simard et al., 2006).
- **Field-satellite synergy**, where in situ measurements (e.g., chlorophyll, LAI, salinity) are used to train and validate machine learning models built from remotely sensed data (Padilla et al., 2014).

Integrated approaches are increasingly enabled by cloud platforms such as Google Earth Engine (GEE) and advances in machine learning, which facilitate the fusion of multi-temporal and multi-sensor datasets into operational mangrove monitoring systems (Singh et al., 2025c) (Fig. 4). In the Sundarbans, several studies have implemented multi-sensor approaches to improve the reliability of ecological assessments. For instance, fusion of Sentinel-1 SAR and Sentinel-2 optical data has enabled fine-scale classification of mangrove zones, overcoming limitations posed by persistent cloud cover (Islam et al., 2018). Likewise, ALOS PALSAR L-band SAR has been integrated with vegetation indices to map above-ground biomass, revealing spatial heterogeneity in forest structure. Emerging applications have incorporated GEDI LiDAR data with multispectral imagery to produce high-resolution canopy height models, supporting carbon stock estimation and habitat quality analysis (Fatoyinbo & Simard, 2013). Ground-truth data from field campaigns, including salinity and chlorophyll measurements, have further enhanced the accuracy of these outputs.

13 Platforms for Remote Sensing

RS platforms are the physical carriers of sensors used to acquire spatial data across varying resolutions, frequencies, and ecological contexts. These platforms—categorized as satellite-based, airborne, UAV-based, and ground-based—differ in spatial coverage, flexibility, cost, and data quality (Fig. 5). In MEs, where access is restricted and cloud cover is frequent, selecting the appropriate platform or

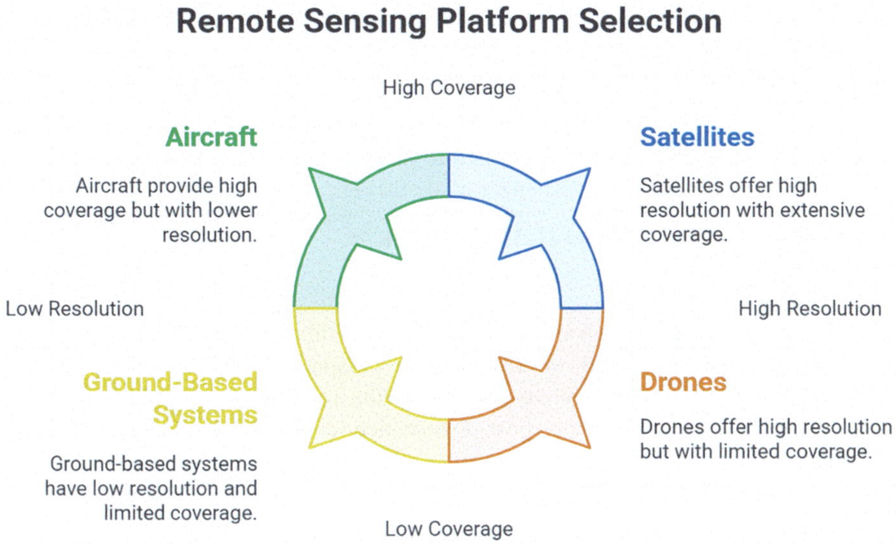

Fig. 5 Comparative framework illustrating the trade-offs between resolution and coverage across remote sensing platforms—satellites, aircraft, drones, and ground-based systems—for optimized monitoring of mangrove trees

combination of platforms is crucial for effective monitoring (Kuenzer et al., 2011; Jensen, 2015).

1. **Satellite-Based Platforms**

- **Benefits**
- Satellite platforms provide global to regional coverage, long-term historical archives, and consistent revisit times. They support multi-temporal analysis and are equipped with diverse sensors (e.g., optical, SAR, LiDAR). Platforms like Landsat, Sentinel-1 and 2, and MODIS have been instrumental in mangrove mapping, biomass estimation, and hydrological studies (Giri et al., 2007; Fatoyinbo & Simard, 2013).
- **Limitations**
- Moderate spatial resolution (10–30 m) may not be suitable for fine-scale applications. Optical sensors are affected by cloud cover—an issue in humid regions like the Sundarbans. Temporal flexibility is limited compared to UAVs (Lucas et al., 2007).

2. **Airborne (Aircraft-Based) Platforms**

- **Benefits**
- Airborne platforms offer customizable flight plans and high-resolution data, ideal for capturing 3D canopy structure and biomass using airborne LiDAR

or hyperspectral sensors. They bridge the gap between satellite coverage and field resolution (Lefsky et al., 2002).

- **Limitations**
- Airborne campaigns are costly, weather-dependent, and require special permissions. Their use is often limited to targeted research missions rather than operational monitoring (Kuenzer et al., 2011).

3. **Unmanned Aerial Vehicles (UAVs/Drones)**

- **Benefits**
- UAVs enable ultra-high spatial resolution (centimeter-level) imaging and can be equipped with RGB, multispectral, hyperspectral, or lightweight LiDAR payloads. They are effective for monitoring restoration plots, species-level vegetation mapping, and post-disturbance assessments (Paneque-Gálvez et al., 2014).
- **Limitations**
- UAVs have short flight durations and limited spatial coverage per mission and are affected by wind and precipitation. Regulatory restrictions and the need for trained operators are additional barriers in protected areas like the Sundarbans (Fatoyinbo & Simard, 2013).

4. **Ground-Based Platforms**

- **Benefits**
- Ground platforms provide precise ecological measurements using spectroradiometers, chlorophyll meters, LAI sensors, and salinity probes. These data are crucial for calibrating and validating satellite- and UAV-derived models (Dahdouh-Guebas et al., 2004).
- **Limitations**
- Despite high accuracy, ground data is spatially limited, labor-intensive, and difficult to collect in tide-influenced, muddy, or cyclone-prone mangrove regions (Rahman et al., 2021).

14 Challenges and Limitations

While RS technologies offer valuable tools for mangrove tree monitoring, several challenges persist in accurately capturing tree-specific attributes in dynamic and complex coastal environments.

14.1 Sensor-Specific and Environmental Challenges

Mangrove trees, with their dense canopies and multi-tiered structure, pose challenges for RS. Optical sensors often cannot penetrate the upper canopy, hindering assessments of understory regeneration and species variation (Lucas et al., 2007). In mature mangrove stands, crown overlap limits traditional vegetation indices in measuring individual tree health or biomass. Additionally, factors like cloud cover and tidal cycles in areas such as the Sundarbans reduce optical image availability and spectral consistency, affecting classification accuracy. Seasonal phenological changes in species like *Avicennia marina* or *Excoecaria agallocha* can also complicate time-series comparisons.

14.2 Data Access, Resolution, and Cost Constraints

Tree-level assessment, especially for canopy height, crown spread, and biomass, requires high-resolution imagery or LiDAR point clouds, which are often inaccessible due to high acquisition costs or limited spatial availability. In the Sundarbans, dense forest cover and logistical constraints limit opportunities for airborne LiDAR surveys or repeat UAV missions. Additionally, hyperspectral sensors capable of distinguishing among mangrove species are largely unavailable in open-access satellite programs, hindering spectral separation of tree species like *Rhizophora mucronata* and *Sonneratia apetala.*

14.3 Regulatory and Operational Barriers

In ecologically sensitive zones like the Indian Sundarbans, drone operations for tree canopy mapping or restoration monitoring require special permits and are often restricted by wildlife protection laws. Field data collection for validating RS outputs—such as measuring tree height or trunk diameter—is frequently hampered by tidal flooding, mudflats, and dangerous wildlife habitats, including tiger corridors.

15 Emerging Trends and Future Directions

To overcome these limitations, emerging technologies and analytical approaches are opening new frontiers in mangrove tree monitoring, offering more granular, real-time, and species-sensitive observations (Fig. 6).

Mangrove Tree Monitoring Technologies

Characteristic	Next-Gen Sensors	AI & Analytics	Participatory Monitoring
Species Classification	Enhanced by NISAR mission	Deep learning distinguishes species	N/A
Biomass Estimation	GEDI provides vertical profiling	AI models automate stress symptom monitoring	N/A
Seedling Monitoring	N/A	OBIA tracks survival and growth	Local stakeholders track planted saplings
Data Enrichment	N/A	N/A	Mobile apps record tree health

Fig. 6 Matrix of emerging technologies for mangrove monitoring, comparing next-generation sensors, AI analytics, and participatory approaches for species classification, biomass estimation, seedling tracking, and data enrichment

15.1 Next-Generation Sensors and Missions

Advanced missions such as GEDI now provide spaceborne LiDAR data that enable detailed vertical profiling of mangrove tree canopies, including metrics like height, crown depth, and canopy cover—critical for estimating tree-specific above-ground biomass. The forthcoming NISAR mission, with its dual-frequency SAR capabilities, is expected to enhance tree structure and moisture content estimation, particularly in flood-prone mangrove zones like the Sundarbans.

15.2 Artificial Intelligence and Tree-Level Analytics

The integration of AI and deep learning with high-resolution RS enables individual tree crown delineation, species classification, and detection of diseases through subtle spectral and structural cues (Singh & Srivastava, 2024). These methods effectively distinguish mangrove species based on leaf pigment and canopy architecture and can be trained with UAV or field spectral libraries. AI-driven models can automate the monitoring of plant stress symptoms using UAV images. Additionally, object-based image analysis (OBIA) shows promise in tracking seedling survival, growth rate, and canopy expansion in restoration areas.

15.3 Participatory Tree Monitoring

Community-based UAV programs are increasingly being piloted to involve local stakeholders in tree-level monitoring. Equipped with smartphones and simple drones, local forest dwellers can assist in capturing canopy images, flagging unusual tree mortality, and tracking planted saplings, contributing to both science and stewardship.

15.4 Future Opportunities for Sundarbans Mangrove Tree Monitoring

(a) AI-Enabled Tree Classification: Deep learning applied to high-resolution satellite and UAV imagery can help differentiate co-occurring mangrove species like *Avicennia*, *Rhizophora*, and *Bruguiera* based on leaf traits, crown morphology, and spectral response, improving biodiversity assessments and carbon stock mapping.
(b) LiDAR-Driven Tree Biomass Estimation: Integration of GEDI LiDAR and UAV LiDAR can enable tree-level canopy height modeling, aiding in species-specific biomass and carbon stock calculations for REDD+ and blue carbon initiatives in the Sundarbans.
(c) Drone-Based Seedling Survival Monitoring: UAVs deployed over reforested or cyclone-affected plots can be used to monitor individual sapling survival, assess planting success, and generate early warnings for replanting needs—especially in community-led afforestation zones.
(d) Mobile-Enabled Citizen Science: Empowering local communities with smartphone applications to record tree health, flowering patterns, and dieback incidents can enrich long-term phenological datasets and support localized ecological modeling.

16 Conclusions

Mangrove trees are crucial components of tropical coastal ecosystems, known for their unique aerial roots, salt tolerance, and carbon sequestration abilities. They play a vital role in ecological stability and shoreline protection. However, threats from human activities, sea-level rise, and extreme weather jeopardize their survival. Remote sensing has emerged as an effective method for monitoring mangrove health across various landscapes. This chapter provides an overview of RS technologies—such as optical, SAR, LiDAR, hyperspectral, thermal infrared, UAV-based, and field-based systems, and their roles in mapping mangrove extent, identifying species differences, estimating biomass, and detecting physiological stress in mangroves. It emphasizes the advantage of multi-sensor approaches for more accurate and consistent assessments. The Sundarbans mangrove forest serves as a case study, showcasing the need for precise monitoring in this biodiversity hotspot. Challenges like cloud cover, tidal interference, and data resolution limitations highlight the importance of integrating technologies like spaceborne LiDAR (i.e. GEDI), synthetic aperture radar (i.e. Sentinel-1, NISAR), and UAV-based hyperspectral imaging, validated by in situ measurements. The future of mangrove tree monitoring lies in the fusion of next-generation sensors, artificial intelligence, and community-driven data collection frameworks. AI-enhanced image analysis, real-time UAV surveillance, and participatory monitoring can together enable precise tree-level mapping of canopy health, seedling survival, and biomass change—even in fragmented or cyclone-affected zones. Such integrative systems are essential not only for local conservation and climate resilience but also for contributing to global initiatives like REDD+, blue carbon accounting, and ecosystem-based disaster risk reduction. Ultimately, mangrove tree monitoring must shift from periodic surveys to real-time, adaptive, and participatory strategies, underpinned by advanced RS technologies. By making trees visible across scales—from satellite mosaics to centimeter-level drone imagery—RS empowers both science and society to safeguard these vital coastal forests into the future.

References

Adam, E., Mutanga, O., & Rugege, D. (2010). Multispectral and hyperspectral remote sensing for identification and mapping of wetland vegetation: A review. *Wetlands Ecology and Management, 18*(3), 281–296.

Alongi, D. M. (2008). Mangrove forests: Resilience, protection from tsunamis, and responses to global climate change. *Estuarine, Coastal and Shelf Science, 76*(1), 1–13.

Barbier, E. B., Hacker, S. D., Kennedy, C., Koch, E. W., Stier, A. C., & Silliman, B. R. (2011). The value of estuarine and coastal ecosystem services. *Ecological Monographs, 81*(2), 169–193.

Campbell, J. B., & Wynne, R. H. (2011). *Introduction to remote sensing* (5th ed.). Guilford Press.

Dahdouh-Guebas, F., Verheyden, A., Kairo, J. G., Jayatissa, L. P., & Koedam, N. (2004). Capacity building in tropical coastal resource monitoring in developing countries: A re-appraisal of mangrove vegetation remote sensing methods. *Environment, Development and Sustainability, 6*(2), 173–197.

Dasgupta, S., Sobhan, M. I., & Wheeler, D. (2019). The impact of climate change on salinity and mangroves in the Sundarbans. *Water Policy, 21*(5), 870–883.

Donato, D. C., Kauffman, J. B., Murdiyarso, D., Kurnianto, S., Stidham, M., & Kanninen, M. (2011). Mangroves are among the most carbon-rich forests in the tropics. *Nature Geoscience, 4*(5), 293–297.

Duke, N. C., Meynecke, J. O., Dittmann, S., Ellison, A. M., Anger, K., Berger, U., & Dahdouh-Guebas, F. (2007). A world without mangroves? *Science, 317*(5834), 41–42.

Fatoyinbo, T. E., & Simard, M. (2013). Height and biomass of mangroves in Africa from ICESat/GLAS and SRTM. *International Journal of Remote Sensing, 34*(2), 668–681.

Friess, D. A., Rogers, K., Lovelock, C. E., Krauss, K. W., Hamilton, S. E., Lee, S. Y., et al. (2019). The state of the world's mangrove forests: Past, present, and future. *Annual Review of Environment and Resources, 44*, 89–115.

Giri, C., Pengra, B., Zhu, Z., Singh, A., & Tieszen, L. L. (2007). Monitoring mangrove forest dynamics of the Sundarbans in Bangladesh and India using multi-temporal satellite data from 1973 to 2000. *Estuarine, Coastal and Shelf Science, 73*(1–2), 91–100.

Giri, C., Ochieng, E., Tieszen, L. L., Zhu, Z., Singh, A., Loveland, T., et al. (2011). Status and distribution of mangrove forests of the world using Earth observation satellite data. *Global Ecology and Biogeography, 20*(1), 154–159.

Govender, M., Chetty, K., & Bulcock, H. (2007). A review of hyperspectral remote sensing and its application in vegetation and water resource studies. *Water SA, 33*(2), 145–151.

Harikesh, Singh, S., Shrivastava, V., & Sharma, V. (2020). *CityGML based 3D modeling of urban area using UAV dataset for estimation of solar potential* (pp. 355–367). https://doi.org/10.1007/978-3-030-37393-1_30

Hazra, S., Mistri, B., & Samanta, K. (2020). Sea level rise and resilience: Sundarbans, India. In *Climate change and community resilience* (pp. 85–110). Springer.

Heumann, B. W. (2011). Satellite remote sensing of mangrove forests: Recent advances and future opportunities. *Progress in Physical Geography, 35*(1), 87–108.

Islam, M. M., Rahman, M. M., & Mustapha, M. A. (2018). Estimating biomass in the Sundarbans using ALOS PALSAR data. *Remote Sensing Letters, 9*(6), 552–561.

Jensen, J. R. (2015). *Introductory digital image processing: A remote sensing perspective* (4th ed.). Pearson.

Kuenzer, C., Bluemel, A., Gebhardt, S., Quoc, T. V., & Dech, S. (2011). Remote sensing of MEs: A review. *Remote Sensing, 3*(5), 878–928.

Lefsky, M. A., Cohen, W. B., Parker, G. G., & Harding, D. J. (2002). Lidar remote sensing for ecosystem studies. *Bioscience, 52*(1), 19–30.

Lucas, R., Mitchell, A., Rosenqvist, A., et al. (2007). Remote sensing of MEs: Methods and applications. *Remote Sensing of Environment, 109*(2), 214–224.

Padilla, F. M., Gallardo, M., Manzano-Agugliaro, F., & Thompson, R. B. (2014). Evaluation of optical sensor measurements of chlorophyll content for predicting nitrogen status and yield of muskmelon. *Precision Agriculture, 15*, 541–557.

Paneque-Gálvez, J., McCall, M. K., Napoletano, B. M., Wich, S. A., & Koh, L. P. (2014). Small drones for community-based forest monitoring: An assessment of their feasibility and potential in tropical areas. *Forest, 5*(6), 1481–1507.

Ponton, S., Dupouey, J. L., Bouchard, M., & Dreyer, E. (2014). Comparison of LAI estimates by Ceptometer and LAI-2000 sensors in broad-leaved forests. *Agricultural and Forest Meteorology, 121*(3–4), 175–186.

Rahman, M. A., Khan, M. H., & Asaduzzaman, M. (2021). Spatiotemporal analysis of mangrove cover change in the Sundarbans of Bangladesh using Landsat imagery. *Environmental Challenges, 3*, 100039.

Schmugge, T. J., Kustas, W. P., Ritchie, J. C., Jackson, T. J., & Rango, A. (2002). Remote sensing in hydrology. *Advances in Water Resources, 25*(8–12), 1367–1385.

Simard, M., Zhang, K., Rivera-Monroy, V. H., Ross, M. S., Ruiz, P. L., Castañeda-Moya, E., et al. (2006). Mapping height and biomass of mangrove forests in Everglades National Park with SRTM elevation data. *Photogrammetric Engineering & Remote Sensing, 72*(3), 299–311.

Singh, H., & Srivastava, S. K. (2024). From firestick to satellites: Technological advancement and indigenous cultural practice in managing forest fires in Australia. *The Historic Environment: Policy & Practice, 16*, 1–24. https://doi.org/10.1080/17567505.2024.2425246

Singh, H., Ang, L.-M., & Srivastava, S. K. (2025a). Active wildfire detection via satellite imagery and machine learning: An empirical investigation of Australian wildfires. *Natural Hazards*. https://doi.org/10.1007/s11069-025-07163-w

Singh, H., Srivastava, P. K., Prasad, R., & Srivastava, S. K. (2025b). Tracking post-fire vegetation regrowth and burned areas using bitemporal Sentinel-1 SAR data: A Google earth engine approach in heath vegetation of Mooloolah River National Park, Queensland, Australia. *Remote Sensing (Basel), 17*, 2031. https://doi.org/10.3390/rs17122031

Singh, H., Ang, L.-M., Paudyal, D., Acuna, M., Srivastava, P. K., & Srivastava, S. K. (2025c). A comprehensive review of empirical and dynamic wildfire simulators and machine learning techniques used for the prediction of wildfire in Australia. *Technology, Knowledge and Learning*. https://doi.org/10.1007/s10758-025-09839-5

Tomlinson, P. B. (2016). *The botany of mangroves* (2nd ed.). Cambridge University Press.

Way Forward and Challenges in Mangrove Monitoring Using Earth Observation

Megha Paul, Smrutisikha Mohanty, Pavan Kumar, Prashant K. Srivastava, and Sanjeev Kumar Srivastava

Abstract Anthropogenic stresses and climate-related stressors are posing a growing danger to MEs, which are essential for biodiversity, carbon sequestration, and coastal protection. In terms of tracking, mapping, and managing these dynamic coastal forests, Earth Observation (EO) technologies have become indispensable due to their high temporal and synoptic coverage. Using cutting-edge technology like LiDAR, unmanned aerial vehicle (UAV) systems, multispectral and radar satellites, and artificial intelligence (AI)-driven data analytics, this chapter examines the changing field of EO applications in mangrove monitoring. Each platform's advantages and disadvantages are emphasized, along with the necessity of multisensor integration and consistent monitoring procedures. Nevertheless, apart from many significant advancements, obstacles still exist that prevent the full operationalization of EO for conservation planning, such as spectrum confusion, cloud cover, data fragmentation, a lack of ground truth data, and institutional limitations. A forward--looking approach that incorporates sensor fusion, cloud-based AI analytics, SmallSat missions, citizen research, and capacity building programs is needed. In order to enable prompt, precise, and inclusive mangrove conservation, these strategies seek to close the gap between EO science and policy.

M. Paul (✉) · P. Kumar
Geospatial Laboratory, College of Horticulture and Forestry, Rani Lakshmi Bai Central Agricultural University, Jhansi, India

S. Mohanty
Department of Life Sciences, School of Natural Sciences, Shiv Nadar Institution of Eminence (Deemed to be University), Tehsil Dadri, India

P. K. Srivastava
Remote Sensing Laboratory, Institute of Environment and Sustainable Development, Banaras Hindu University, Varanasi, India

Geospatial Analytics for Conservation and Management, School of Science Technology and Engineering (SSTE), University of the Sunshine Coast, Sippy Downs, QLD, Australia

S. K. Srivastava
Geospatial Analytics for Conservation and Management, School of Science Technology and Engineering (SSTE), University of the Sunshine Coast, Sippy Downs, QLD, Australia

S. K. Srivastava et al. (eds.), *Earth Observation for Monitoring Mangrove Ecosystems*, https://doi.org/10.1007/978-3-032-16668-5_16

Keywords Mangrove monitoring · Earth observation · Remote sensing · Coastal ecosystems · Climate change impacts · Geospatial analysis

1 Introduction

Mangrove habitats constitute some of the most efficient and biologically significant environments on Earth, situated where terrestrial and marine biomes converge. These salt-tolerant forests, which cover tropical and subtropical coastlines, offer a variety of ecosystem services as breeding grounds for many marine species as well as coastal protection and nutrient cycling (Giri et al., 2011). In addition to their ecological significance, mangroves sustain millions of livelihoods of individuals living along the coast by providing resources including fish, lumber, and traditional remedies. The mangrove ecosystems have been steadily deteriorating during the last few decades, despite their distinctive value. Anthropogenic factors, such as port construction, vacationing, growing urbanization, and aquaculture expansion, also caused the destruction and division of habitat (Friess et al., 2020). According to Ward et al. (2016), climate-induced disturbances such temperature discrepancies, saltwater invasion, accelerated ocean level expand and along with hurricanes also worsen the risks for mangrove sensitivity as well as geographical distribution. The decline of mangroves has outpaced the decline of upland tropical forests in many regions, leading to severe consequences that necessitate prompt restoration and surveillance actions.

Earth Observation (EO) techniques are now vital resources for visualization, surveillance, and handling the ecosystems of mangroves in the constantly shifting climatic and geopolitical situation. The acquisition of synoptic, multi-temporal, even multispectral images over vast and often inaccessible coastlines is made possible by EO's array of equipment, including remote sensing satellites, unmanned aerial vehicles (UAVs), and on-the-ground sensors (Kuenzer et al., 2012). These innovations offer an adaptable and cost-effective alternative to traditional field-based inquiries, which are typically constrained by logistical, financial, and temporal issues. Historically, the majority of EO systems for mangrove surveillance rely on the SPOT and Landsat intermediate resolution optical satellite images. Finding long--term trends in the shifting patterns of recovery following disasters and degradation hotspots has been made possible owing in large part to such databases. It is now considerably simpler to determine the diversity of species as well as morphological characteristics of mangrove due to radar-based systems (such as Sentinel-1 along with ALOS PALSAR) and higher-resolution detectors (e.g., Sentinel-2 as well as Planet Scope) (Lucas et al., 2021). Furthermore, the development of multidimensional forest measurements has been made possible by recent developments in LiDAR with hyperspectral photography, which have enhanced the optical segregation of species of mangroves with nearby land uses.

Mangrove surveillance has advanced because of the integration of EO information with contemporary analysis, particularly through machine learning (ML), an approach associated with computational intelligence. For more accurate and efficient automated classification, change detection, and biomass estimation, algorithms like convolutional neural networks (CNNs), support vector machines, and random forests are being utilized more and more (Mandal et al., 2020). Simultaneously, cloud-based services including Google Earth Engine (GEE) and Amazon Web Services (AWS) have made large EO datasets and high-performance computer infrastructure more accessible to the general public (Gorelick et al., 2017). However, EO operationalization for mangrove conservation is not at all simple. There are still several institutional, environmental, and technical obstacles to overcome. Technically, there are still many obstacles to overcome, including spectral confusion, which is especially problematic when identifying mangrove species or mixed vegetation types. The intricacy of radar backscatter signals in tidal zones, cloud contamination in optical datasets, and variations in image quality all make analysis more difficult (Proisy et al., 2018). Furthermore, EO-based evaluations are not as reproducible or comparable across locations due to the absence of established categorization techniques and ground truth datasets.

The environment adds another level of complication. Mangrove Ecosystems (MEs) are dynamic systems that are impacted by seasonal salinity gradients, cyclones, floods, and tidal inundation. High-frequency observations are necessary to accurately capture these temporal dynamics, but they may not always be accessible, especially in areas with limited resources. Furthermore, the spatial resolution of the majority of satellite derived digital elevation models (DEMs) frequently falls short of fine-scale elevation changes, which are essential for evaluating vulnerability to sea level rise. These changes adopt more accurate but costly alternatives, such as airborne LiDAR. The broad use of EO techniques for mangrove management is further hampered by issues with institutions and policies. The administrative frameworks, financial resources, and technological know-how necessary to successfully execute EO-based monitoring programs are lacking in many poor nations with substantial mangrove cover. Data management issues like organizational disintegration, license constraints, and accessibility constraints also contribute to inadequate use. Moreover, discrepancies between assessments based on geospatial data as well as organizational frontiers or legal identities could give rise to policy stagnation or conflicting views of land utilization development (Webb et al., 2014).

Despite these constraints, EO has tremendous potential to observe mangroves. An environment that is conducive to growing EO usage has been created by technological innovation, emerging information exchange channels, with worldwide preservation agreements (such as the United Nations Decade on restoring ecosystems and its Goals of the Achievement of Sustainability). Mangrove Monitor Globally (Bunting et al., 2018), which maps the worldwide distribution of mangroves using multisensory satellite imagery, is a prime instance of the efficacy of cooperative, science-driven surveillance strategies. A systematic amalgamation of multisource EO the information, based on artificial intelligence analytics, field-based approval, including active surveillance is necessary to address the intricate problems of mangrove preservation. To render EO concepts more applicable in practice and relevant to regulations, macroeconomic and political elements might be incorporated. In the long term, bridging the disparity among EO science along with quaternary coastline management techniques will prove critical to the sustainable development and cultural importance of these ecosystems.

2 State of the Art in Earth Observation-Based Mangrove Monitoring

2.1 Radar and Optical Sensors

Over the past several decades, Earth Observation (EO) technological advances have been increasingly important for monitoring mangroves, assessing their state of health, and detecting changes. This has been utilized heavily on satellite-based optical including radar devices. Accurate surveillance of mangrove changes over both space and time has been rendered possible by spectral imagery using optical sensors such as Sentinel-2, MODIS, especially the Landsat range. Using vegetation measures, mangrove physiologic state is being recorded, including leaf amount of chlorophyll and crown density, along with availability of moisture (Giri et al., 2018). Constant precipitation limits the usefulness of optical sensors, especially in tropical coastal regions where mangroves constitute the predominant vegetation. Synthetic Aperture Radar (SAR) sensors, such as Sentinel-1, ALOS PALSAR, and RADARSAT-2, have developed into supplementary devices that can get around this restriction. Night and day images that can penetrate clouds are rendered feasible by SAR's operation in the microwave frequency domain. Canopy order, roughness of the surface, as well as amount of moisture are variables that prove particularly crucial during mangrove cartography along with estimation of biomass since they all affect radar scattering features (Simard et al., 2019). Dual as well as tetra-oriented SAR data further facilitate mapping in regions with extensive canopy or submerged terrain and enhance species-level discrimination.

2.2 UAV and LiDAR Systems

UAVs along with airplanes have implemented LiDAR technologies to achieve higher altitude as well as spatial accuracy. LiDAR creates three-dimensional point cloud images that accurately represent ground altitude and crown elevation including layering—all crucial factors for determining above-ground biomass, carbon stocks, as well as complexity of structures, claim Fatoyinbo et al. (2018). LiDAR-generated precise DEMs enable ongoing assessments of mangroves' vulnerability to erosion along the coast along with sea level increase. UAV-mounted detectors including RGB, multispectral, along with thermal imaging sensors provide adaptable, ultra-high-resolution data acquisition at user-specified periods. These platforms are particularly effective in small-scale probes, recovery sites, and places that are difficult to access. UAVs facilitate rapid post-disaster assessment and phenological surveillance, along with ecological suitability studies, especially when paired using observation in the field and imagery from satellites (Jeronimo et al., 2018). Integrating UAV photos using LiDAR and satellite imagery enhances the precision of classification, reduces ambiguity, and accelerates up making decisions in adaptive control systems.

2.3 *Novel Strategies*

Recent advancements in machine learning (artificial intelligence (AI)) with computing via the cloud have greatly expanded the statistical possibilities of EO-based mangrove management. Systems including Google Earth Engine (GEE) as well as Amazon Web Services (AWS) have rendered it feasible to conduct massive, continuous observation by allowing investigators to organize and evaluate a petabyte information in virtually real time (Gorelick et al., 2017). These technologies equalize the availability of worldwide EO information and reduce technical obstacles in gaining access, especially in developing countries. In segmentation along with identification, approaches based on machine learning (ML), such as random forests and support vector machines, are now standard. However, the advent of deep learning, particularly CNNs, has resulted in a notable breakthrough. CNNs have demonstrated exceptional accuracy for recognizing species ecosystems, determining mangrove restrictions, and detecting subtle shifts in woodland cover beneath difficult ecological circumstances (Mandal et al., 2020). These algorithms operate especially effectively especially trained using large, structured data sets that are paired with exceptionally detailed images from multiple detectors. These cutting-edge EO technologies, which include optical, radar, LiDAR, and UAV platforms, provide mangrove monitoring with previously unheard of spatial, temporal, and thematic resolution. As they continue to develop, mangrove conservation policies around the world should become more accurate, effective, and responsive (Table 1).

3 Limitations in Monitoring Mangroves Employing Earth Observation

Earth Observation (EO) technologies have advanced significantly, but a combination of institutional, environmental, and technical obstacles still make it difficult to use them for mangrove monitoring. These restrictions limit the accuracy, consistency, and usefulness of insights generated by EO, especially in areas with complex

Table 1 A contrasting evaluation of EO technologies for monitoring mangroves

Technology/ platform	Sensor type	Key applications	Strengths	Limitations
UAV (RGB, multispectral)	Drone-based optical	Restoration sites, disaster mapping	Ultra-high resolution, flexible deployment	Small coverage, high operational effort
MODIS	Moderate-- resolution	Phenological trends, large scale dynamics	High temporal resolution	Coarse spatial resolution
LiDAR (airborne/UAV)	Active laser scanning	Biomass, canopy height, topography (DEMs)	High vertical precision, 3D modeling	Expensive, limited temporal frequency
Landsat, Sentinel-2	Multispectral optical	Long-term mapping, health indices (NDVI, EVI)	Free, global, consistent temporal coverage	Cloud obstruction in tropical zones
Sentinel-1, ALOS PALSAR	Synthetic Aperture Radar (SAR)	Canopy structure, flooded terrain mapping	All-weather, night-time capability	Complex preprocessing, limited spectral info

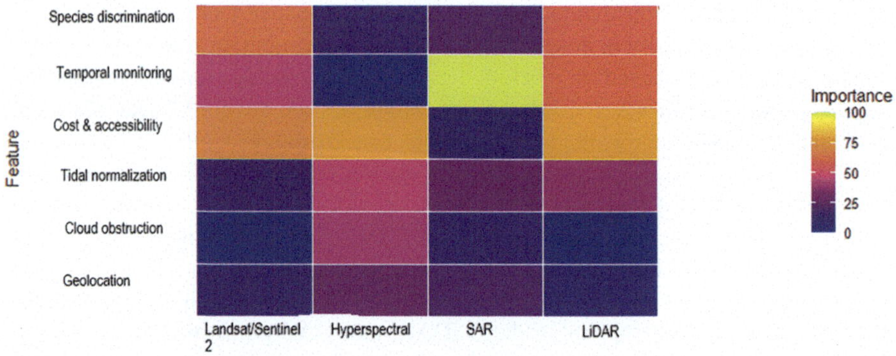

Fig. 1 Sensor performance with the reduction of challenges

ecosystems and dispersed administrative systems. Figure 1 illustrates how expensive LiDAR is and has a restricted time frame, despite its great accuracy for species recognition and boundary mapping. Despite their accessibility and temporal dependability, optical sensors like Sentinel-2 and Landsat have trouble distinguishing between various species and cloud cover. SAR all weather characteristics allow it to perform exceptionally well in tidal and cloud prone areas. All things considered, the table highlights the necessity of a multi sensor strategy to successfully handle the various monitoring limitations.

3.1 Discrimination Through Species and Spectral Confusion

The challenge of species-level classification utilizing remote sensing data is one of the most enduring technical challenges in mangrove monitoring. It is difficult to distinguish between distinct mangrove species using traditional multispectral sensors because they share comparable spectral reflectance properties in the visible and near-infrared (VNIR) spectrum (Heumann, 2011). In areas with high species richness or in transitional zones where mangroves meet seagrass beds, salt marshes, or other coastal vegetation, this issue is exacerbated. Furthermore, by combining spectral data from several land cover types into a single pixel, mixed pixels, which are typical in moderate resolution photography, significantly impair classification accuracy (Kuenzer et al., 2012). Although LiDAR and hyperspectral data provide better species discrimination, their restricted availability and expensive cost prevent their routine usage in many areas.

3.2 Temporal Inconsistencies and Cloud Obstruction

Most MEs are found in humid tropical and subtropical regions, where continuous cloud cover severely impairs the utility of optical satellite data. In addition to reducing the number of cloud free images available, this difficulty creates temporal gaps that make it more difficult to analyze time series and identify seasonal trends (Huang et al., 2020). Even while cloud cover has no effect on radar based sensors like Sentinel-1 and ALOS PALSAR, processing and interpreting their data requires sophisticated knowledge, particularly when distinguishing between vegetation structures under various tidal and hydrological circumstances (Simard et al., 2019).

3.3 Difficulties in Geolocation and Delineation of Enclosures

Geolocation errors frequently limit the ability to map mangroves with high spatial accuracy, particularly in small estuary regions or narrow, fractured coastal strips. Misclassification may arise from positional inaccuracies in satellite data, especially along dynamic coastal margins where mangrove limits change periodically as a result of sediment deposition and tidal inundation (Blasco et al., 1997). Therefore, advanced methods like multidate compositing, tidal normalization, or data fusion are necessary for accurate border delineation but are not yet commonly used or standardized in many monitoring programs.

3.4 Combining Field Constraints and Ground Truth Integration

The EO precision is dependent on the availability of ground truth information for calibration and validation, even if it offers broad insights. However, field access is hazardous or logistically challenging in many mangrove ecosystems due to their remoteness, ecological fragility, or political sensitivity. Ground surveys are therefore frequently sparse, geographically skewed, or nonexistent. Furthermore, interoperability and the creation of regionally adaptive EO models are hindered by the lack of standardized, publicly available field datasets (d'Andrimont et al., 2020). Table 2 shows the areas that require investment and have the worst EO ground integration.

3.5 Interoperability and Data Fragmentation

Datasets have proliferated as a result of the growing availability of EO platforms, which range from private systems like PlanetScope and ICEYE to public satellites like Landsat and Sentinel. Nevertheless, the lack of consistency among spectral bands, format of files, and geographic resolutions, along with metadata requirements, complicates cooperation and interaction (Bunting et al., 2018). These differences impede empirical comparisons, particularly in the case of international systems of mangroves or ongoing tracking that requires constant chronological baselines.

3.6 Regulations and Institutional Boundaries

Organizational difficulties continue to be a major barrier to converting EO-generated information toward efficient preservation activity. The primary barriers to EO implementations across numerous developing nations are a shortage of skilled experience, insufficient funding, including disparity in structure among ecological,

Table 2 Challenges of field data in the mangrove hotspot

Region	Ground access difficulty	Field frequency	Open field datasets	Political/security constraints
Sundarbans	High	Low	Few	Moderate
Niger delta	Very high	Very low	None	High
Mekong delta	Low	Moderate	Some	Low
Amazon coast	Moderate	Moderate	Few	Low
PNG islands	High	Low	None	Moderate

forest products, including coastal administration authorities (Friess et al., 2016). Disputes among EO-generated maps and official administrative documents causing systems of land tenure often lead to disputes or inefficiencies in the implementation of policy. EO-based detection faces the potential of continuing a merely technological operation with limited practical ramifications if established procedures are not in place for information confirmation, approval of usage, and integration into the process of decision-making.

4 A Route to Advancement

Global environmental obligations, technological developments, and an increasing urgency to preserve coastlines are driving Earth Observation's (EO) mangrove observation path into a critical phase. While much progress has been made, a fundamental approach is still required to translate the promise of EO into realistic, equitable, and empirically grounded conservation efforts. In order to advance EO-based mangrove monitoring into the next generation of decision support systems, this section lays out strategic routes for overcoming present obstacles. For policy briefings or strategic planning, Table 3 provides a clear link between each forward-looking effort and its associated technology, benefit, obstacle, and readiness level.

Table 3 Evaluation of EO developments and their concrete implications

Strategic direction	Core technology	Main challenges addressed	Key benefits
Sensor fusion	Optical, SAR, LiDAR, Hyperspectral	Spectral confusion, boundary accuracy	High accuracy
Data standardization	GMW, metadata protocols	Data fragmentation	Interoperability, global consistency
SmallSat integration	PlanetScope, ICEYE	Temporal gaps, storm or illegal activity mapping	High revisit, early detection of rapid changes
Citizen science	Mobile apps, participatory GIS	Field data limitations, trust building	Ground validation, stakeholder involvement
Capacity building	Training, curriculum, institutional networks	Technical skills, policy integration	Sustainable EO adoption, equitable access
AI and cloud analytics	Deep learning, Google Earth Engine	Big data handling, data scarcity	Scalable processing, near real-time analysis

4.1 Enhancing Strategies for Sensor Fusion

Mangrove habitats are too complicated to be fully captured by a single sensor type. A reliable solution is provided by sensor fusion, which combines optical, radar, LiDAR, and hyperspectral data while utilizing the complementing advantages of each modality. LiDAR records exact canopy height and vertical complexity, radar allows cloud penetrating structural insights, and optical sensors offer vegetation indices and spectral richness. Combining these modalities with sophisticated algorithms like wavelet transform, Bayesian data fusion, or deep ensemble learning can significantly improve hydrological evaluations, biomass estimation, and species recognition (Zhu et al., 2020). Mainstreaming multisource EO solutions into operational workflows will depend heavily on the creation of interoperable, sensor agnostic platforms.

4.2 Cloud Analytics and Artificial Intelligence Integration

Analytical frameworks that are scalable are necessary because of the exponential growth of EO data. In mangrove environments, artificial intelligence (AI), especially deep learning and transfer learning, has shown promising results in envision classification, segmentation, and anomaly detection (Mandal et al., 2020). Since many tropical regions lack labeled datasets, transfer learning enables pretrained models to adjust to region-specific conditions with little training data. These models, when combined with cloud-based geospatial engines including Google Earth Engine and Amazon SageMaker, can provide decision assistance, alert systems, and near real-time monitoring with little strain on the infrastructure (Gorelick et al., 2017). Further enhancing model interpretability and confidence in automated decision-making will be the use of explainable AI technologies.

4.3 Frequency Occurring, SmallSat Missions

The emergence of Small Satellite (SmallSat) constellations, such as those available from ICEYE, Capella Space, and PlanetScope, has transformed EO's temporal resolution. The monitoring of swift ecological occurrences including storm damage, illicit logging, and phenological changes is made possible by these satellites daily to weekly revisit rates (Pusat et al., 2016). Their frequently commercially available high-resolution photography fills in the gaps in data availability and enhances conventional large-scale sensors. Monitoring pipelines that incorporate SmallSat inputs are more responsive, enabling early intervention and adaptive management in dynamic coastal environments.

4.4 Monitoring via Collaboration and Citizen Science

Mangrove monitoring can benefit from the ethical and practical application of citizen science. According to Danielsen et al. (2014), local populations, who are frequently the most impacted stakeholders, have extensive ecological knowledge and are able to offer in situ observations via participatory GIS platforms and mobile apps. Such involvement promotes EO efforts' social legitimacy, transparency, and data democratization. Furthermore, data provided by citizens can support the validation of EO findings, especially in areas where formal ground-truthing is difficult or impossible because of the terrain or conflict.

4.5 The Frameworks Promoting Open Data and Standardization

For EO instruments to be comparable and scalable, unified data protocols must keep up with their proliferation. Standardized metadata templates, classification methods, and accuracy evaluation procedures are necessary for cross-regional analysis. Under initiatives comprise the United Nations Mission on Ecosystem Restoration; the Sustainable Development Goals (SDGs), expanding these frameworks and promoting national adoption might hasten the global mangrove conservation targets.

4.6 Building Ability and Strengthening Organizations

Without institutional and human capacity, technology cannot bring about revolutionary change on its own accord. To establish lon-term capability in environmental governance and geospatial analysis, it is imperative to invest in training initiatives, interdisciplinary curriculum development, and institutional collaborations (Friess et al., 2016). Governments, space agencies, nongovernmental organizations, and academic institutions can work together to create collaborative networks that facilitate data exchange, co-design of EO technologies, and policy integration. Collaborations between the North and South should be prioritized in order to reduce capability inequalities and advance fair access to conservation provided by EO.

5 Conclusion

Technologies for Earth Observation (EO) have completely changed ways mangrove ecosystems are evaluated, tracked, and maintained. From AI-driven analysis and community-based observation to advanced imagery from satellites and radar systems, EO techniques currently offer unparalleled knowledge on the spatial as well as temporal shifts of mangrove ecosystems. By facilitating the identification of forest destruction, the estimation of atmospheric carbon inventory, and the provision for warning signs for natural disruptions, these advancements have increased the adaptability of conservation initiatives. Spectral disorientation, sensor compatibility, and cloud interference are technical limitations that persist with institutional shortcomings such as data disintegration, inadequate ground confirmation, and inadequate policy convergence. To move above these barriers, a thorough and well-coordinated strategy that incorporates technology developments alongside cross-sector collaboration, capacity building, and democratic leadership is essential. Consistent observation practices, accessible data platforms, as well and broad involvement of stakeholders must be the cornerstones of EO-enabled mangrove preservation. Finally, maintaining mangroves which are critical for ecosystems, protection of the coast, and climate regulation requires incorporating EO within a robust, equity-centered paradigm. This would greatly enhance worldwide climate and environmental goals while also increasing the efficacy of preservation.

References

Blasco, M. A., Lee, H. W., Hande, M. P., Samper, E., Lansdorp, P. M., DePinho, R. A., & Greider, C. W. (1997). Telomere shortening and tumor formation by mouse cells lacking telomerase RNA. *Cell, 91*(1), 25–34.

Bunting, L., Davidson, G., McCartan, C., Hanratty, J., Bywaters, P., Mason, W., & Steils, N. (2018). The association between child maltreatment and adult poverty—A systematic review of longitudinal research. *Child Abuse & Neglect, 77*, 121–133.

d'Andrimont, R., Yordanov, M., Martinez-Sanchez, L., Eiselt, B., Palmieri, A., Dominici, P., et al. (2020). Harmonised LUCAS in-situ land cover and use database for field surveys from 2006 to 2018 in the European Union. *Scientific Data, 7*(1), 352.

Danielsen, F., Jensen, P. M., Burgess, N. D., Altamirano, R., Alviola, P. A., Andrianandrasana, H., et al. (2014). A multicountry assessment of tropical resource monitoring by local communities. *Bioscience, 64*(3), 236–251.

Fatoyinbo, T., Feliciano, E. A., Lagomasino, D., Lee, S. K., & Trettin, C. (2018). Estimating mangrove aboveground biomass from airborne LiDAR data: A case study from the Zambezi River delta. *Environmental Research Letters, 13*(2), 025012.

Friess, D. A., Thompson, B. S., Brown, B., Amir, A. A., Cameron, C., Koldewey, H. J., et al. (2016). Policy challenges and approaches for the conservation of mangrove forests in Southeast Asia. *Conservation Biology, 30*(5), 933–949.

Friess, D. A., Yando, E. S., Abuchahla, G. M., Adams, J. B., Cannicci, S., Canty, S. W., et al. (2020). Mangroves give cause for conservation optimism, for now. *Current Biology, 30*(4), R153–R154.

Giri, C., Ochieng, E., Tieszen, L. L., Zhu, Z., Singh, A., Loveland, T., et al. (2011). Status and distribution of mangrove forests of the world using earth observation satellite data. *Global Ecology and Biogeography, 20*(1), 154–159.

Giri, S. S., Yun, S., Jun, J. W., Kim, H. J., Kim, S. G., Kang, J. W., et al. (2018). Therapeutic effect of intestinal autochthonous Lactobacillus reuteri P16 against waterborne lead toxicity in Cyprinus carpio. *Frontiers in Immunology, 9*, 1824.

Gorelick, P. B., Furie, K. L., Iadecola, C., Smith, E. E., Waddy, S. P., Lloyd-Jones, D. M., et al. (2017). Defining optimal brain health in adults: A presidential advisory from the American Heart Association/American Stroke Association. *Stroke, 48*(10), e284–e303.

Heumann, B. W. (2011). Satellite remote sensing of mangrove forests: Recent advances and future opportunities. *Progress in Physical Geography, 35*(1), 87–108.

Huang, H. Y., Lin, Y. C. D., Li, J., Huang, K. Y., Shrestha, S., Hong, H. C., et al. (2020). miRTarBase 2020: Updates to the experimentally validated microRNA–target interaction database. *Nucleic Acids Research, 48*(D1), D148–D154.

Jeronimo, S. M., Kane, V. R., Churchill, D. J., McGaughey, R. J., & Franklin, J. F. (2018). Applying LiDAR individual tree detection to management of structurally diverse forest landscapes. *Journal of Forestry, 116*(4), 336–346.

Kuenzer, C., Zhang, J., Sun, Y., Jia, Y., & Dech, S. (2012). Coal fires revisited: The Wuda coal field in the aftermath of extensive coal fire research and accelerating extinguishing activities. *International Journal of Coal Geology, 102*, 75–86.

Lucas, R., Otero, V., Van De Kerchove, R., Lagomasino, D., Satyanarayana, B., Fatoyinbo, T., & Dahdouh-Guebas, F. (2021). Monitoring Matang's Mangroves in Peninsular Malaysia through Earth observations: A globally relevant approach. *Land Degradation & Development, 32*(1), 354–373.

Mandal, S., Bhatnagar, T., Arinaminpathy, N., Agarwal, A., Chowdhury, A., Murhekar, M., et al. (2020). Prudent public health intervention strategies to control the coronavirus disease 2019 transmission in India: A mathematical model-based approach. *Indian Journal of Medical Research, 151*(2–3), 190–199.

Proisy, C., Viennois, G., Sidik, F., Andayani, A., Enright, J. A., Guitet, S., et al. (2018). Monitoring mangrove forests after aquaculture abandonment using time series of very high spatial resolution satellite images: A case study from the Perancak estuary, Bali, Indonesia. *Marine Pollution Bulletin, 131*, 61–71.

Pusat, S., Akkoyunlu, M. T., Pekel, E., Akkoyunlu, M. C., Özkan, C., & Kara, S. S. (2016). Estimation of coal moisture content in convective drying process using ANFIS. *Fuel Processing Technology, 147*, 12–17.

Simard, M., Fatoyinbo, L., Smetanka, C., Rivera-Monroy, V. H., Castañeda-Moya, E., Thomas, N., & Van der Stocken, T. (2019). Mangrove canopy height globally related to precipitation, temperature and cyclone frequency. *Nature Geoscience, 12*(1), 40–45.

Ward, M. M., Deodhar, A., Akl, E. A., Lui, A., Ermann, J., Gensler, L. S., et al. (2016). American College of Rheumatology/Spondylitis Association of America/Spondyloarthritis Research and Treatment Network 2015 recommendations for the treatment of ankylosing spondylitis and nonradiographic axial spondyloarthritis. *Arthritis and Rheumatology, 68*(2), 282–298.

Webb, E. L., Jachowski, N. R., Phelps, J., Friess, D. A., Than, M. M., & Ziegler, A. D. (2014). Deforestation in the Ayeyarwady Delta and the conservation implications of an internationally-engaged Myanmar. *Global Environmental Change, 24*, 321–333.

Zhu, Z., Cai, T., Fan, L., Lou, K., Hua, X., Huang, Z., & Gao, G. (2020). Clinical value of immune-inflammatory parameters to assess the severity of coronavirus disease 2019. *International Journal of Infectious Diseases, 95*, 332–339.

Index

S. K. Srivastava et al. (eds.), *Earth Observation for Monitoring Mangrove Ecosystems*, https://doi.org/10.1007/978-3-032-16668-5

The manufacturer's authorised representative in the EU is Springer Nature Customer Service Centre GmbH, Europaplatz 3, 69115 Heidelberg, Germany. If you have any concerns regarding our products, please contact ProductSafety@springernature.com

Printed and bound by CPI Group (UK) Ltd, Croydon, CR0 4YY
07/10/2026
02229350-0007